EIGHTH EDITION

Laboratory Manual

HOLE'S ESSENTIALS OF HUMAN ANATOMY AND PHYSIOLOGY

DAVID SHIER • JACKIE BUTLER • RICKI LEWIS

TERRY R. MARTIN

Kishwaukee College

McGraw Hill

Boston Burr Ridge, IL Dubuque, IA Madison, WI New York San Francisco St. Louis
Bangkok Bogotá Caracas Kuala Lumpur Lisbon London Madrid Mexico City
Milan Montreal New Delhi Santiago Seoul Singapore Sydney Taipei Toronto

McGraw-Hill Higher Education

A Division of The ***McGraw-Hill*** *Companies*

LABORATORY MANUAL FOR HOLE'S ESSENTIALS OF HUMAN ANATOMY AND PHYSIOLOGY, EIGHTH EDITION

Published by McGraw-Hill, a business unit of The McGraw-Hill Companies, Inc., 1221 Avenue of the Americas, New York, NY 10020.

This book is printed on acid-free paper.

1 2 3 4 5 6 7 8 9 0 WEB/WEB 0 9 8 7 6 5 4 3 2

ISBN 0-07-235120-9

Publisher: *Martin J. Lange*
Sponsoring editor: *Michelle Watnick*
Senior developmental editor: *Patricia Hesse*
Director of development: *Kristine Tibbetts*
Project manager: *Rose Koos*
Senior production supervisor: *Sandy Ludovissy*
Media project manager: *Sandra M. Schnee*
Senior media technology producer: *Barbara R. Block*
Designer: *K. Wayne Harms*
Interior designer: *Kathleen Theis*
Cover image: *Doug Pensinger/Allsport*
Senior photo research coordinator: *John C. Leland*
Photo research: *Mary Reeg*
Compositor: *Precision Graphics*
Typeface: *10/12 Garamond Book*
Printer: *Webcrafters, Inc.*

The credits section for this book begins on page 372 and is considered an extension of the copyright page.

Some of the laboratory experiments included in this text may be hazardous if materials are handled improperly or if procedures are conducted incorrectly. Safety precautions are necessary when you are working with chemicals, glass test tubes, hot water baths, sharp instruments, and the like, or for any procedures that generally require caution. Your school may have set regulations regarding safety procedures that your instructor will explain to you. Should you have any problems with materials or procedures, please ask your instructor for help.

www.mhhe.com

Contents

Preface

This laboratory manual was prepared to supplement the textbook *Hole's Essentials of Human Anatomy and Physiology,* 8th edition, by David Shier, Jackie Butler, and Ricki Lewis. As in the case of the textbook, the laboratory manual is designed for students with minimal backgrounds in the physical and biological sciences who are pursuing careers in allied health fields.

The laboratory manual contains forty-eight laboratory exercises and reports, which are closely integrated with the chapters of the textbook. The exercises are planned to illustrate and review the anatomical and physiological facts and principles presented in the textbook and to help students investigate some of these ideas in greater detail.

Often the laboratory exercises are short or are divided into several separate procedures. This allows an instructor to select those exercises or parts of exercises that will best meet the needs of a particular program. Also, exercises requiring a minimal amount of laboratory equipment have been included.

The laboratory exercises include a variety of special features that are designed to stimulate interest in the subject matter, to involve students in the learning process, and to guide them through the planned activities. These special features include the following:

MATERIALS NEEDED

This section lists the laboratory materials that are required to complete the exercise and to perform the demonstrations and optional activities.

SAFETY

A list of safety guidelines is included inside the front cover. Each lab session that requires special safety guidelines has a safety section following "Materials Needed." Your instructor might require some modifications of these guidelines.

INTRODUCTION

The introduction briefly describes the subject of the exercise or the ideas that will be investigated.

PURPOSE OF THE EXERCISE

The purpose provides a statement concerning the intent of the exercise—that is, what will be accomplished.

LEARNING OBJECTIVES

The learning objectives list in general terms what a student should be able to do after completing the exercise.

PROCEDURE

The procedure provides a set of detailed instructions for accomplishing the planned laboratory activities. Usually these instructions are presented in outline form so that a student can proceed through the exercise in stepwise fashion. Frequently, the student is referred to particular sections of the textbook for necessary background information or for review of subject matter presented previously.

The procedures include a wide variety of laboratory activities and, from time to time, direct the student to complete various tasks in the laboratory reports.

LABORATORY REPORTS

A laboratory report to be completed by the student immediately follows each exercise. These reports include various types of review activities, spaces for sketches of microscopic objects, tables for recording observations and experimental results, and questions dealing with the analysis of such data.

It is hoped that as a result of these activities, students will develop a better understanding of the structural and functional characteristics of their bodies and will increase their skills in gathering information by observation and experimentation. Some of the exercises also include demonstrations, optional activities, and useful illustrations.

DEMONSTRATIONS

Demonstrations appear in separate boxes. They describe specimens, specialized laboratory equipment, or other materials of interest that an instructor may want to display to enrich the student's laboratory experience.

OPTIONAL ACTIVITIES

Optional activities also appear in separate boxes. They encourage students to extend their laboratory experiences. Some of these activities are open-ended in that they suggest the student plan an investigation or experiment and carry it out after receiving approval from the laboratory instructor.

THE USE OF ANIMALS IN BIOLOGY EDUCATION*

The National Association of Biology Teachers (NABT) believes that the study of organisms, including nonhuman animals, is essential to the understanding of life on Earth. NABT recommends the prudent and responsible use of animals in the life science classroom. NABT believes that biology teachers should foster a respect for life. Biology teachers also should teach about the interrelationship and interdependency of all things.

Classroom experiences that involve nonhuman animals range from observation to dissection. NABT supports these experiences so long as they are conducted within the long-established guidelines of proper care and use of animals, as developed by the scientific and educational community.

As with any instructional activity, the use of nonhuman animals in the biology classroom must have sound educational objectives. Any use of animals, whether for observation or dissection, must convey substantive knowledge of biology. NABT believes that biology teachers are in the best position to make this determination for their students.

NABT acknowledges that no alternative can substitute for the actual experience of dissection or other use of animals and urges teachers to be aware of the limitations of alternatives. When the teacher determines that the most effective means to meet the objectives of the class do not require dissection, NABT accepts the use of alternatives to

* Adopted by the Board of Directors in October 1995. This policy supersedes and replaces all previous NABT statements regarding animals in biology education.

dissection, including models and the various forms of multimedia. The Association encourages teachers to be sensitive to substantive student objections to dissection and to consider providing appropriate lessons for those students where necessary.

To implement this policy, NABT endorses and adopts the "Principles and Guidelines for the Use of Animals in Precollege Education" of the Institute of Laboratory Animals Resources (National Research Council). Copies of the "Principles and Guidelines" may be obtained from the ILAR (2101 Constitution Avenue, NW, Washington, DC 20418; 202 334-2590).

ILLUSTRATIONS

Diagrams from the textbook and diagrams similar to those in the textbook often are used as aids for reviewing subject matter. Other illustrations provide visual instructions for performing steps in procedures or are used to identify parts of instruments or specimens. Micrographs are included to help students identify microscopic structures or to evaluate student understanding of tissues.

In some exercises, the figures include line drawings that are suitable for students to color with colored pencils. This activity may motivate students to observe the illustrations more carefully and help them to locate the special features represented in the figures. Students can check their work by referring to the corresponding full-color illustrations in the textbook.

FEATURES OF THIS EDITION

This new edition of the laboratory manual has been made more user-friendly. Many of the changes are a result of evaluations and suggestions from anatomy and physiology students. Numerous suggestions from users and reviewers of the seventh edition have been incorporated. Some features include the following:

1. To meet the need for clearer and more definite safety guidelines, a safety list is located inside the front cover, and safety sections are found in appropriate labs.
2. A section called "Study Skills for Anatomy and Physiology" is located in the front material. This section was written by students enrolled in a Human Anatomy and Physiology course.
3. The "Materials Needed" section is located at the beginning of the laboratory exercise to enable greater ease in laboratory preparations.
4. New art has been created for the textbook and the lab manual.
5. To clarify whether a label refers to a general area or a specific structure, "clue" words in parentheses have been added to some figures to direct students in their answers. The first example is figure 2.1.
6. A list of terms is provided to assist the labeling of many figures; the first example is figure 11.3.
7. Computer literacy is integrated to  relevant laboratory exercises. "Web Quest" activities are found at the end of most exercises. Discover the answers to many scientific questions at www.mhhe.com/shieress8 Here you'll find links to help you with your quest.
8. "Critical Thinking Applications"  are incorporated within most of the laboratory exercises to enhance valuable critical thinking skills that students need throughout their lives.
9. References for laboratory exercise  correlations to *The Virtual Physiology Lab CD-ROM* are indicated by the icon in the appropriate lab following "Materials Needed." The ten animal-based physiological experiments can be used for prelab preparation, for an actual lab alternative or supplement, or for postlab review.
10. The *Instructor's Manual for Laboratory Manual to Accompany Hole's Essentials of Human Anatomy and Physiology* has been revised and updated. It is located online as part of the Instructor Resources of the Online Learning Center. www.mhhe.com/shieress8
11. Two assessment tools (rubrics) for laboratory reports are included in Appendix 2.
12. A supplement of four computerized physiology labs with laboratory reports using Intelitool products is available. The title is *Intelitool Supplement to Accompany Anatomy and Physiology Labs* (0-697-33368-X).

REVIEWERS

I would like to express my sincere gratitude to all users of the laboratory manual who provided suggestions for its improvement. I am especially grateful for the contributions of the reviewers who kindly reviewed the manual and examined the manuscript of the new edition and gave their thoughtful comments and valuable suggestions. They include the following:

Cynthia A. Bottrell
Scott Community College

Shirley A. Colvin
Gadsden State Community College

Ann Henninger
Wartburg College

Mark Jonasson
Crafton Hills College

Eunice R. Knouse
Spartanburg Methodist College

Vanessa M. Montgomery
Paradise Valley Community College

Tricia Moore
Northern Oklahoma College

Tony E. Morris
Fairmont State College

Sheila Owen
Augusta Prep School

ACKNOWLEDGMENTS

I value all the support and encouragement by the staff at McGraw-Hill, including Michael Lange, Kristine Tibbetts, Marty Lange, Pat Hesse, Rose Koos, and Darlene Schueller. Special recognition is granted to Colin Wheatley for his insight, confidence, wisdom, warmth, and friendship.

I am grateful for the professional talent of David Shier, Jackie Butler, and Ricki Lewis for all of their coordination efforts and guidance with the laboratory manual. The deepest gratitude is extended to John W. Hole, Jr., for his years of dedicated effort in previous editions of this classic work and for the opportunity to revise his established laboratory manual.

I am particularly thankful to Dr. David Louis, president of Kishwaukee College, who remains a supporter of my endeavors. I am appreciative for the expertise of J. Womack Photography. The professional reviews of the nursing procedures were provided by Kathy Schnier. I am also grateful to Michele Dukes, Troy Hanke, Jenifer Holtzclaw, Angele Myska, Sparkle Neal, Robert Stockley, Shatina Thompson, Jana Voorhis, and DeKalb Clinic Chartered for their contributions. There have been valuable contributions from my students who have supplied thoughtful suggestions and assisted in clarification of details.

To my son Ross, I owe gratitude for his keen eye and creative suggestions. Foremost, I am appreciative to Sherrie Martin, my spouse and best friend, for advice, understanding, and devotion throughout the writing and revising.

Terry R. Martin
Kishwaukee College
21193 Malta Road
Malta, IL 60150

ABOUT THE AUTHOR

This eighth edition is the third revision by Terry R. Martin of Kishwaukee College. Terry's teaching experience of over thirty years, his interest in students and love for college instruction, and his innovative attitude and use of technology-based learning enhance the solid tradition of John Hole's laboratory manual. Among Terry's awards are the 1972 Kishwaukee College Outstanding Educator, 1977 Phi Theta Kappa Outstanding Instructor Award, 1989 Kishwaukee College ICCTA Outstanding Educator Award, 1996 Who's Who Among America's Teachers, 1996 Kishwaukee College Faculty Board of Trustees Award of Excellence, and 1998 Continued Excellence Award for Phi Theta Kappa Advisors. Terry's professional memberships include the National Association of Biology Teachers, Illinois Association of Community College Biologists, Human Anatomy and Physiology Society, Chicago Area Anatomy and Physiology Society (founding member), Phi Theta Kappa (honorary member), DeKalb County Prairie Stewards, and Nature Conservancy. In addition to writing many publications, he co-produced with Hassan Rastegar a videotape entitled *Introduction to the Human Cadaver and Prosection,* published by Wm. C. Brown Publishers in 1989. Terry revised the *Laboratory Manual to Accompany Hole's Human Anatomy and Physiology,* 9th edition, and authored *Human Anatomy and Physiology Laboratory Manual, Fetal Pig Dissection,* 1st edition. Terry is a contributing author and advisor for BioCourse.com, which is a McGraw-Hill Internet resource site for biology

faculty and students. During 1994, Terry was a faculty exchange member in Ireland. He has also been involved in his community, most notably as a Commissioner for Boy Scouts of America. We are pleased to have Terry continue the tradition of John Hole's laboratory manual.

The Editor

TO THE STUDENT

The exercises in this laboratory manual will provide you with opportunities to observe various anatomical parts and to investigate certain physiological phenomena. Such experiences should help you relate specimens, models, microscope slides, and your own body to what you have learned in the lecture and read about in the textbook.

The following list of suggestions may help to make your laboratory activities more effective and profitable.

1. Prepare yourself before attending the laboratory session by reading the assigned exercise and reviewing the related sections of the textbook. It is important to have some understanding of what will be done in the laboratory before you come to class.
2. Bring your laboratory manual and textbook to each laboratory session. These books are closely integrated and will help you complete most of the exercises.
3. Be on time. During the first few minutes of the laboratory meeting, the instructor often will provide verbal instructions. Make special note of any changes in materials to be used or procedures to be followed. Also listen carefully for information concerning special techniques to be used and precautions to be taken.
4. Keep your work area clean and your materials neatly arranged so that you can locate needed items quickly. This will enable you to proceed efficiently and will reduce the chances of making mistakes.
5. Pay particular attention to the purpose of the exercise, which states what you are to accomplish in general terms, and to the learning objectives, which list what you should be able to do as a result of the laboratory experience. Then, before you leave the class, review the objectives and make sure that you can meet them.
6. Precisely follow the directions in the procedure and proceed only when you understand them clearly. Do not improvise procedures unless you have the approval of the laboratory instructor. Ask questions if you do not understand exactly what you are supposed to do and why you are doing it.
7. Handle all laboratory materials with care. These materials often are fragile and expensive to replace. Whenever you have questions about the proper treatment of equipment, ask the instructor.
8. Treat all living specimens humanely and try to minimize any discomfort they might experience.
9. Although at times you might work with a laboratory partner or a small group, try to remain independent when you are making observations, drawing conclusions, and completing the activities in the laboratory reports.

10. Record your observations immediately after making them. In most cases, such data can be entered in spaces provided in the laboratory reports.
11. Read the instructions for each section of the laboratory report before you begin to complete it. Think about the questions before you answer them. Your responses should be based on logical reasoning and phrased in clear and concise language.
12. At the end of each laboratory period, clean your work area and the instruments you have used. Return all materials to their proper places and dispose of wastes, including glassware or microscope slides that have become contaminated with human blood or body fluids, as directed by the laboratory instructor. Wash your hands thoroughly before leaving the laboratory.

STUDY SKILLS FOR ANATOMY AND PHYSIOLOGY

My students have found that certain study skills worked well for them while enrolled in Human Anatomy and Physiology. Although each individual has a somewhat different learning style, there are techniques that work well for the majority of students. Utilizing some of the skills listed here could make your course more enjoyable and rewarding.

1. **Note taking:** Look for the main ideas and briefly express them in your own words. Organize, edit, and review your notes soon after the lecture. Add textbook information to your notes as you reorganize them. Underline or highlight with different colors the important points, major headings, and key terms. Study your notes daily, as they provide sequential building blocks of the course content.
2. **Chunking:** Organize information into logical groups or categories. Study and master one chunk of information at a time. For example, study the bones of the upper limb, lower limb, trunk, and head as separate study tasks.
3. **Mnemonic devices:** An *acrostic* is a combination of association and imagery to aid your memory. It is often in the form of a poem, rhyme, or jingle in which the first letter of each word corresponds to the first letters of the words you need to remember. **S**o **L**ong **T**op **P**art, **H**ere **C**omes **T**he **T**humb is an example of such a mnemonic device to remember the eight carpals in the correct sequence. *Acronyms* are words that are formed by the first letters of the items to remember. *IPMAT* is an example of this type of mnemonic device to help remember the phases of the cell cycle in the correct sequence. Try some of your own.
4. **Study groups:** Small study groups that meet periodically to review course material and compare notes have helped and encouraged many students. However, keep the group on the task at hand. Work as a team and alternate leaders. This group often becomes a support group.
5. **Recording and recitation:** An auditory learner can benefit by recording lectures and review sessions with a cassette recorder. Many students listen to the taped sessions as they drive or just before going to bed. Reading your notes aloud can help also. Explain the material to anyone (even if there are no listeners). Talk about anatomy and physiology in everyday conversations.
6. **Note cards/flash cards:** Make your own. Add labels and colors to enhance the material. Keep them with you in your pocket or purse. Study them frequently and for short periods of time. Concentrate on a small number of cards at one time. Shuffle your cards and have someone quiz you on their content. As you become familiar with the material, you can set aside cards that don't require additional mastery.
7. **Time management:** Prepare monthly, weekly, and daily schedules. Include dates of quizzes, exams, and projects on the calendar. On your daily schedule, budget several short study periods. Daily repetition alleviates cramming for exams. Prioritize your time so that you still have time for work and leisure activities. Find an appropriate study atmosphere with minimum distractions.

Best wishes on your anatomy and physiology endeavor.

Correlation of Textbook Chapters and Laboratory Exercises

Textbook Chapters		*Related Laboratory Exercises*	
Chapter 1	Introduction to Human Anatomy and Physiology	Exercise 1	Scientific Method and Measurements
		Exercise 2	Body Organization and Terminology
Chapter 2	Chemical Basis of Life		
Chapter 3	Cells	Exercise 3	Care and Use of the Compound Microscope
		Exercise 4	Cell Structure and Function
		Exercise 5	Movements Through Cell Membranes
		Exercise 6	The Cell Cycle
Chapter 4	Cellular Metabolism		
Chapter 5	Tissues	Exercise 7	Epithelial Tissues
		Exercise 8	Connective Tissues
		Exercise 9	Muscle and Nervous Tissues
Chapter 6	Skin and the Integumentary System	Exercise 10	Integumentary System
Chapter 7	Skeletal System	Exercise 11	Structure of Bone
		Exercise 12	Organization of the Skeleton
		Exercise 13	The Skull
		Exercise 14	Vertebral Column and Thoracic Cage
		Exercise 15	Pectoral Girdle and Upper Limb
		Exercise 16	Pelvic Girdle and Lower Limb
		Exercise 17	The Joints
Chapter 8	Muscular System	Exercise 18	Skeletal Muscle Structure
		Exercise 19	Muscles of the Face, Head, and Neck
		Exercise 20	Muscles of the Chest, Shoulder, and Upper Limb
		Exercise 21	Muscles of the Abdominal Wall and Pelvic Outlet
		Exercise 22	Muscles of the Hip and Lower Limb
Chapter 9	Nervous System	Exercise 23	Nervous Tissue and Nerves
		Exercise 24	The Reflex Arc and Reflexes
		Exercise 25	The Meninges and Spinal Cord
		Exercise 26	The Brain and Cranial Nerves
		Exercise 27	Dissection of the Sheep Brain
Chapter 10	Somatic and Special Senses	Exercise 28	The Ear and Hearing
		Exercise 29	The Eye
		Exercise 30	Visual Tests and Demonstrations
Chapter 11	Endocrine System	Exercise 31	Endocrine System
Chapter 12	Blood	Exercise 32	Blood Cells
		Exercise 33	Blood Testing—A Demonstration
		Exercise 34	Blood Typing
Chapter 13	Cardiovascular System	Exercise 35	Structure of the Heart
		Exercise 36	The Cardiac Cycle
		Exercise 37	Blood Vessels
		Exercise 38	Pulse Rate and Blood Pressure
		Exercise 39	Major Arteries and Veins
Chapter 14	Lymphatic System and Immunity	Exercise 40	Lymphatic System

Laboratory Exercise 1

Scientific Method and Measurements

MATERIALS NEEDED

Meterstick
Calculator
Human skeleton

Scientific investigation involves a series of logical steps to arrive at explanations for various biological phenomena. This technique, called the *scientific method,* is used in all disciplines of science. It allows scientists to draw logical and reliable conclusions about phenomena.

The scientific method begins with *observations* related to the topic under investigation. This step commonly involves the accumulation of previously acquired information and/or your own observations of the phenomenon. These observations are used to formulate a tentative explanation known as the *hypothesis.* An important attribute of a hypothesis is that it must be testable. The testing of the hypothesis involves performing a carefully controlled experiment to obtain data that can be used to support, disprove, or modify the hypothesis. An *analysis of data* is conducted using all of the information collected during the experiment. Data analysis includes organization and presentation of data as tables, graphs, and drawings. From the interpretation of the data analysis, *conclusions* are drawn. The final presentation of the information is made from the conclusions. Results and conclusions are presented to the scientific community for evaluation through peer reviews, presentations at professional meetings, and published articles. If numerous investigators working independently can validate the hypothesis by arriving at the same conclusions, the explanation becomes a **theory.** A theory that is verified continuously over a period of time and accepted by the scientific community becomes known as a **scientific law** or **principle.** A scientific law serves as the standard explanation for an observation unless it is disproved by new information. The five components of the scientific method are summarized as

Observations
↓
Hypothesis
↓
Experiment
↓
Analysis of data
↓
Conclusions

Metric measurements are characteristic tools of scientific investigations. Because the English system of measurements is often used in the United States, the investigator must make conversions from the English system to the metric system. A reference table for the conversion of English units of measure to metric units for length, mass, volume, time, and temperature is located inside the back cover of the laboratory manual.

PURPOSE OF THE EXERCISE

To become familiar with the scientific method of investigation, to learn how to formulate sound conclusions, and to provide opportunities to use the metric system of measurements.

LEARNING OBJECTIVES

After completing this exercise, you should be able to

1. list in the correct order and describe all steps of the scientific method;
2. use the scientific method to test the validity of a hypothesis concerning the direct, linear relationship between human height and upper limb length;
3. make conversions from English measurements to the metric system, and vice versa;
4. formulate a hypothesis and test it using the scientific method.

Figure 1.1 Measurement of upper limb length.

PROCEDURE A—USING THE STEPS OF THE SCIENTIFIC METHOD

1. Many people have observed a correlation between the length of the upper and lower limbs and the height (height for this lab means overall height of the subject) of an individual. For example, a person who has long upper limbs (the arm, forearm, and hand combined) tends to be tall. Make some visual observations of other people in your class to observe a possible correlation.
2. From such observations, the following hypothesis is formulated: The length of a person's upper limb is equal to 0.4 (40%) of the height of the person. Test this hypothesis by performing the following experiment.
3. In this experiment, use a meterstick to measure an upper limb length of ten subjects. For each measurement, place the meterstick in the axilla (armpit) and record the length in centimeters to the end of the longest finger (fig. 1.1). Obtain the height of each person in centimeters by measuring them without shoes against a wall (fig. 1.2). The height of each person can be calculated by multiplying each individual's height in inches by 2.54 to obtain his/her height in centimeters. Record all your measurements in Part A of Laboratory Report 1.

Figure 1.2 Measurement of height.

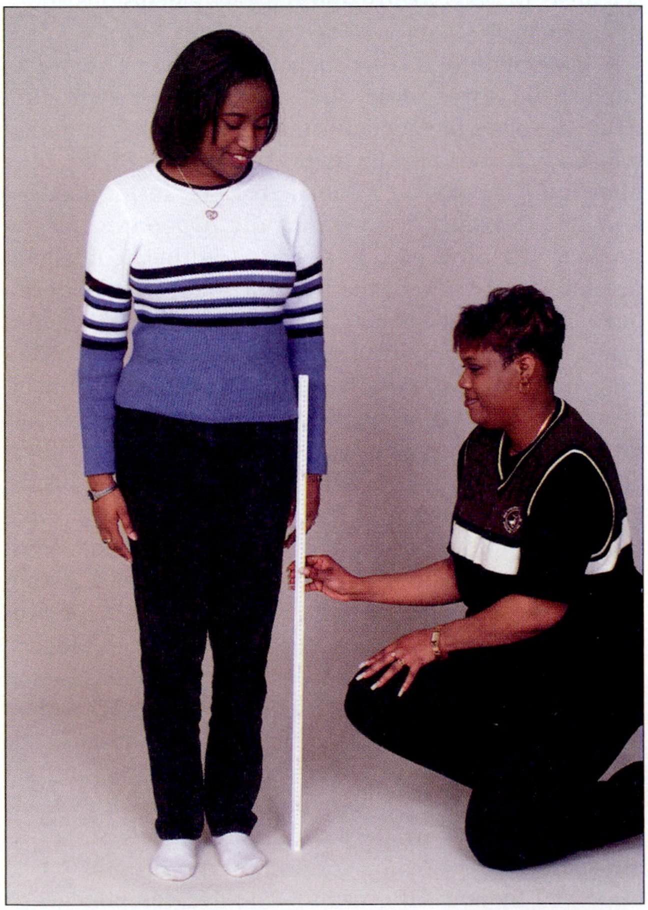

4. The data collected from all of the measurements can now be analyzed. The expected correlation between upper limb length and height is determined using the following equation:

   Height $\times$ 0.4 = expected upper limb length

   The observed correlation to be used to test the hypothesis is determined by

   Length of upper limb/height
   = actual % of height

5. A graph is an excellent method to obtain a visual representation of the data. Plot the subjects' data in Part A of the laboratory report. Plot the upper limb length of each subject on the x-axis and the height of each person on the y-axis. A line is already located on the graph that represents a hypothetical relationship of 0.4 (40%) upper limb length compared to height. This is a graphic representation of the original hypothesis.
6. Compare the distribution of all of the points (actual height and upper limb length) that you placed on the graph with the distribution of the expected correlation represented by the hypothesis.
7. Complete Part A of the laboratory report.

PROCEDURE B—DESIGN AN EXPERIMENT

Critical Thinking Application

You have probably concluded that there is some correlation to the length of body parts and height. Often when a skeleton is found, it is not complete, especially when paleontologists discover a skeleton. It is occasionally feasible to use the length of a single bone to estimate the height of an individual. Observe human skeletons and locate the radius bone in the forearm. Use your observations to identify a mathematical relationship between the length of the radius and height. Formulate a hypothesis that can be tested. Make measurements, analyze data, and develop a conclusion from your experiment. Complete Part B of the laboratory report.

Web Quest

Examine additional information about the scientific method at www.mhhe.com/shieress8

Laboratory Report 1

Name ______________________

Date ______________________

Section ______________________

Scientific Method and Measurements

PART A

1. Record measurements for height and the upper limb length of ten subjects. Use a calculator to determine the expected upper limb length and the actual percentage (as a decimal or a percentage) of the height for the ten subjects. Record your results in the following table.

Subject	Height (cm)	Measured Upper Limb Length (cm)	Height x 0.4 = Expected Upper Limb Length (cm)	Actual % of Height = Upper Limb Length (cm)/Height (cm)
1.				
2.				
3.				
4.				
5.				
6.				
7.				
8.				
9.				
10.				

2. Plot the distribution of data (upper limb length and height) collected for the ten subjects on the following graph. The line located on the graph represents the **expected** 0.4 (40%) upper limb length compared to measured height (the original hypothesis). (Note that the x-axis represents upper limb length and the y-axis represents height.) Draw a line of *best fits* through the distribution of points of the plotted data of the ten subjects. Compare the two distributions (expected line and the distribution line drawn for the ten subjects).

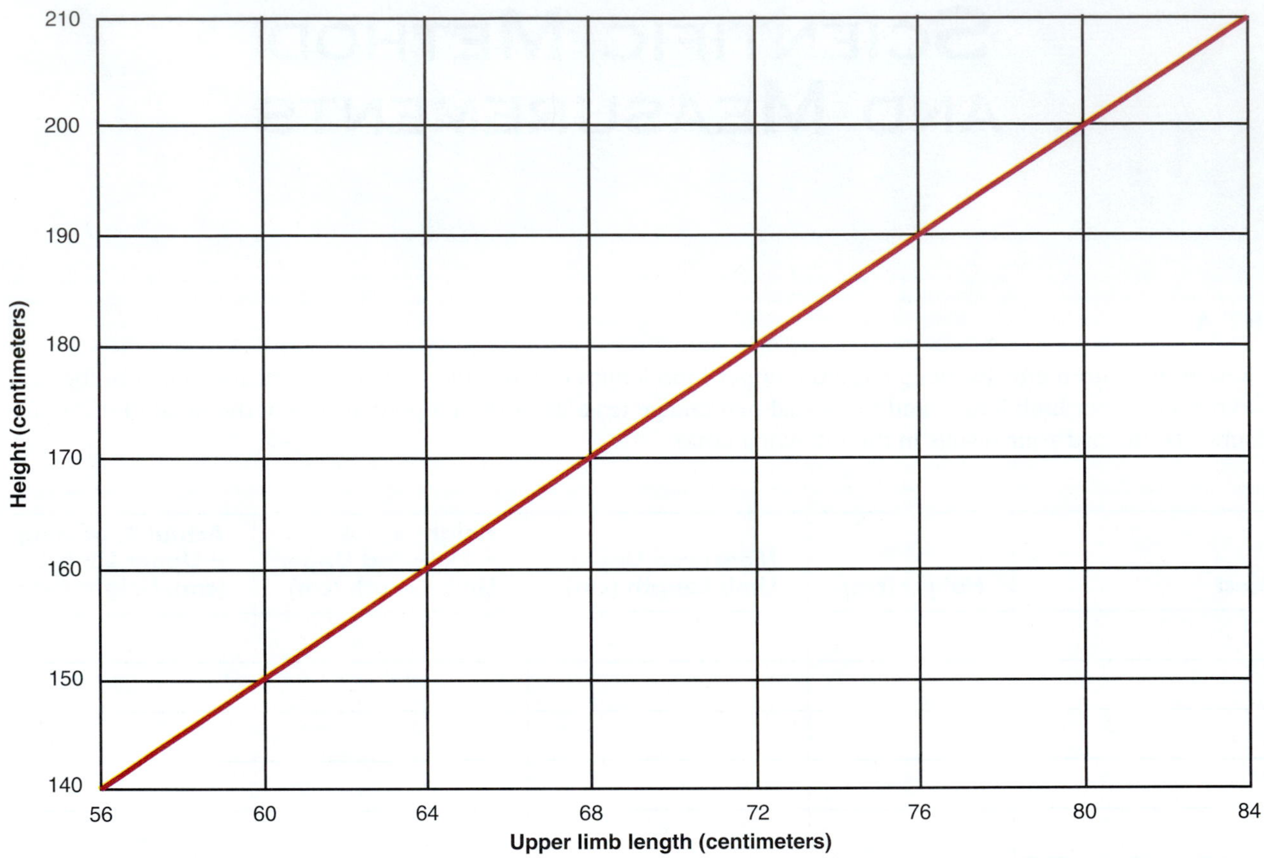

3. Does the distribution of the ten subjects' measured upper limb length support or disprove the original hypothesis? ___________ Explain your answer.

PART B

1. Describe your observations of a possible correlation between the radius length to height.

2. Write a hypothesis based on your observations.

3. Describe the design of the experiment that you devised to test your hypothesis.

4. Place your analysis of the data in this space in the form of a table and a graph.

5. Based from an analysis of your data, what conclusions can you make? Did these conclusions confirm or refute your original hypothesis?

6. Discuss your results and conclusions with other classmates. What common conclusion can the class formulate concerning the correlation between radius length to height?

BODY ORGANIZATION AND TERMINOLOGY

MATERIALS NEEDED

Textbook
Dissectible torso (manikin)
Variety of specimens or models sectioned along various planes

For Optional Activity:

Colored pencils

The major features of the human body include certain cavities, a set of membranes associated with these cavities, and a group of organ systems composed of related organs. In order to communicate effectively with each other about the body, scientists have devised names to describe these body features. They also have developed terms to represent the relative positions of body parts, imaginary planes passing through these parts, and body regions.

PURPOSE OF THE EXERCISE

To review the organizational pattern of the human body, to review its organ systems and the organs included in each system, and to become acquainted with the terms used to describe the relative position of body parts, body sections, and body regions.

LEARNING OBJECTIVES

After completing this exercise, you should be able to

1. locate and name the major body cavities and identify the membranes associated with each cavity;
2. name the organ systems of the human organism;
3. list the organs included within each system and locate the organs in a dissectible torso;
4. describe the general functions of each system;
5. define the terms used to describe the relative positions of body parts;
6. define the terms used to identify body sections and identify the plane along which a particular specimen is cut;
7. define the terms used to identify body regions.

PROCEDURE A—BODY CAVITIES AND MEMBRANES

1. Review the sections entitled "Body Cavities" and "Thoracic and Abdominopelvic Membranes" in chapter 1 of the textbook.
2. As a review activity, label figures 2.1, 2.2, and 2.3.
3. Locate the following features on the reference plates near the end of chapter 1 of the textbook and on the dissectible torso:

dorsal cavity
- cranial cavity
- vertebral canal (spinal cavity)

ventral cavity
- thoracic cavity
  - mediastinum
  - pleural cavity
- abdominopelvic cavity
  - abdominal cavity
  - pelvic cavity

diaphragm

smaller cavities within the head
- oral cavity
- nasal cavity with connected sinuses
- orbital cavity
- middle ear cavity

membranes and cavities
- pleural cavity
  - parietal pleura
  - visceral pleura
- pericardial cavity
  - parietal pericardium (covered by fibrous pericardium)
  - visceral pericardium (epicardium)

Figure 2.1 Label the major body cavities.

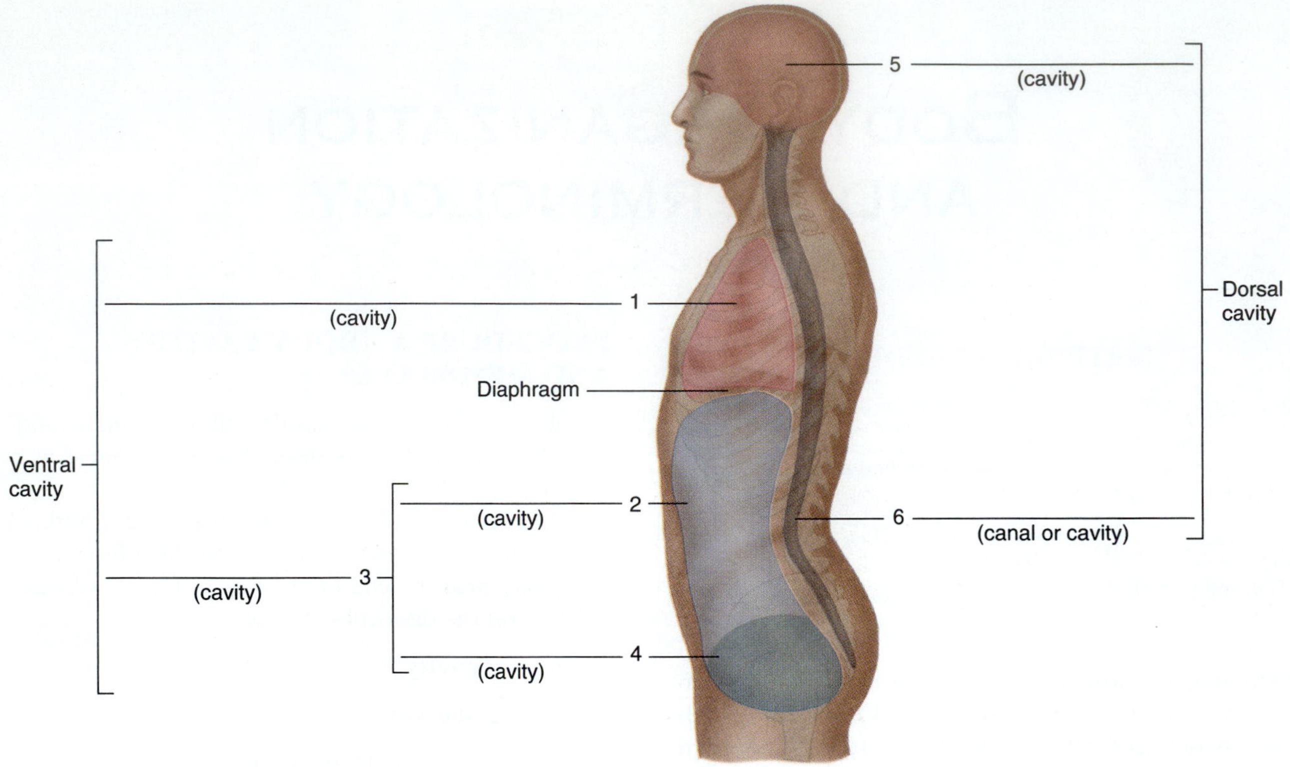

peritoneal cavity

parietal peritoneum

visceral peritoneum

4. Complete Parts A and B of Laboratory Report 2.

PROCEDURE B—ORGAN SYSTEMS

1. Review the section entitled "Organ Systems" in chapter 1 of the textbook.
2. Use the reference plates near the end of chapter 1 of the textbook and the dissectible torso to locate the following organs:

integumentary system

skin

accessory organs such as hair and nails

skeletal system

bones

ligaments

cartilages

muscular system

skeletal muscles

tendons

nervous system

brain

spinal cord

nerves

endocrine system

pituitary gland

thyroid gland

adrenal glands

pancreas

ovaries

testes

thymus gland

Figure 2.2 Label the smaller cavities and sinuses within the head.

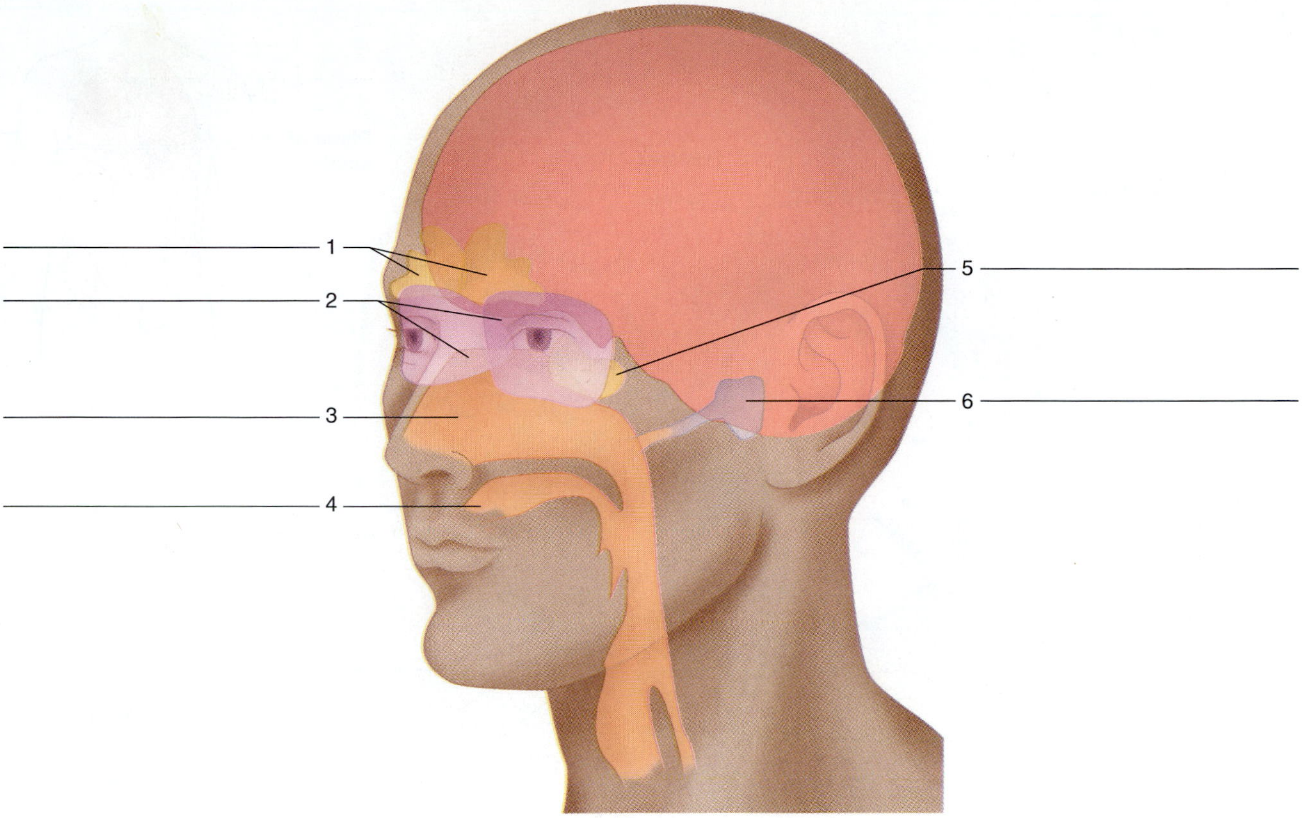

cardiovascular system

heart

arteries

veins

lymphatic system

lymphatic vessels

lymph nodes

thymus gland

spleen

digestive system

mouth

tongue

teeth

salivary glands

pharynx

esophagus

stomach

liver

gallbladder

pancreas

small intestine

large intestine

respiratory system

nasal cavity

pharynx

larynx

trachea

bronchi

lungs

urinary system

kidneys

ureters

urinary bladder

urethra

Figure 2.3 Label the thoracic membranes and cavities in (*a*) and the abdominopelvic membranes and cavity in (*b*) as shown in these superior views of transverse sections.

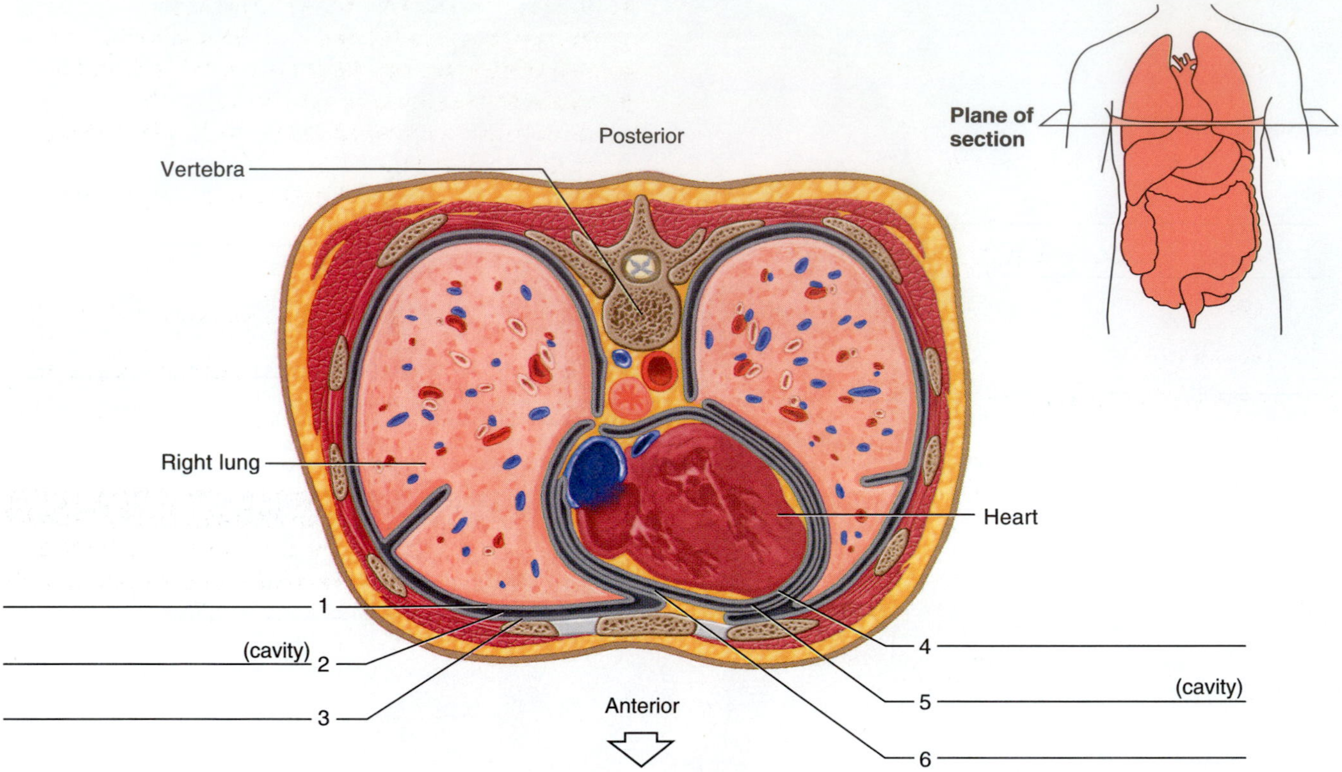

(a)

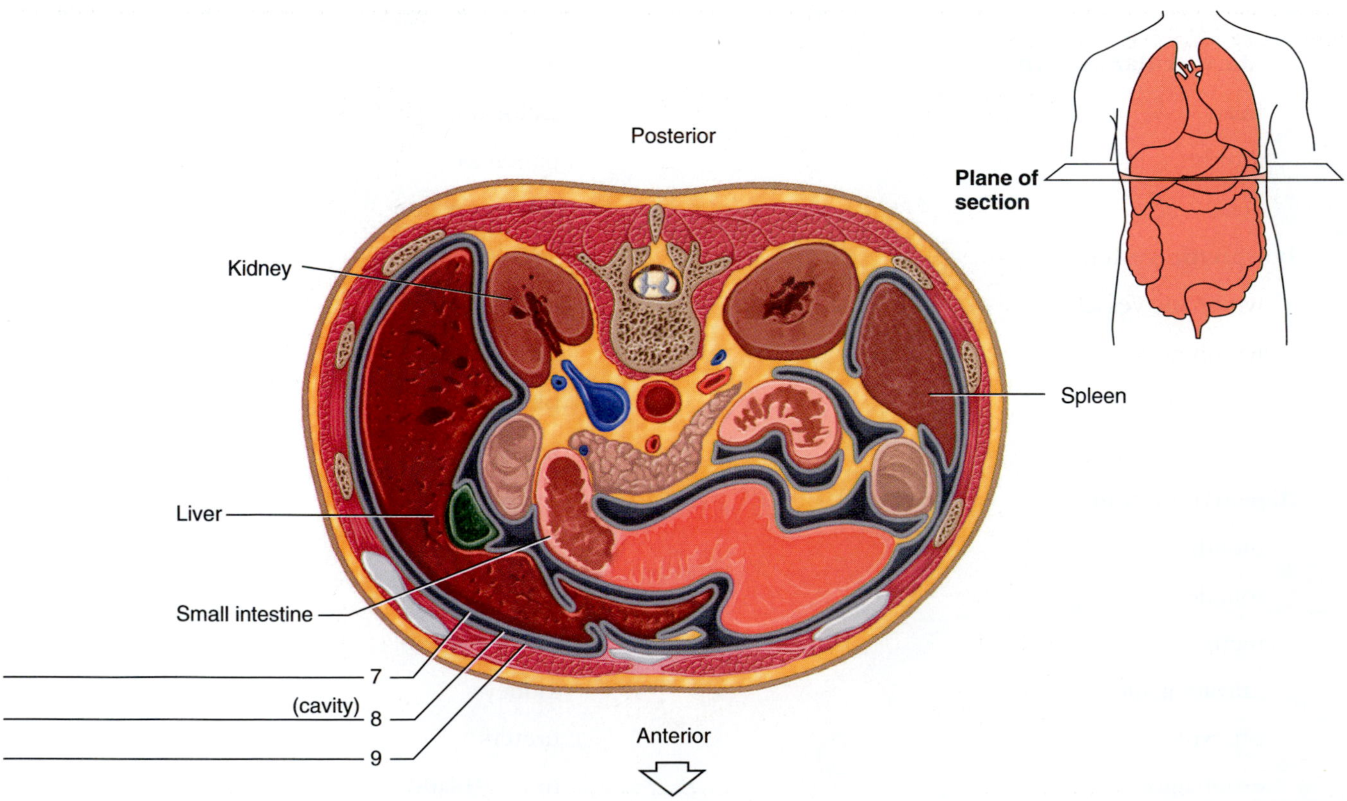

(b)

male reproductive system

scrotum

testes

penis

urethra

female reproductive system

ovaries

uterine tubes (oviducts; fallopian tubes)

uterus

vagina

3. Complete Part C of the laboratory report.

PROCEDURE C—RELATIVE POSITIONS, PLANES, SECTIONS, AND REGIONS

1. Review the section entitled "Anatomical Terminology" in chapter 1 of the textbook.
2. As a review activity, label figures 2.4, 2.5, and 2.6.
3. Examine the sectioned specimens on the demonstration table and identify the plane along which each is cut.
4. Complete Parts D, E, F, and G of the laboratory report.

OPTIONAL ACTIVITY

Use different colored pencils to distinguish the body regions in figure 2.6.

Web Quest

Identify organs in the three planes of the body from digitized sections of a human body. Search these at www.mhhe.com/shieress8

Figure 2.4 Label the planes represented in this illustration.

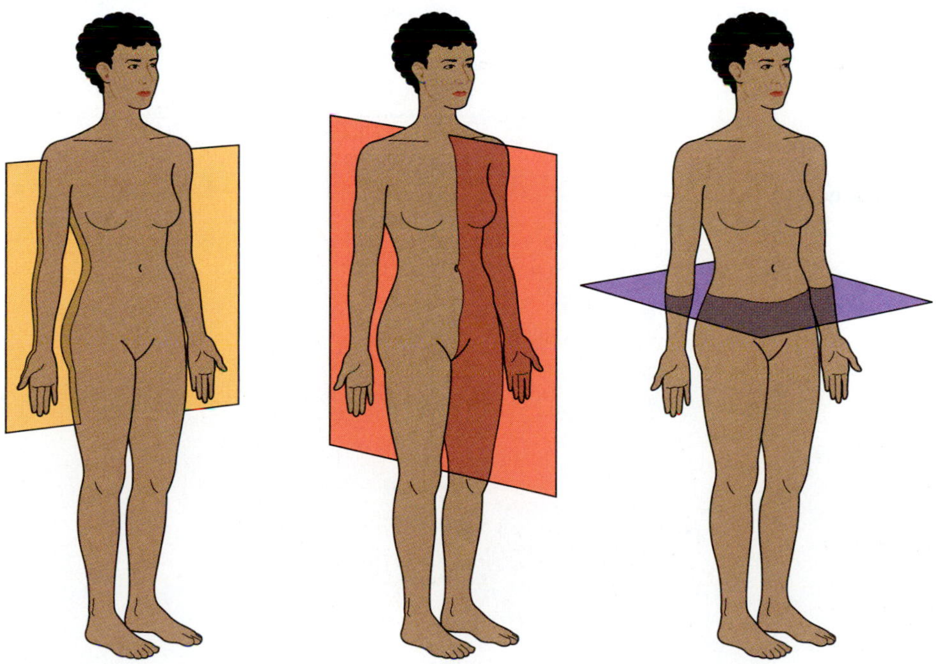

(1) ____________ (2) ____________ (3) ____________

Figure 2.5 Label (*a*) the regions and compare them to (*b*) the quadrants of the abdominal area.

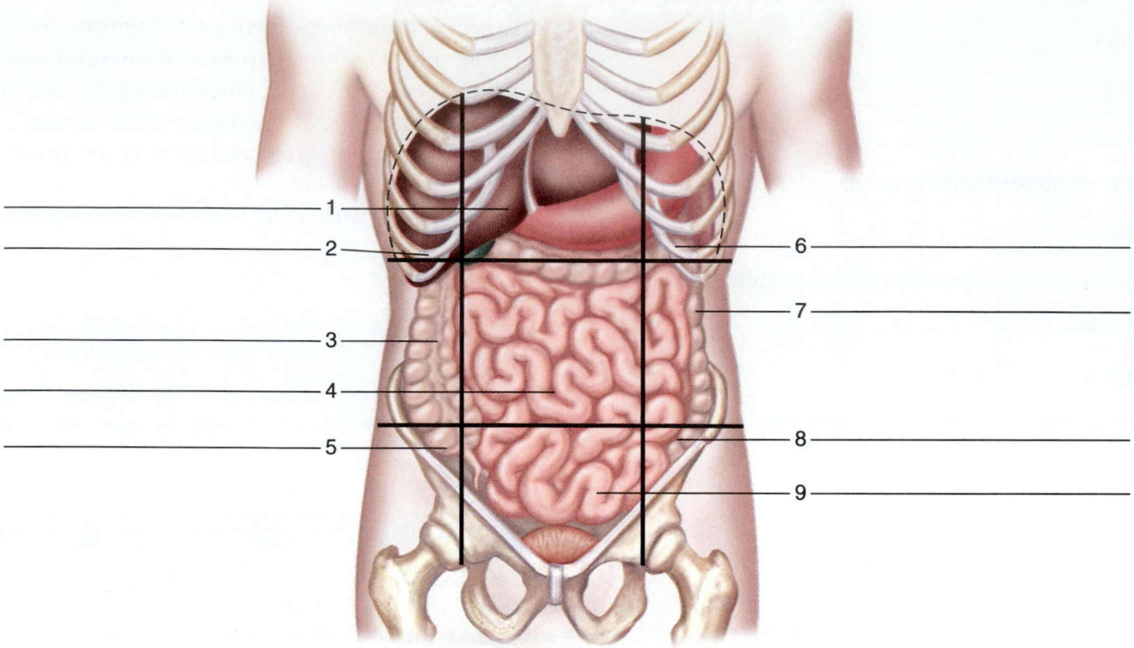

(a)

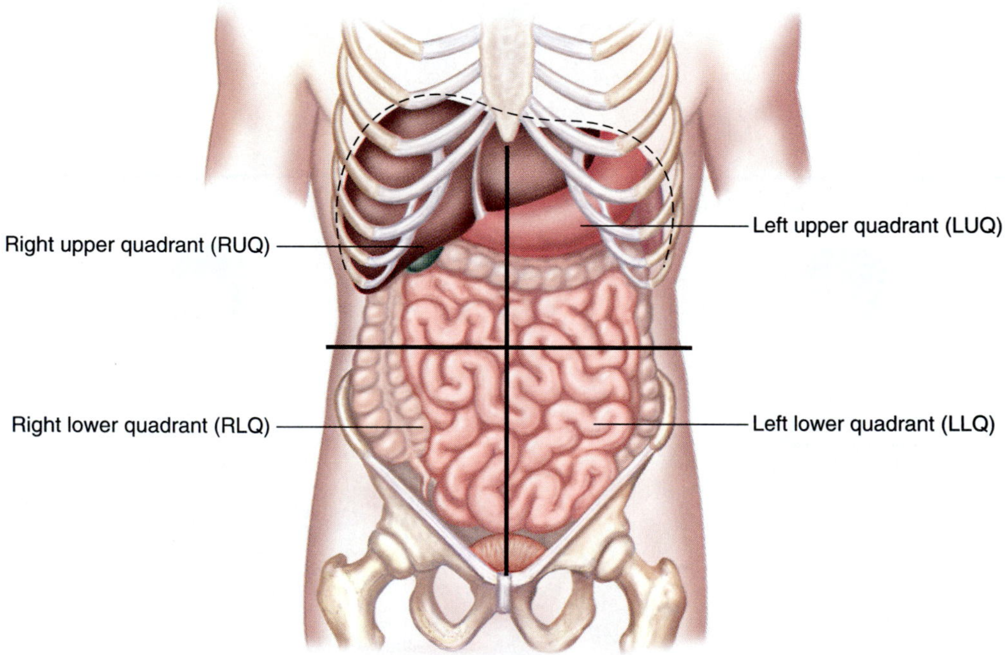

(b)

Figure 2.6 Label these diagrams with terms used to describe body regions: (*a*) anterior regions; (*b*) posterior regions.

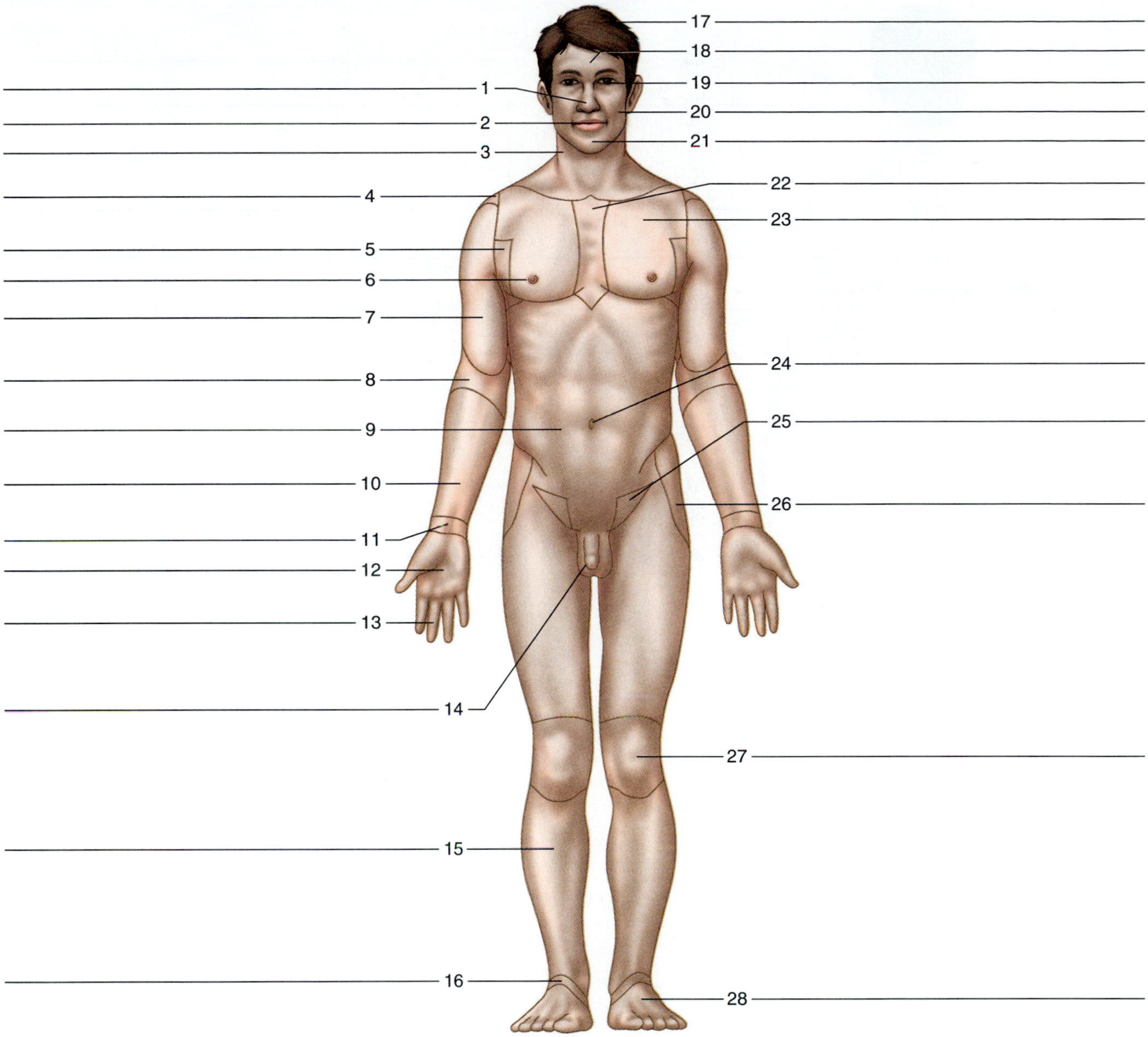

(a)

Figure 2.6 Continued.

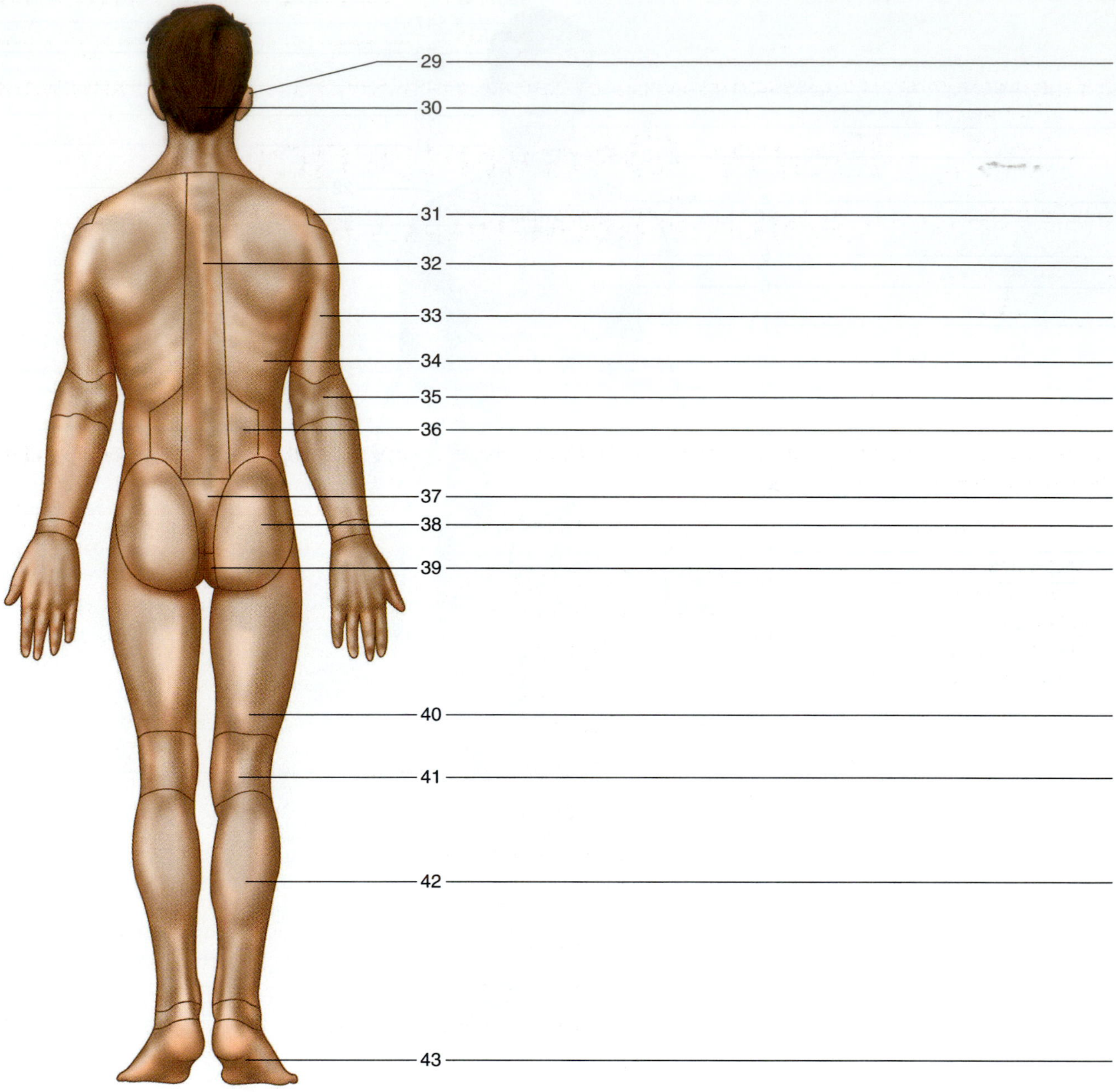

(b)

Laboratory Report **2**

Name ______________________________

Date ______________________________

Section ______________________________

Body Organization and Terminology

PART A

Match the cavities in column A with the organs contained in the cavities in column B. Place the letter of your choice in the space provided.

Column A

a. Abdominal cavity
b. Cranial cavity
c. Pelvic cavity
d. Thoracic cavity
e. Vertebral canal (spinal cavity)

Column B

_______ 1. Liver
_______ 2. Lungs
_______ 3. Spleen
_______ 4. Stomach
_______ 5. Brain
_______ 6. Trachea and esophagus
_______ 7. Gallbladder
_______ 8. Urinary bladder
_______ 9. Small intestine
_______ 10. Spinal cord
_______ 11. Thymus gland
_______ 12. Internal reproductive organs
_______ 13. Heart
_______ 14. Mediastinum

PART B

Complete the following statements:

1. The membrane on the surface of the lung is called the ______________________________ .
2. The membrane on the surface of the heart is called the ______________________________ .
3. The membrane that lines the wall of the abdominopelvic cavity is called the ______________________________ .
4. The membrane on the surface of the stomach (in the abdominal cavity) is called the ______________________________ .
5. The region between the lungs that separates the thoracic cavity into two compartments is called the ______________________________ .
6. The muscular structure that separates the thoracic and abdominopelvic cavities is called the ______________________________ .

PART C

Match the organ systems in column A with the general functions in column B. Place the letter of your choice in the space provided.

Column A	Column B
a. Cardiovascular system	_______ 1. Main system that secretes hormones
b. Digestive system	_______ 2. Provides an outer covering
c. Endocrine system	_______ 3. Produces a new organism
d. Integumentary system	_______ 4. Stimulates muscles to contract and interprets information from sensory units
e. Lymphatic system	_______ 5. Provides a framework for soft tissues and produces blood cells in red marrow
f. Muscular system	_______ 6. Exchanges gases between air and blood
g. Nervous system	_______ 7. Transports excess fluid from tissues to blood
h. Reproductive system	_______ 8. Maintains posture and generates most body heat
i. Respiratory system	_______ 9. Removes liquid wastes from blood and transports them to outside
j. Skeletal system	_______ 10. Converts food molecules into forms that are absorbed
k. Urinary system	_______ 11. Transports nutrients, wastes, and gases throughout the body

PART D

Indicate if each of the following sentences makes correct or incorrect usage of the word in boldface type (assume that the body is in the anatomical position). If the sentence is incorrect, supply a term that will make it correct in the space provided.

1. The mouth is **superior** to the nose. _______________
2. The stomach is **inferior** to the diaphragm. _______________
3. The trachea is **anterior** to the spinal cord. _______________
4. The larynx is **posterior** to the esophagus. _______________
5. The heart is **medial** to the lungs. _______________
6. The kidneys are **inferior** to the adrenal glands. _______________
7. The hand is **proximal** to the elbow. _______________
8. The knee is **proximal** to the ankle. _______________
9. Blood in **deep** blood vessels gives color to the skin. _______________
10. A **peripheral** nerve passes from the spinal cord into the limbs. _______________
11. The dermis is the **superficial** layer of the skin. _______________

Figure 2.7 Name the sections represented in these diagrams.

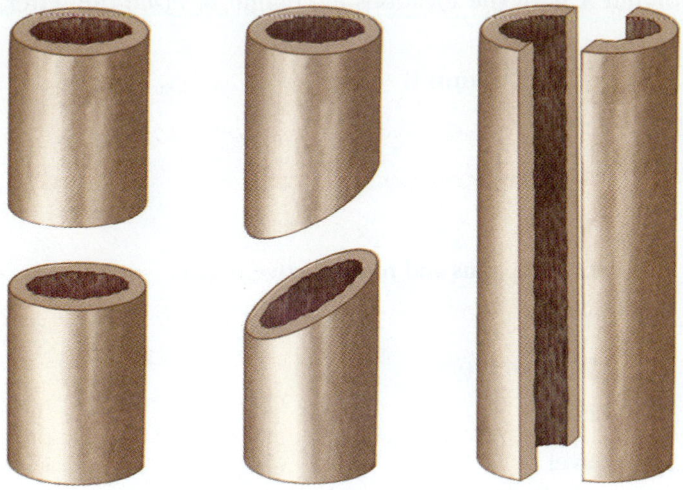

PART E

Name each of the sections represented in figure 2.7.

1. ______________________________
2. ______________________________
3. ______________________________

PART F

Match the body regions in column A with the body parts in column B. Place the letter of your choice in the space provided.

Column A		Column B
a. Antebrachial	______	1. Wrist
b. Antecubital	______	2. Ribs
c. Axillary	______	3. Reproductive organs
d. Brachial	______	4. Armpit
e. Buccal	______	5. Elbow
f. Carpal	______	6. Forehead
g. Cephalic	______	7. Buttocks
h. Cervical	______	8. Forearm
i. Costal	______	9. Back
j. Crural	______	10. Neck
k. Cubital	______	11. Arm
l. Dorsal	______	12. Cheek
m. Frontal	______	13. Leg
n. Genital	______	14. Head
o. Gluteal	______	15. Space in front of elbow

PART G

Match the body regions in column A with the locations in column B. Place the letter of your choice in the space provided.

Column A

a. Inguinal
b. Lumbar
c. Mammary
d. Mental
e. Occipital
f. Otic
g. Palmar
h. Pectoral
i. Pedal
j. Pelvic
k. Perineal
l. Plantar
m. Popliteal
n. Sternal
o. Umbilical

Column B

_______ 1. Pelvis
_______ 2. Breasts
_______ 3. Ear
_______ 4. Between anus and reproductive organs
_______ 5. Sole
_______ 6. Middle of thorax
_______ 7. Chest
_______ 8. Navel
_______ 9. Chin
_______ 10. Behind knee
_______ 11. Foot
_______ 12. Lower posterior region of head
_______ 13. Abdominal wall near thigh
_______ 14. Lower back
_______ 15. Palm

PART H

Critical Thinking Application

State the quadrant of the abdominopelvic cavity where the pain or sound would be located for each of the six common conditions listed. In some cases, there may be more than one correct answer, and pain is sometimes referred to another region. This phenomenon, called *referred pain,* occurs when pain is interpreted as originating from some area other than the parts being stimulated.

1. Stomach ulcer _______________
2. Appendicitis _______________
3. Bowel sounds _______________
4. Gallbladder attack _______________
5. Kidney stone in left ureter _______________
6. Ruptured spleen _______________

CARE AND USE OF THE COMPOUND MICROSCOPE

MATERIALS NEEDED

Compound microscope
Lens paper
Microscope slides
Coverslips
Transparent plastic millimeter ruler
Slide of three colored threads
Medicine dropper
Dissecting needle (needle probe)
Specimen examples for wet mounts
Methylene blue (dilute) or iodine-potassium-iodide stain

For Demonstrations:

Micrometer scale
Stereomicroscope (dissecting microscope)

Because the human eye is unable to perceive objects less than 0.1 mm in diameter, a microscope is an essential tool for the study of small structures such as cells. The microscope usually used for this purpose is the *compound microscope.* It is called compound because it utilizes two sets of lenses: an eyepiece lens system and an objective lens system. The eyepiece lens system magnifies, or compounds, the image reaching it after being magnified by the objective lens system. Such an instrument can magnify images of small objects up to about one thousand times.

PURPOSE OF THE EXERCISE

To become familiar with the major parts of a compound microscope and their functions and to make use of the compound microscope to observe small objects.

LEARNING OBJECTIVES

After completing this exercise, you should be able to

1. locate and identify the major parts of a compound microscope;
2. describe the functions of these parts;
3. calculate the total magnification produced by various combinations of eyepiece and objective lenses;
4. prepare a simple microscope slide;
5. make proper use of the microscope to observe small objects.

PROCEDURE

1. Familiarize yourself with the following list of rules for care of the microscope:
   a. Handle the microscope with great care. It is an expensive and delicate instrument. To move it or carry it, hold it by its *arm* with one hand and support its *base* with the other hand (fig. 3.1).
   b. Keep the microscope under its *dustcover* and in a cabinet when it is not being used.
   c. To clean the lenses, rub them gently with *lens paper* or a high-quality cotton swab. If the lenses need additional cleaning, follow the directions in the "Lens-Cleaning Technique" section that follows.
   d. If the microscope has a substage lamp, be sure the electric cord does not hang off the laboratory table where someone might trip over it. The bulb life can be extended if the lamp is cool before the microscope is moved.
   e. Never drag the microscope across the laboratory table after its placement.
   f. Never remove parts of the microscope or try to disassemble the eyepiece or objective lenses.
   g. If your microscope is not functioning properly, report the problem to your laboratory instructor immediately.

LENS-CLEANING TECHNIQUE

1. Moisten one end of a high-quality cotton swab with one drop of Kodak lens cleaner. Keep the other end dry.
2. Clean the optical surface with the wet end. Dry it with the other end, using a circular motion.
3. Use a hand aspirator to remove lingering dust particles.
4. Start with the scanning objective and work upward in magnification, using a new cotton swab for each objective.
5. When cleaning the eyepiece, do not open the lens unless it is absolutely necessary.
6. Use alcohol for difficult cleaning, and only as a last resort use xylene. Regular use of xylene will destroy lens coatings.

Figure 3.1 Major parts of a compound microscope with a monocular body and a mechanical stage. Some compound microscopes are equipped with a binocular body.

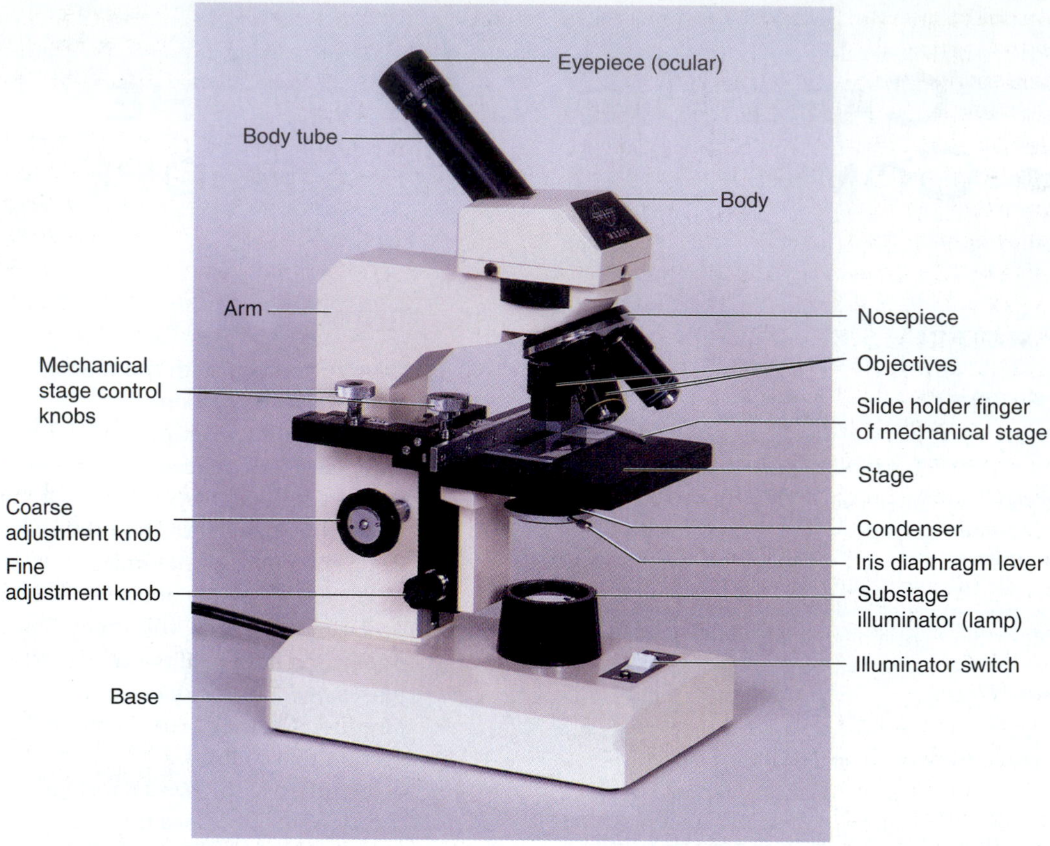

Figure 3.2 The powers of this 10× eyepiece (*a*) and this 40× objective (*b*) are marked in the metal. DIN is an international optical standard on quality optics. The 0.65 on the 40× objective is the numerical aperture, which is a measure of the light-gathering capabilities.

(a)

(b)

2. Observe a compound microscope and study figure 3.1 to learn the names of its major parts. Note that the lens system of a compound microscope includes three parts—the condenser, objective lens, and eyepiece.

   Light enters this system from a *substage illuminator* (*lamp*) or *mirror* and is concentrated and focused by a *condenser* onto a microscope slide or specimen placed on the *stage.* The condenser, which contains a set of lenses, usually is kept in its highest position possible.

   The *iris diaphragm,* which is located between the light source and the condenser, can be used to increase or decrease the intensity of the light entering the condenser. Locate the lever that operates the iris diaphragm beneath the stage and move it back and forth. Note how this movement causes the size of the opening in the diaphragm to change. (Some microscopes have a revolving plate called a disc diaphragm beneath the stage instead of an iris diaphragm. Disc diaphragms have different-sized holes to admit varying amounts of light.) Which way do you move the diaphragm to increase the light intensity?________________ Which way to decrease it?________________________

   After light passes through a specimen mounted on a microscope slide, it enters an *objective lens system.* This lens projects the light upward into the *body tube,* where it produces a magnified image of the object being viewed.

   The *eyepiece (ocular) lens* system then magnifies this image to produce another image, which is seen by the eye. Typically, the eyepiece lens magnifies the image ten times (10×). Look for the number in the metal of the eyepiece that indicates its power (fig. 3.2). What is the eyepiece power of your microscope?__________________

   The objective lenses are mounted in a revolving *nosepiece* so that different magnifications can be achieved by rotating any one of several objective lenses into position above the specimen. Commonly, this set of lenses includes a scanning objective (4×), a low-power objective (10×), and a high-power objective, also called a high-dry-power objective (about 40×). Sometimes an oil immersion objective (about 100×) is present. Look for the number printed on each objective that indicates its power. What are the objective lens powers of your microscope? ________________

   To calculate the *total magnification* achieved when using a particular objective, multiply the power of the eyepiece by the power of the objective used. Thus, the 10× eyepiece and the 40× objective produce a total magnification of 10×40, or 400×.
3. Complete Part A of Laboratory Report 3.
4. Turn on the substage illuminator and look through the eyepiece. You will see a lighted circular area called the *field of view.*

   You can measure the diameter of this field of view by focusing the lenses on the millimeter scale of a transparent plastic ruler. To do this, follow these steps:
   a. Place the ruler on the microscope stage in the spring clamp of a slide holder finger on a mechanical stage or under the stage (slide) clips. (*Note:* If your microscope is equipped with a mechanical stage, it may be necessary to use a short section cut from a transparent plastic ruler. The section should be several millimeters long and can be mounted on a microscope slide for viewing.)
   b. Center the millimeter scale in the beam of light coming up through the condenser and rotate the scanning objective into position.
   c. While you watch from the side to prevent the lens from touching anything, lower the objective until it is as close to the ruler as possible, using the *coarse adjustment knob* and then using the *fine adjustment knob* (fig. 3.3).

Figure 3.3 When you focus using a particular objective, you can prevent it from touching the specimen by watching from the side.

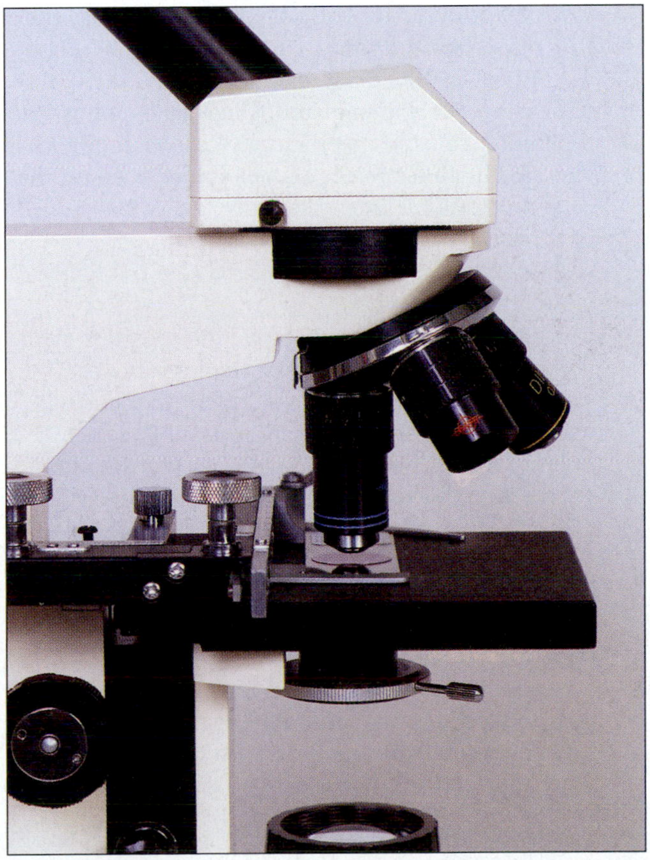

(*Note:* The adjustment knobs on some microscopes move the stage upward and downward for focusing.)

d. Look into the eyepiece and use the coarse adjustment knob to raise the objective lens until the lines of the millimeter scale come into sharp focus.

e. Adjust the light intensity by moving the *iris diaphragm lever* so that the field of view is brightly illuminated but comfortable to your eye. At the same time, take care not to overilluminate the field because transparent objects tend to disappear in very bright light.

f. Position the millimeter ruler so that its scale crosses the greatest diameter of the field of view. Also, move the ruler so that one of the millimeter marks is against the edge of the field of view.

g. Measure the distance across the field of view in millimeters.

5. Complete Part B of the laboratory report.

6. Most microscopes are designed to be *parfocal.* This means that when a specimen is in focus with a lower-power objective, it will be in focus (or nearly so) when a higher-power objective is rotated into position. Always center the specimen in the field of view before changing to higher objectives.

   Rotate the low-power objective into position, and then look at the millimeter scale of the transparent plastic ruler. If you need to move the low-power objective to sharpen the focus, use the *fine adjustment knob.*

   Adjust the iris diaphragm so that the field of view is properly illuminated. Once again, adjust the millimeter ruler so that the scale crosses the field of view through its greater diameter, and position the ruler so that a millimeter mark is against one edge of the field.

   Try to measure the distance across the field of view in millimeters.

7. Rotate the high-power objective into position while you watch from the side, and then observe the millimeter scale on the plastic ruler. All focusing using high-power magnification should be done only with the fine adjustment knob. If you use the coarse adjustment knob with the high-power objective, you can accidently force the objective into the coverslip. This is because the *working distance* (the distance from the objective lens to the slide on the stage) is much shorter when using higher magnifications.

   Adjust the iris diaphragm for proper illumination. Usually more illumination when using higher magnifications will help to view the objects more clearly. Try to measure the distance across the field of view in millimeters.

8. Locate the numeral 4 (or 9) on the plastic ruler and focus on it using the scanning objective. Note how the number appears in the field of view. Move the plastic ruler to the right and note which way the image moves. Slide the ruler away from you and again note how the image moves.

9. Examine the slide of the three colored threads using the low-power objective and then the high-power objective. Focus on the location where the three threads cross. By using the fine adjustment knob, determine the order from top to bottom by noting which color is in focus at different depths. The other colored threads will still be visible, but they will be blurred. Be sure to notice whether the stage or the body tube moves up and down with the adjustment knobs of the microscope that is being used for this depth determination. The vertical depth of a specimen that is clearly in focus is called the *depth of field (focus).* Whenever specimens are examined, continue to use the fine adjustment focusing knob to determine relative depths of structures that are clearly in focus within cells, giving a three-dimensional perspective. It should be noted that the depth of field is less at higher magnifications.

Critical Thinking Application

What was the sequence of the three colored threads from top to bottom?

____

____

____

10. Complete Parts C and D of the laboratory report.

DEMONSTRATION

A compound microscope is sometimes equipped with a micrometer scale mounted in the eyepiece. Such a scale is subdivided into fifty to one hundred equal divisions (fig. 3.4). These arbitrary divisions can be calibrated against the known divisions of a micrometer slide placed on the microscope stage. Once the values of the divisions are known, the length and width of a microscopic object can be measured by superimposing the scale over the magnified image of the object.

Observe the micrometer scale in the eyepiece of the demonstration microscope. Focus the low-power objective on the millimeter scale of a micrometer slide (or a plastic ruler) and measure the distance between the divisions on the micrometer scale in the eyepiece. What is the distance between the finest divisions of the scale in micrometers? ______________________

Figure 3.4 The divisions of a micrometer scale in an eyepiece can be calibrated against the known divisions of a micrometer slide. (Courtesy of Swift Instruments, Inc., San Jose, California)

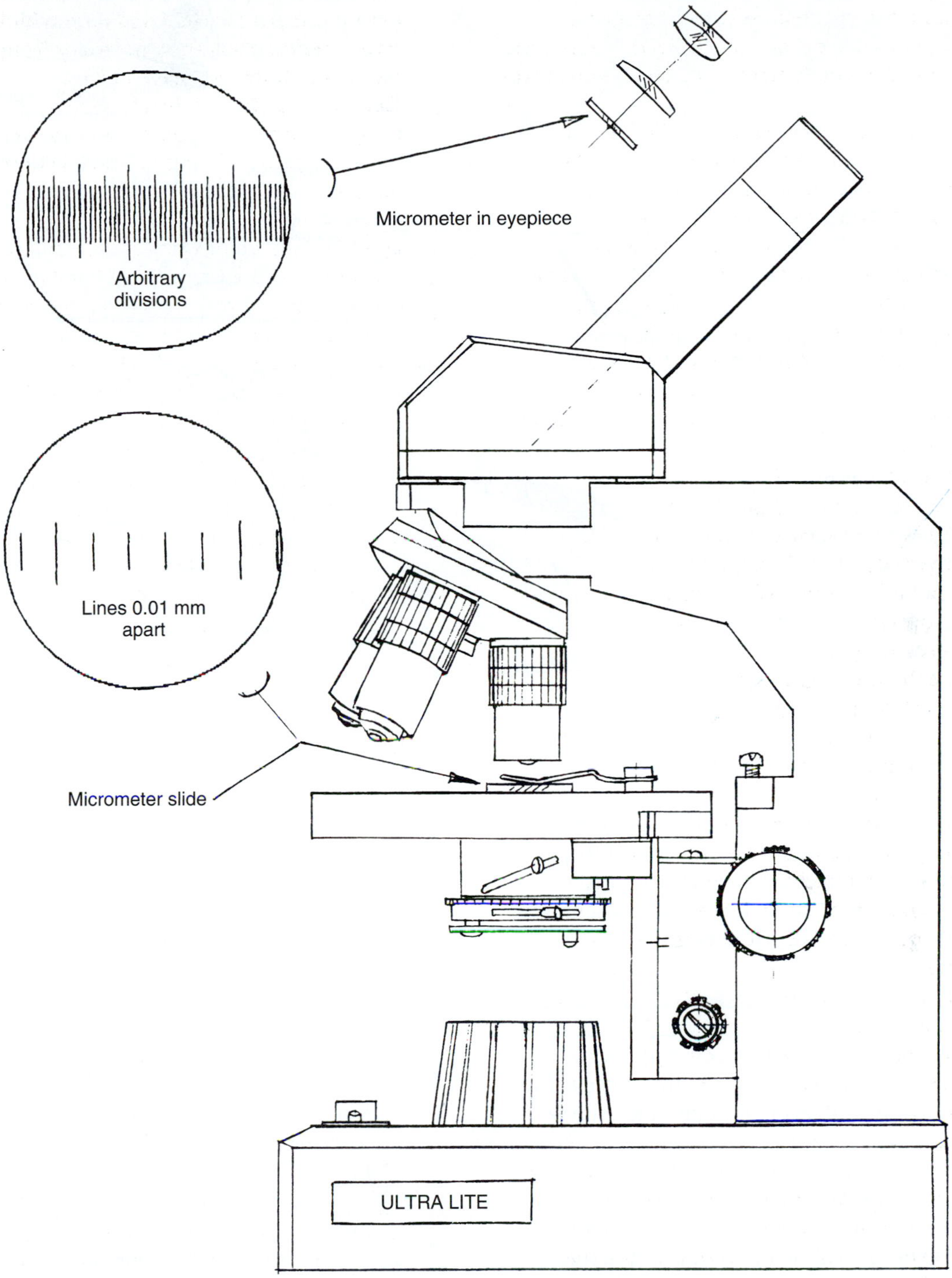

11. Prepare several temporary *wet mounts,* using any small, transparent objects of interest, and examine the specimens using the low-power objective and then a high-power objective to observe their details. To prepare a wet mount, follow these steps (fig. 3.5):

   a. Obtain a precleaned microscope slide.
   b. Place a tiny, thin piece of the specimen you want to observe in the center of the slide, and use a medicine dropper to put a drop of water over it. Consult with your instructor if a drop of stain might enhance the image of any cellular

Figure 3.5 Steps in the preparation of a wet mount.

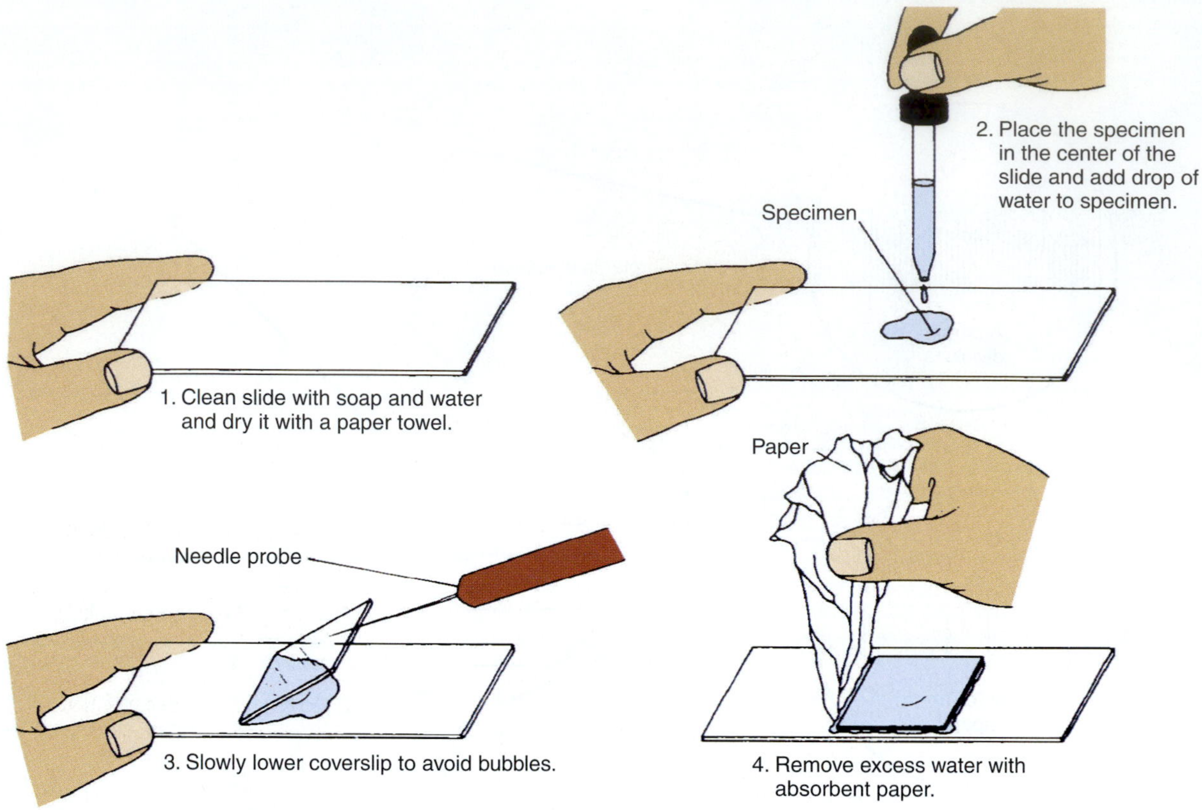

structures of your specimen. If the specimen is solid, you might want to tease some of it apart with dissecting needles. In any case, the specimen must be thin enough so that light can pass through it. Why is it necessary for the specimen to be so thin?

____

____

c. Cover the specimen with a coverslip. Try to avoid trapping bubbles of air beneath the coverslip by slowly lowering it at an angle into the drop of water.
d. Remove any excess water from the edge of the coverslip with absorbent paper. If your microscope has an inclination joint, do not tilt the microscope while observing wet mounts.
e. Place the slide under the stage (slide) clips or in the slide holder on a mechanical stage and position the slide so that the specimen is centered in the light beam passing up through the condenser.
f. Focus on the specimen using the scanning objective first. Next focus using the low-power objective, and then examine it with the high-power objective.

12. If an oil immersion objective is available, use it to examine the specimen. To use the oil immersion objective, follow these steps:
    a. Center the object you want to study under the high-power field of view.
    b. Rotate the high-power objective away from the microscope slide, place a small drop of immersion oil on the coverslip, and swing the oil immersion objective into position. To achieve sharp focus, use the fine adjustment knob only.
    c. You will need to open the iris diaphragm more fully for proper illumination. More light is needed because the oil immersion objective covers a very small lighted area of the microscope slide.
    d. Because the oil immersion objective must be very close to the coverslip to achieve sharp focus, care must be taken to avoid breaking the coverslip or damaging the objective lens. For this reason, never lower the objective when you are looking into the eyepiece. Instead, always raise the objective to achieve focus, or prevent the objective from touching the coverslip by watching the microscope slide and coverslip from the side if the objective needs to be lowered.

Figure 3.6 A stereomicroscope, which is also called a dissecting microscope.

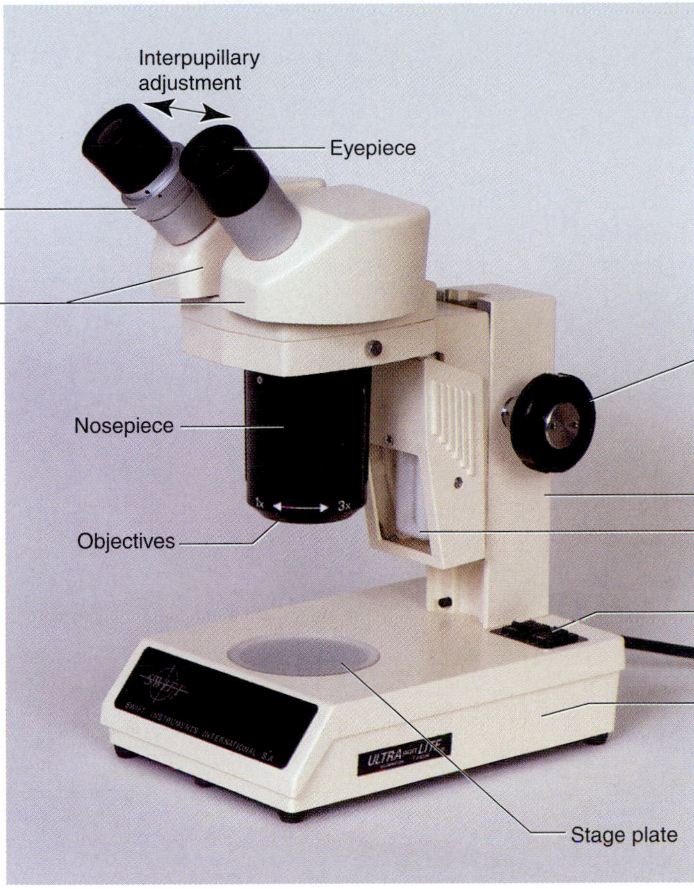

13. When you have finished working with the microscope, remove the microscope slide from the stage and wipe any oil from the objective lens with lens paper or a high-quality cotton swab. Swing the scanning objective or the low-power objective into position. Wrap the electric cord around the base of the microscope and replace the dustcover.
14. Complete Part E of the laboratory report.

DEMONSTRATION

A stereomicroscope (dissecting microscope) (fig. 3.6) is useful for observing the details of relatively large, opaque specimens. Although this type of microscope achieves less magnification than a compound microscope, it has the advantage of producing a three-dimensional image rather than the flat, two-dimensional image of the compound microscope. In addition, the image produced by the stereomicroscope is positioned in the same manner as the specimen, rather than being reversed and inverted as it is by the compound microscope.

Observe the stereomicroscope. Note that the eyepieces can be pushed apart or together to fit the distance between your eyes. Focus the microscope on the end of your finger. Which way does the image move when you move your finger to the right? ____________ When you move it away?____________________

If the instrument has more than one objective, change the magnification to higher power. Use the instrument to examine various small, opaque objects available in the laboratory.

Web Quest

Explore additional explanations on proper microscope use at www.mhhe.com/shieress8

Laboratory Report **3**

Name ______________________________

Date ______________________________

Section ____________________________

Care and Use of the Compound Microscope

PART A

Complete the following:

1. What total magnification will be achieved if the 10× eyepiece and the 10× objective are used? ____________
2. What total magnification will be achieved if the 10× eyepiece and the 100× objective are used? ____________

PART B

Complete the following:

1. Sketch the millimeter scale as it appears under the scanning objective magnification. (The circle represents the field of view through the microscope.)
2. What is the diameter of the scanning field of view in millimeters? ____________
3. Microscopic objects often are measured in *micrometers.* A micrometer equals 1/1,000 of a millimeter and is symbolized by µm. What is the diameter of the scanning power field of view in micrometers? ____________
4. If a circular object or specimen extends halfway across the scanning field, what is its diameter in millimeters? ____________
5. What is its diameter in micrometers? ____________

PART C

Complete the following:

1. Sketch the millimeter scale as it appears using the low-power objective.
2. What do you estimate the diameter of this field of view to be in millimeters? ____________
3. How does the diameter of the scanning power field of view compare with that of the low-power field?

____

4. Why is it more difficult to measure the diameter of the high-power field of view than the low-power field?

____

____

5. What change occurred in the light intensity of the field of view when you exchanged the low-power objective for the high-power objective? ____

6. Sketch the numeral 4 (or 9) as it appears through the scanning objective of the compound microscope.

7. What has the lens system done to the image of the numeral? (Is it right side up, upside down, or what?) _______________________

8. When you moved the ruler to the right, which way did the image move? __________

9. When you moved the ruler away, which way did the image move? __________

PART D

Match the names of the microscope parts in column A with the descriptions in column B. Place the letter of your choice in the space provided.

Column A

a. Adjustment knob
b. Arm
c. Condenser
d. Eyepiece (ocular)
e. Field of view
f. Iris diaphragm
g. Nosepiece
h. Objective lens system
i. Stage
j. Stage (slide) clip

Column B

_______ 1. Increases or decreases the light intensity
_______ 2. Platform that supports a microscope slide
_______ 3. Concentrates light onto the specimen
_______ 4. Causes objective lens (or stage) to move upward or downward
_______ 5. After light passes through specimen, it next enters this lens system
_______ 6. Holds a microscope slide in position
_______ 7. Contains a lens at the top of the body tube
_______ 8. Serves as a handle for carrying the microscope
_______ 9. Part to which the objective lenses are attached
_______ 10. Circular area seen through the eyepiece

PART E

Prepare sketches of the objects you observed using the microscope. For each sketch, include the name of the object, the magnification you used to observe it, and its estimated dimensions in millimeters and micrometers.

CELL STRUCTURE AND FUNCTION

MATERIALS NEEDED

Textbook
Animal cell model
Clean microscope slides
Coverslips
Flat toothpicks
Medicine dropper
Methylene blue (dilute) or iodine-potassium-iodide stain
Prepared microscope slides of human tissues
Compound microscope

For Optional Activities:

Single-edged razor blade
Plant materials such as leaves, soft stems, fruits, onion peel, and vegetables
Cultures of *Amoeba* and *Paramecium*

SAFETY

- Clean laboratory surfaces before and after laboratory procedures.
- Wear disposable gloves for the wet-mount procedures of the cells lining the inside of the cheek.
- Dispose of laboratory gloves, slides, coverslips, and toothpicks as instructed.
- Wash your hands before leaving the laboratory.

Cells are the "building blocks" from which all parts of the human body are formed. They account for the shape, organization, and construction of the body and are responsible for carrying on its life processes. Under the light microscope, with a properly applied stain to make structures visible, the **cell (plasma) membrane,** the **cytoplasm,** and a **nucleus** are easily seen. The cytoplasm is composed of a clear fluid, the *cytosol,* and numerous *cytoplasmic organelles* suspended in the cytosol.

PURPOSE OF THE EXERCISE

To review the structure and functions of major cellular components and to observe examples of human cells.

LEARNING OBJECTIVES

After completing this exercise, you should be able to

1. name and locate the components of a cell on a model or diagram and describe the general functions of these components;
2. prepare a wet mount of cells lining the inside of the cheek, stain the cells, and identify the major components of these cells;
3. locate cells on prepared slides of human tissues and identify their major components.

PROCEDURE

1. Review the section entitled "Composite Cell" in chapter 3 of the textbook.
2. Observe the animal cell model and identify its major parts.
3. As a review activity, label figures 4.1 and 4.2.
4. Complete Parts A and B of Laboratory Report 4.
5. Prepare a wet mount of cells lining the inside of the cheek. To do this, follow these steps:
   a. Gently scrape (force is not necessary and should be avoided) the inner lining of your cheek with the broad end of a flat toothpick.
   b. Stir the toothpick in a drop of water on a clean microscope slide and dispose of the toothpick as directed by your instructor.
   c. Cover the drop with a coverslip.
   d. Observe the cheek cells by using the microscope. Compare your image with figure 4.3. To report what you observe, sketch a single cell in the space provided in Part C of the laboratory report.
6. Prepare a second wet mount of cheek cells, but this time, add a drop of dilute methylene blue or iodine-potassium-iodide stain to the cells. Cover the liquid with a coverslip and observe the cells with the microscope. Add to your sketch any additional structures you observe in the stained cells.
7. Answer the questions in Part C of the laboratory report.

Figure 4.1 Label the structures of this composite cell.

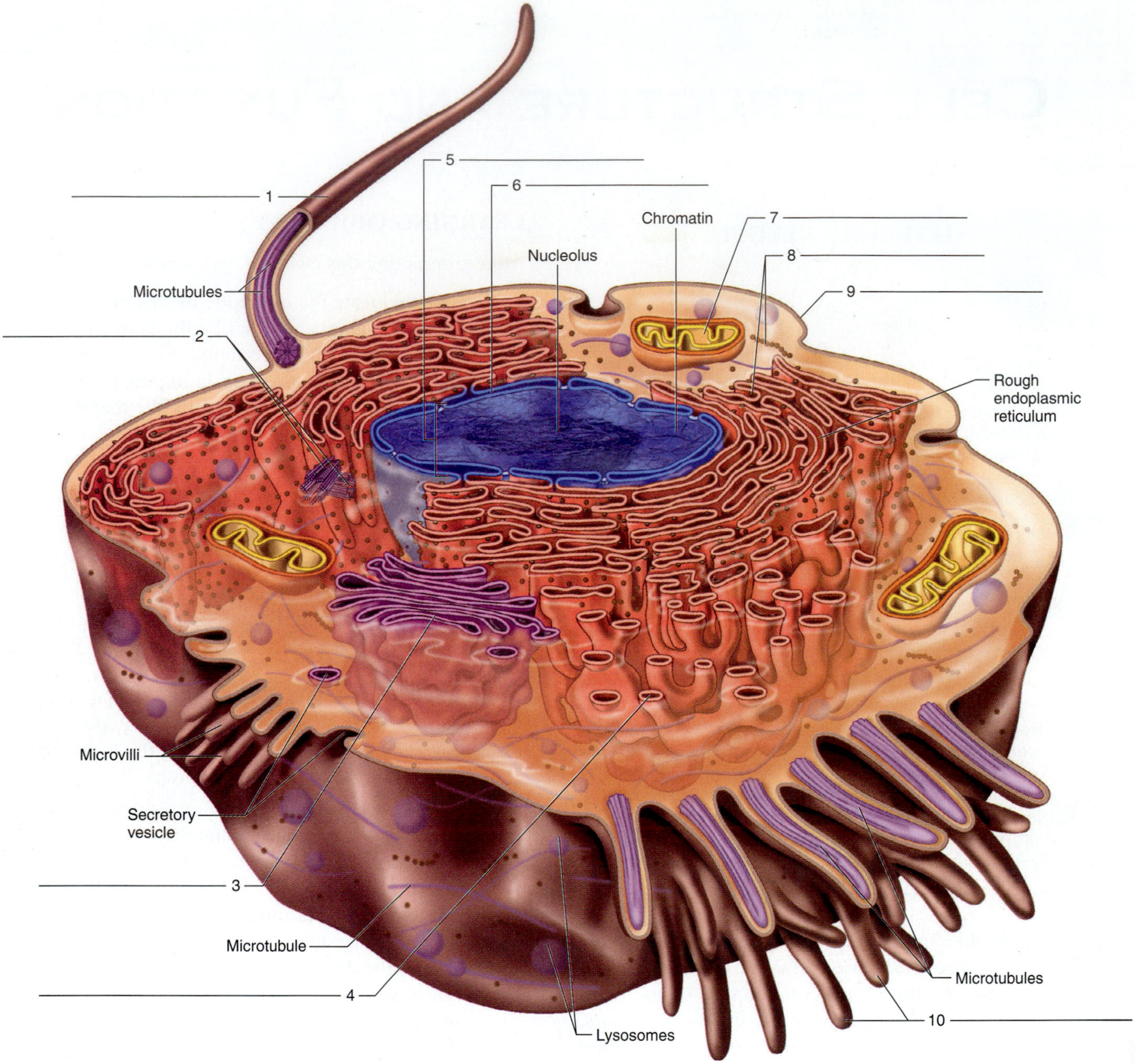

Figure 4.2 Label the structures of the cell membrane.

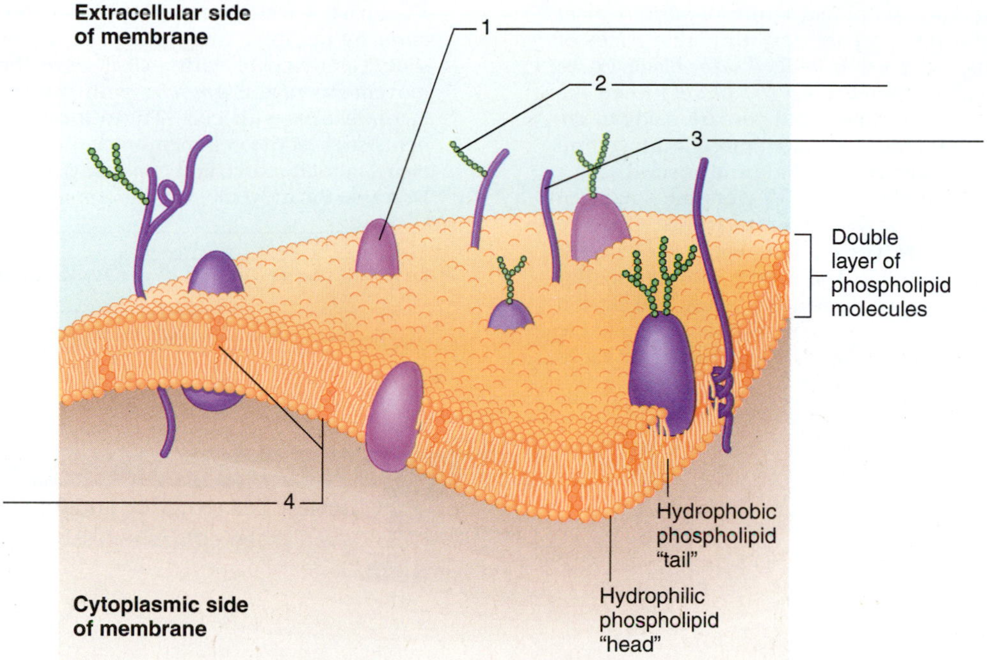

Figure 4.3 Stained cell lining the inside of the cheek as viewed through the compound microscope using the high-power objective (400×).

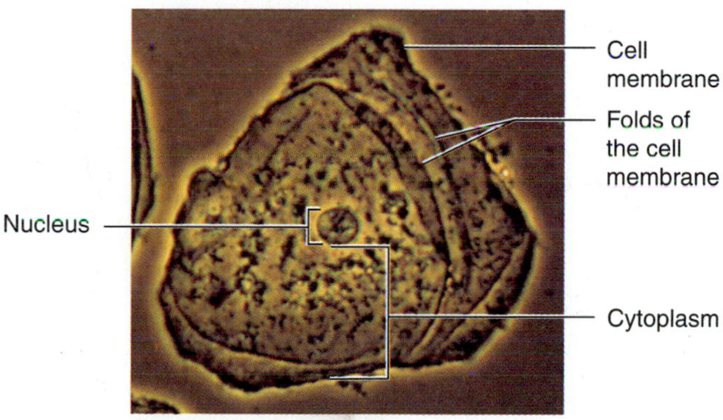

8. Using the microscope, observe each of the prepared slides of human tissues. To report what you observe, sketch a single cell of each type in the space provided in Part D of the laboratory report.
9. Answer the questions in Part D of the laboratory report.

Critical Thinking Application

The cells lining the inside of the cheek are frequently removed for making observations of basic cell structure. The cells are from stratified squamous epithelium. Explain a reason these cells are used instead of outer body surface tissue. Why was the removal of inside cheek cells painless and lacked any blood loss?

____

____

____

____

____

____

____

OPTIONAL ACTIVITY

Investigate the microscopic structure of various plant materials. To do this, prepare very tiny, thin slices of plant specimens, using a single-edged razor blade. (*Take care not to injure yourself with the blade.*) Keep the slices in a container of water until you are ready to observe them. To observe a specimen, place it into a drop of water on a clean microscope slide and cover it with a coverslip. Use the microscope and view the specimen using low- and high-power magnifications. Observe near the edges where your section of tissue is most likely to be one cell thick. Add a drop of dilute methylene blue or iodine-potassium-iodide stain, and note if any additional structures become visible. How are the microscopic structures of the plant specimens similar to the human tissues you observed? ____________________

____

How are they different? ____________________

____

OPTIONAL ACTIVITY

Prepare a wet mount of the *Amoeba* and *Paramecium* by putting a drop of the culture on a clean glass slide. Gently cover with a clean coverslip. Observe the movements of the *Amoeba* with pseudopodia and the *Paramecium* with cilia. Try to locate cellular components such as the cell membrane, nuclear envelope, nucleus, mitochondria, and contractile vacuoles.

Describe the movement of the *Amoeba.*

____

____

Describe the movement of the *Paramecium.*

____

____

Web Quest

Review cell structure and function at www.mhhe.com/shieress8

Laboratory Report **4**

Name ______________________

Date ______________________

Section ______________________

CELL STRUCTURE AND FUNCTION

PART A

Match the cellular components in column A with the descriptions in column B. Place the letter of your choice in the space provided.

Column A

a. Centrosome
b. Chromatin
c. Cilia
d. Cytoplasm
e. Endoplasmic reticulum
f. Flagellum
g. Golgi apparatus
h. Lysosome
i. Microfilament
j. Microtubule
k. Mitochondrion
l. Nuclear envelope
m. Nucleolus
n. Nucleus
o. Ribosome
p. Vesicle/vacuole

Column B

_______ 1. Loosely coiled fiber containing DNA and protein within nucleus
_______ 2. Location of the energy release from food molecules
_______ 3. Nonmembranous structure that contains the centrioles
_______ 4. Small RNA-containing particle for synthesis of proteins
_______ 5. Membranous sac formed by pinching off of pieces of cell membrane
_______ 6. Dense body of RNA within nucleus
_______ 7. Slender tubes that provide movement in cilia and flagella
_______ 8. Organelle composed of membrane-bound canals
_______ 9. Occupies space between cell membrane and nucleus
_______ 10. Flattened membranous sacs that package a secretion
_______ 11. Numerous, short processes for movement
_______ 12. Tiny rod in meshworks or bundles that help cell to shorten
_______ 13. Membranous sac that contains digestive enzymes
_______ 14. Separates nuclear contents from cytoplasm
_______ 15. Spherical organelle that contains chromatin and nucleolus
_______ 16. Process for movement that is usually single and long

PART B

Complete the following:

1. The cell membrane is composed mainly of ______________________ .
2. The basic framework of a cell membrane can be described as ______________________ .
3. A cell membrane is relatively impermeable to substances that are ______________________ .
4. Molecules of ______________________ are responsible for special functions of cell membranes.
5. Chemically, membrane pores consist of ______________________ .

PART C

Complete the following:

1. Sketch a single cell from inside the cheek that has been stained. Label the cellular components you recognize. (The circle represents the field of view through the microscope.)

Magnification ___ ×

2. After comparing the wet mount and the stained cheek cells, describe the advantage that was gained by staining.
____

3. Are the stained cheek cells nearly the same size and shape? ______________________________
Explain your answer. ____

PART D

Complete the following:

1. Sketch a single cell of each kind you observed in the prepared slides of human tissues. Name the tissue and label the cellular components you recognize.

______________ ×

Tissue ______________

______________ ×

Tissue ______________

2. What do the various kinds of cells in these tissues have in common? ______________________
____
____

3. What are the main differences you observed among these cells? ________________________
____
____

Laboratory Exercise 5

Movements Through Cell Membranes

MATERIALS NEEDED

For Procedure A—Diffusion:

Textbook
Petri dish
White paper
Forceps
Potassium permanganate crystals
Millimeter ruler (transparent)

For Procedure B—Osmosis:

Thistle tube
Molasses (or Karo dark corn syrup)
Selectively permeable (semipermeable) membrane (presoaked dialysis tubing of $1\frac{5}{16}$″ or greater diameter)
Ring stand and clamp
Beaker
Rubber band
Millimeter ruler

For Procedure C—Hypertonic, Hypotonic, and Isotonic Solutions:

Test tubes
Marking pen
Test-tube rack
10-mL graduated cylinder
Medicine dropper
Uncoagulated animal blood
Distilled water
0.9% NaCl (aqueous solution)
3.0% NaCl (aqueous solution)
Clean microscope slides
Coverslips
Microscope

For Procedure D—Filtration:

Glass funnel
Filter paper
Ring stand and ring
Beaker
Powdered charcoal
1% glucose (aqueous solution)
1% starch (aqueous solution)
Test tubes
10-mL graduated cylinder
Water bath (boiling water)
Benedict's solution
Iodine-potassium-iodide solution
Medicine dropper

For Alternative Osmosis Activity:

Fresh chicken egg
Beaker
Laboratory balance
Spoon
Vinegar
Corn syrup (Karo)

Virtual Physiology Lab: 9, Diffusion, Osmosis, and Tonicity

SAFETY

- Clean laboratory surfaces before and after laboratory procedures.
- Wear disposable gloves when handling chemicals and animal blood.
- Wear safety glasses when using chemicals.
- Dispose of laboratory gloves and blood-contaminated items as instructed.
- Wash your hands before leaving the laboratory.

A cell membrane functions as a gateway through which chemical substances and small particles may enter or leave a cell. These substances move through the membrane by physical processes such as diffusion, osmosis, and filtration, or by physiological processes such as active transport, phagocytosis, or pinocytosis.

PURPOSE OF THE EXERCISE

To demonstrate some of the physical processes by which substances move through cell membranes.

LEARNING OBJECTIVES

After completing this exercise, you should be able to

1. define *diffusion* and identify examples of diffusion;
2. define *osmosis* and identify examples of osmosis;

3. distinguish among hypertonic, hypotonic, and isotonic solutions and observe the effects of these solutions on animal cells;
4. define *filtration* and identify examples of filtration.

PROCEDURE A—DIFFUSION

1. Review the section entitled "Diffusion" in chapter 3 of the textbook.
2. To demonstrate diffusion, follow these steps:
   a. Place a petri dish half filled with water on a piece of white paper that has a millimeter ruler positioned on the paper. Wait until the water surface is still. Allow approximately 3 minutes.
   b. Using forceps, place one crystal of potassium permanganate near the center of the petri dish and near the millimeter ruler (fig. 5.1).
   c. Measure the radius of the purple circle at 1-minute intervals for 10 minutes.
3. Complete Part A of Laboratory Report 5.

OPTIONAL ACTIVITY

Repeat the demonstration of diffusion using a petri dish filled with ice-cold water and a second dish filled with very hot water. At the same moment, add a crystal of potassium permanganate to each dish and observe the circle as before. What difference do you note in the rate of diffusion in the two dishes? How do you explain this difference? ________________________________

____

____

____

Figure 5.1 To demonstrate diffusion, place one crystal of potassium permanganate in the center of a petri dish containing water. Place the crystal near the millimeter ruler (positioned under the petri dish).

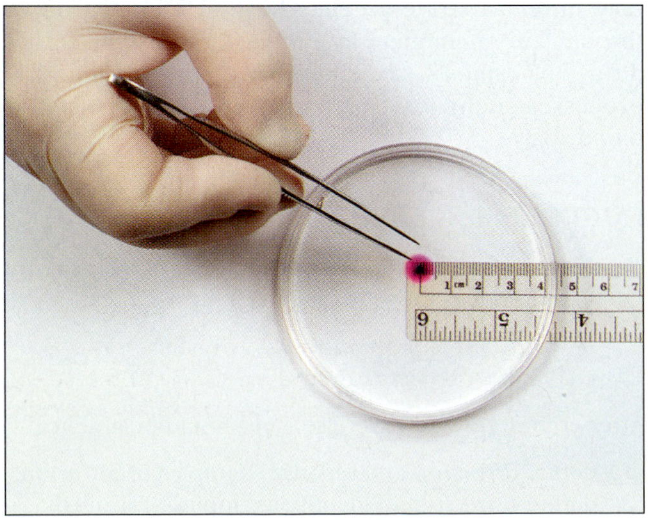

PROCEDURE B—OSMOSIS

1. Review the section entitled "Osmosis" in chapter 3 of the textbook.
2. To demonstrate osmosis, refer to figure 5.2 as you follow these steps:
   a. One person plugs the tube end of a thistle tube with a finger.
   b. Another person then fills the bulb with molasses until it is about to overflow at the top of the bulb. Note that air remains trapped in the stem.
   c. Cover the bulb opening with a single thickness piece of moist selectively permeable (semipermeable) membrane. Dialysis tubing that has been soaked for 30 minutes can easily be cut open because it becomes pliable.
   d. Tightly secure the membrane in place with several wrappings of a rubber band.
   e. Immerse the bulb end of the tube in a beaker of water. If leaks are noted, repeat the procedures.
   f. Support the upright portion of the tube with a clamp on a ring stand. Folded paper under the clamp will protect the thistle tube stem from breakage.
   g. Mark the meniscus level of the molasses in the tube. *Note:* The best results will occur if the mark of the molasses is a short distance up the stem of the thistle tube when the experiment starts.
   h. Measure the level changes after 10 minutes and 30 minutes.
3. Complete Part B of the laboratory report.

ALTERNATIVE PROCEDURE

Eggshell membranes possess selectively permeable properties. To demonstrate osmosis using a natural membrane, soak a fresh chicken egg in vinegar for about 24 hours to remove the shell. Use a spoon to carefully handle the delicate egg. Place the egg in a hypertonic solution (corn syrup) for about 24 hours. Remove the egg, rinse it, and using a laboratory balance, weigh the egg to establish a baseline weight. Place the egg in a hypotonic solution (distilled water). Remove the egg and weigh it every 15 minutes for an elapsed time of 75 minutes. Explain any weight changes that were noted during this experiment.

____

____

____

____

____

____

Figure 5.2 (*a*) Fill the bulb of the thistle tube with molasses; (*b*) tightly secure a piece of selectively permeable (semipermeable) membrane over the bulb opening; and (*c*) immerse the bulb in a beaker of water. *Note:* These procedures require participation of two people.

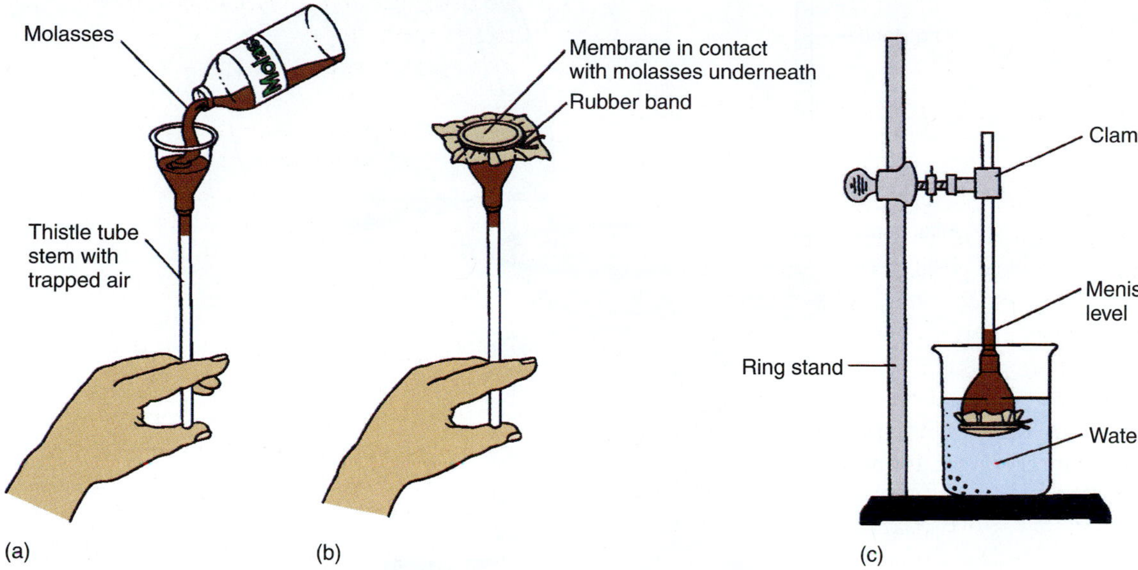

PROCEDURE C—HYPERTONIC, HYPOTONIC, AND ISOTONIC SOLUTIONS

1. Review the section entitled "Osmosis" in chapter 3 of the textbook.
2. To demonstrate the effect of hypertonic, hypotonic, and isotonic solutions on animal cells, follow these steps:
   a. Place three test tubes in a rack and mark them as *tube 1, tube 2,* and *tube 3.* (*Note:* One set of tubes can be used to supply test samples for the entire class.)
   b. Using 10-mL graduated cylinders, add 3 mL of distilled water to tube 1; add 3 mL of 0.9% NaCl to tube 2; and add 3 mL of 3.0% NaCl to tube 3.
   c. Place three drops of fresh uncoagulated animal blood into each of the tubes and gently mix the blood with the solutions. Wait 5 minutes.
   d. Using three separate medicine droppers, remove a drop from each tube and place the drops on three separate microscope slides marked *1, 2,* and *3.*
   e. Cover the drops with coverslips and observe the blood cells, using the high power of the microscope.
3. Complete Part C of the laboratory report.

ALTERNATIVE PROCEDURE

Various substitutes for blood can be used for Procedure C. Onion, cucumber, or cells lining the inside of the cheek represent three possible options.

PROCEDURE D—FILTRATION

1. Review the section entitled "Filtration" in chapter 3 of the textbook.
2. To demonstrate filtration, follow these steps:
   a. Place a glass funnel in the ring of a ring stand over an empty beaker. Fold a piece of filter paper in half and then in half again. Open one thickness of the filter paper to form a cone, wet the cone, and place it in the funnel. The filter paper is used to demonstrate how movement across membranes is limited by size of the molecules, but it does not represent a working model of biological membranes.
   b. Prepare a mixture of 5 cc (approximately 1 teaspoon) powdered charcoal and equal amounts of 1% glucose solution and 1% starch solution in a beaker. Pour some of the mixture into the funnel until it nearly reaches the top of the filter-paper cone. Care should be taken to prevent the mixture from spilling over the top of the filter paper. Collect the filtrate in the beaker below the funnel (fig. 5.3).
   c. Test some of the filtrate in the beaker for the presence of glucose. To do this, place 1 mL of filtrate in a clean test tube and add 1 mL of Benedict's solution. Place the test tube in a water bath of boiling water for 2 minutes and then allow the liquid to cool slowly. If the color of the solution changes to green, yellow, or red, glucose is present (fig. 5.4).

Figure 5.3 Apparatus used to illustrate filtration.

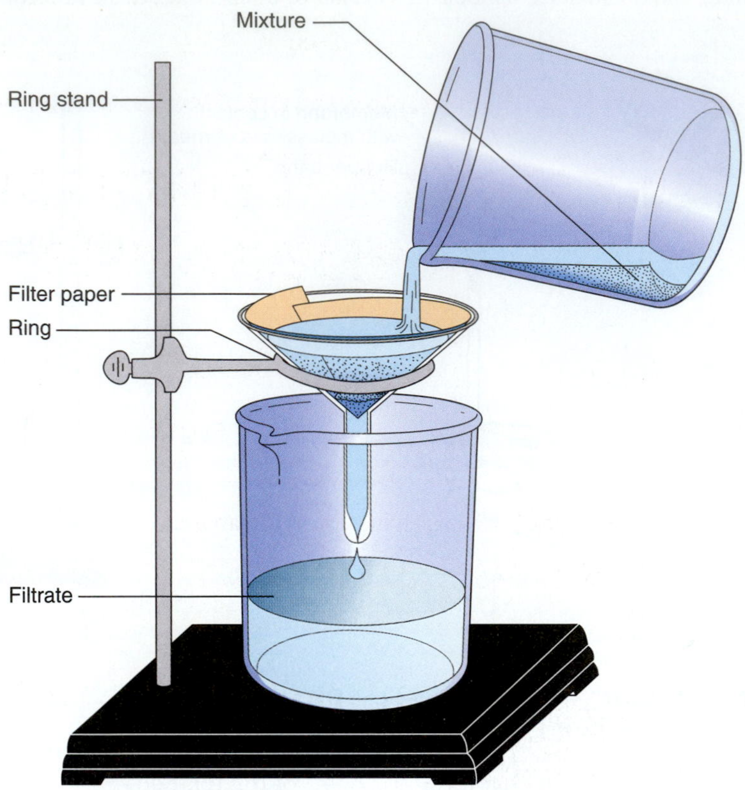

Figure 5.4 Heat the filtrate and Benedict's solution in a boiling water bath for 2 minutes.

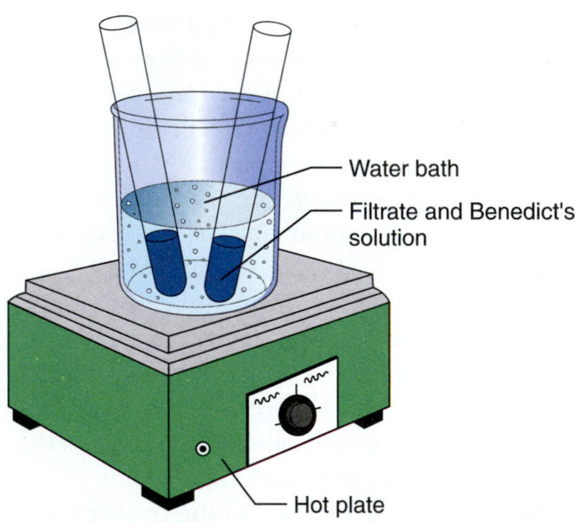

d. Test some of the filtrate in the beaker for the presence of starch. To do this, place a few drops of filtrate in a test tube and add a few drops of iodine-potassium-iodide solution. If the color of the solution changes to blue-black, starch is present.
e. Observe any charcoal in the filtrate.

3. Complete Part D of the laboratory report.

Laboratory Report **5**

Name ______________________________

Date ______________________________

Section ______________________________

MOVEMENTS THROUGH CELL MEMBRANES

PART A

Complete the following:

1. Enter data for changes in the movement of the potassium permanganate.

Elapsed Time	Radius of Purple Circle in Millimeters
initial	______________
1 minute	______________
2 minutes	______________
3 minutes	______________
4 minutes	______________
5 minutes	______________
6 minutes	______________
7 minutes	______________
8 minutes	______________
9 minutes	______________
10 minutes	______________

2. Prepare a graph that illustrates the diffusion distance of potassium permanganate in 10 minutes.

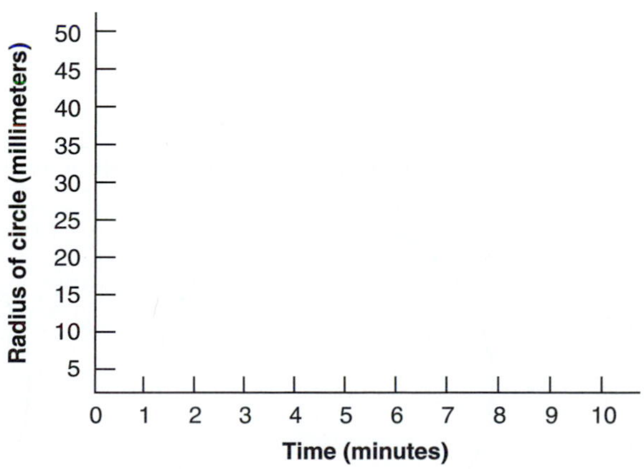

3. Explain your graph. ____

____

4. Briefly define *diffusion.* ____

____

Critical Thinking Application

By answering yes or no, indicate which of the following provides an example of diffusion.

1. A perfume bottle is opened, and soon the odor can be sensed in all parts of the room.__________
2. A sugar cube is dropped into a cup of hot water, and, without being stirred, all of the liquid becomes sweet tasting.__________
3. Water molecules move from a faucet through a garden hose when the faucet is turned on.__________
4. A person blows air molecules into a balloon by exhaling forcefully.__________
5. A crystal of blue copper sulfate is placed in a test tube of water. The next day, the solid is gone, but the water is evenly colored.__________

PART B

Complete the following:

1. What was the change in the level of molasses in 10 minutes? ______________________________
2. What was the change in the level of molasses in 30 minutes? ______________________________
3. How do you explain this change? ____

____

4. Briefly define *osmosis.* ____

____

Critical Thinking Application

By answering yes or no, indicate which of the following involves osmosis.

1. A fresh potato is peeled, weighed, and soaked in a strong salt solution. The next day, it is discovered that the potato has lost weight.__________
2. Garden grass wilts after being exposed to dry chemical fertilizer.__________
3. Air molecules escape from a punctured tire as a result of high pressure inside. __________
4. Plant seeds soaked in water swell and become several times as large as before soaking.__________
5. When the bulb of a thistle tube filled with water is sealed by a selectively permeable membrane and submerged in a beaker of molasses, the water level in the tube falls.__________

PART C

Complete the following:

1. In the spaces, sketch a few blood cells from each of the test tubes, and indicate the magnification.

Tube 1 (distilled water)	Tube 2 (0.9% NaCl)	Tube 3 (3.0% NaCl)
________ ×	________ ×	________ ×

2. Based on your results, which tube contained a solution that was hypertonic to the blood cells? ____________

   Give the reason for your answer. ____________

   ____

3. Which tube contained a solution that was hypotonic to the blood cells? ____________

   Give the reason for your answer. ____________

   ____

4. Which tube contained a solution that was isotonic to the blood cells? ____________

   Give the reason for your answer. ____________

   ____

PART D

Complete the following:

1. Which of the substances in the mixture you prepared passed through the filter paper into the filtrate? ____________

   ____

   ____

2. What evidence do you have for your answer to question 1? ____________

   ____

   ____

3. What force was responsible for the movement of substances through the filter paper? ____________

   ____

4. What substances did not pass through the filter paper? ____________

   ____

5. What factor prevented these substances from passing through? ____________

   ____

   ____

6. Briefly define *filtration*. ____________

   ____

   ____

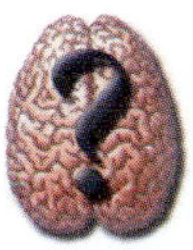

Critical Thinking Application

By answering yes or no, indicate which of the following involves filtration.

1. Oxygen molecules move into a cell and carbon dioxide molecules leave a cell because of differences in the concentrations of these substances on either side of the cell membrane.____________
2. Blood pressure forces water molecules from the blood outward through the thin wall of a blood capillary.____________
3. Urine is forced from the urinary bladder through the tubular urethra by muscular contractions.____________
4. Air molecules enter the lungs through the airways when air pressure is greater outside these organs than inside.____________
5. Coffee is made using a coffeemaker (not instant).____________

Laboratory Exercise 6

The Cell Cycle

MATERIALS NEEDED

Textbook
Models of animal mitosis
Microscope slides of whitefish mitosis (blastula)
Compound microscope

For Demonstration:

Microscope slide of human chromosomes from leukocytes in mitosis
Oil immersion objective

The cell cycle consists of the series of changes a cell undergoes from the time it is formed until it reproduces. Typically, a newly formed cell grows to a certain size and then divides to form two new cells (*daughter cells*). This reproductive process involves two major steps: (1) division of the cell's nuclear parts, which is called *mitosis,* and (2) division of the cell's cytoplasm (cytokinesis). Before the cell divides, it must synthesize biochemicals and other contents. This period of preparation is called *interphase.* The extensive period of interphase is divided into three phases. The S phase, when DNA synthesis occurs, is between two gap phases (G_1 and G_2), when cell growth occurs and cytoplasmic organelles duplicate.

PURPOSE OF THE EXERCISE

To review the stages in the cell cycle and to observe cells in various stages of their life cycles.

LEARNING OBJECTIVES

After completing this exercise, you should be able to

1. describe the cell cycle;
2. identify the stages in the life cycle of a particular cell;
3. arrange into a correct sequence a set of models or drawings of cells in various stages of their life cycles.

PROCEDURE

1. Review the section entitled "The Cell Cycle" in chapter 3 of the textbook.
2. As a review activity, study the various stages of the cell's life cycle represented in figure 6.1 and label the structures indicated in figure 6.2.
3. Observe the animal mitosis models and review the major events in a cell's life cycle represented by each of them. Be sure you can arrange these models in correct sequence if their positions are changed. The acronym IPMAT can help you arrange the correct order of phases in the cell cycle. This includes interphase followed by the four phases of mitosis.
4. Complete Part A of Laboratory Report 6.
5. Obtain a slide of the whitefish mitosis (blastula).
   a. Examine the slide using the high-power objective of a microscope. The tissue on this slide was obtained from a developing embryo (blastula) of a fish, and many of the embryonic cells are undergoing mitosis. Note that the chromosomes of these reproducing cells are darkly stained (fig. 6.3).
   b. Search the tissue for cells in various stages of reproduction. Note that there are several sections on the slide. If you cannot locate different stages in one section, examine the cells of another section since the stages occur at random.
   c. Each time you locate a cell in a different stage, sketch it in an appropriate circle in Part B of the laboratory report.

Critical Thinking Application

Which stage (phase) of the cell cycle was the most numerous in the blastula? ____________
Explain your answer.______________________________
____
____

6. Complete Parts C, D, and E of the laboratory report.

Figure 6.1 The cell cycle: interphase, mitosis, and cytokinesis.

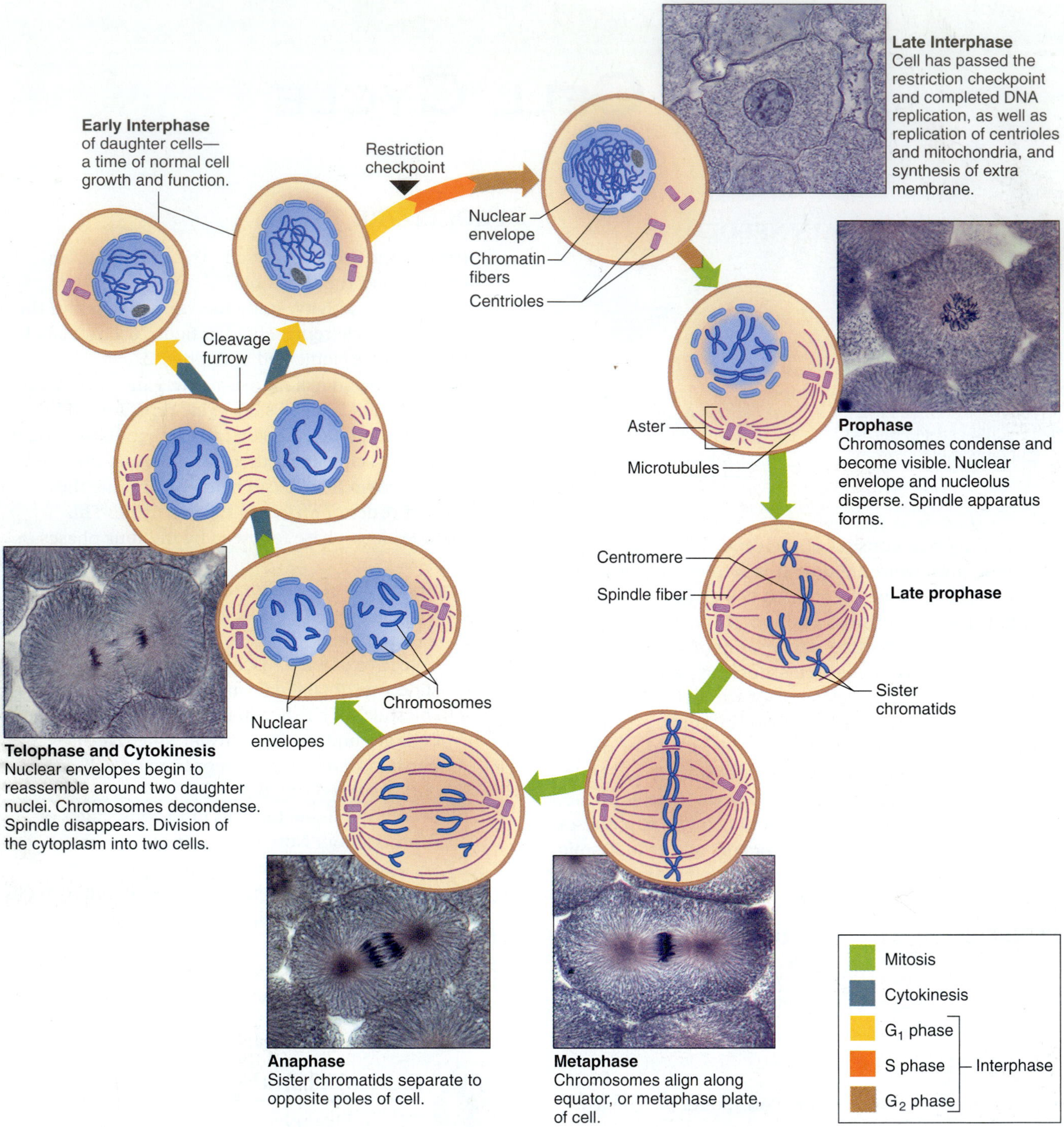

Figure 6.2 Label the structures indicated in the dividing cell.

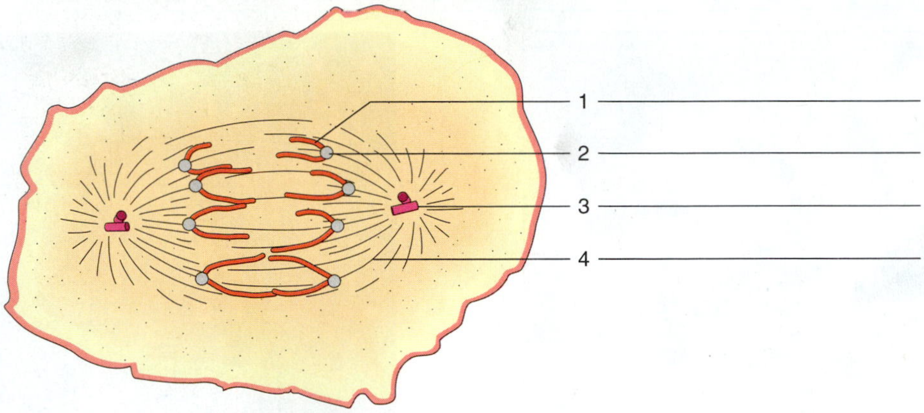

Figure 6.3 Cell in prophase (250× micrograph enlarged to 1,000×).

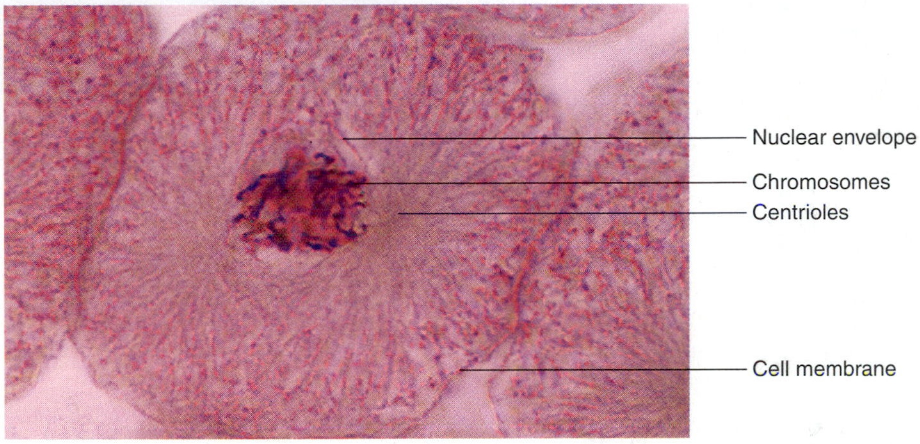

DEMONSTRATION

Using the oil immersion objective of a microscope, see if you can locate some human chromosomes by examining a prepared slide of human chromosomes from leukocytes. The cells on this slide were cultured in a special medium and were stimulated to undergo mitosis. The mitotic process was arrested in metaphase by exposing the cells to a chemical called colchicine, and the cells were caused to swell osmotically. As a result of this treatment, the chromosomes were spread apart. A complement of human chromosomes should be visible when they are magnified about 1,000×. Note that each chromosome is double-stranded and consists of two chromatids joined by a common centromere (fig. 6.4).

Web Quest

Review the phases of the cell cycle and the events during each phase. Search these at www.mhhe.com/shieress8

Figure 6.4 (*a*) A complement of human chromosomes (2,700×). (*b*) A *karyotype* can be constructed by arranging the homologous chromosome pairs together in a chart. The completed karyotype indicates a normal male.

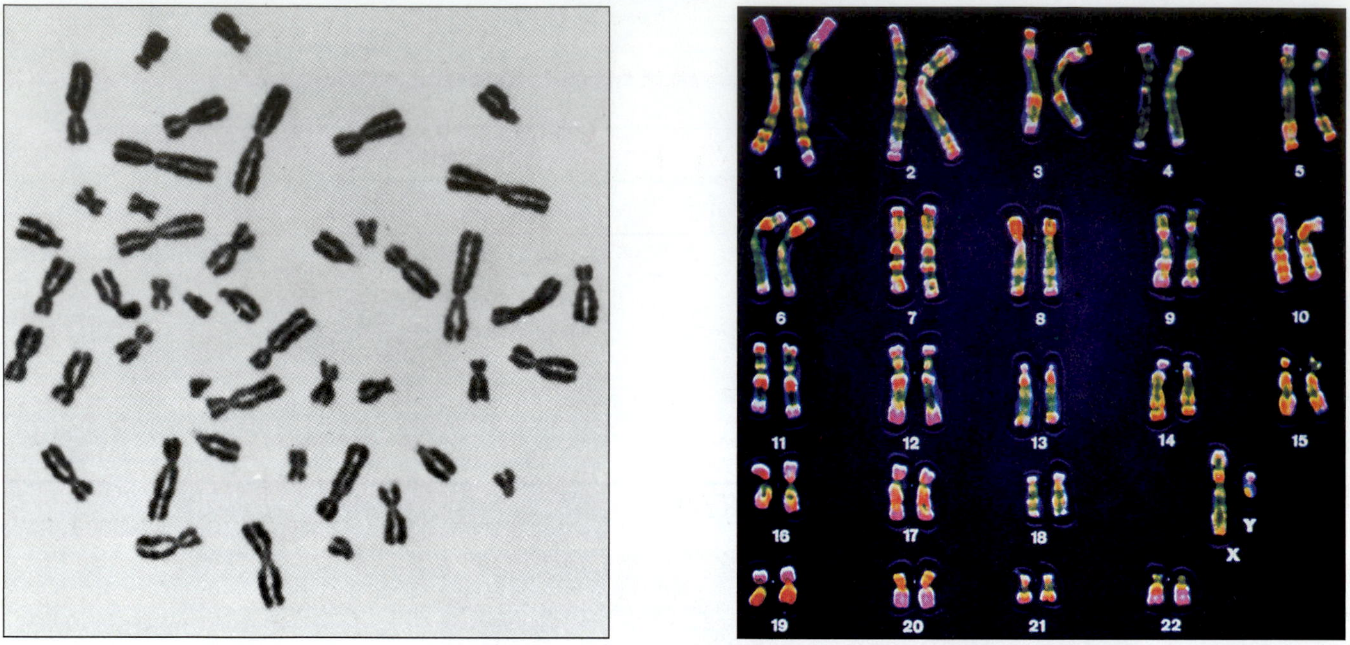

(a)

(b)

Laboratory Report **6**

Name ____________________

Date ____________________

Section ____________________

THE CELL CYCLE

PART A

Complete the table.

Stage	Major Events Occurring
Interphase (G_1, S, and G_2)	
Mitosis (karyokinesis) Prophase	
Metaphase	
Anaphase	
Telophase	
Cytoplasmic division (cytokinesis)	

PART B

Sketch an interphase cell and cells in different stages of mitosis to illustrate the whitefish cell's life cycle. Label the major cellular structures represented in the sketches and indicate cytokinesis locations. (The circles represent fields of view through the microscope.)

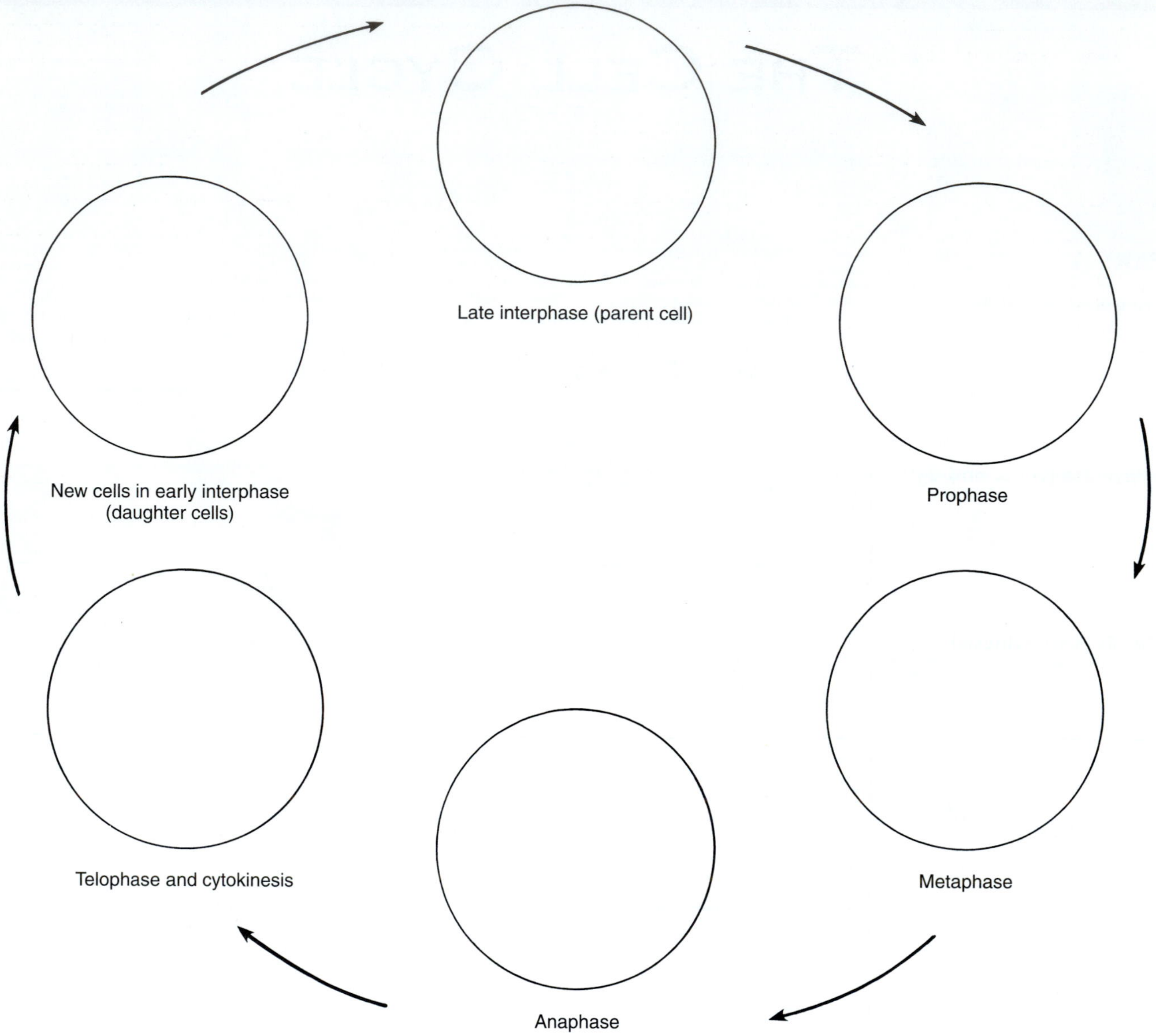

Magnification _____________ ×

PART C

Complete the following:

1. In what ways are the new cells (daughter cells), which result from a cell cycle, similar? ____________________

2. How do the new cells differ slightly? ____________________

3. Distinguish between mitosis (karyokinesis) and cytoplasmic division (cytokinesis). ____________________

PART D

Identify the mitotic stage represented by each of the micrographs in figure 6.5(*a–d*).

a. ____________________

b. ____________________

c. ____________________

d. ____________________

Figure 6.5 Identify the mitotic stage of the cell in each of these micrographs of the whitefish blastula (250×).

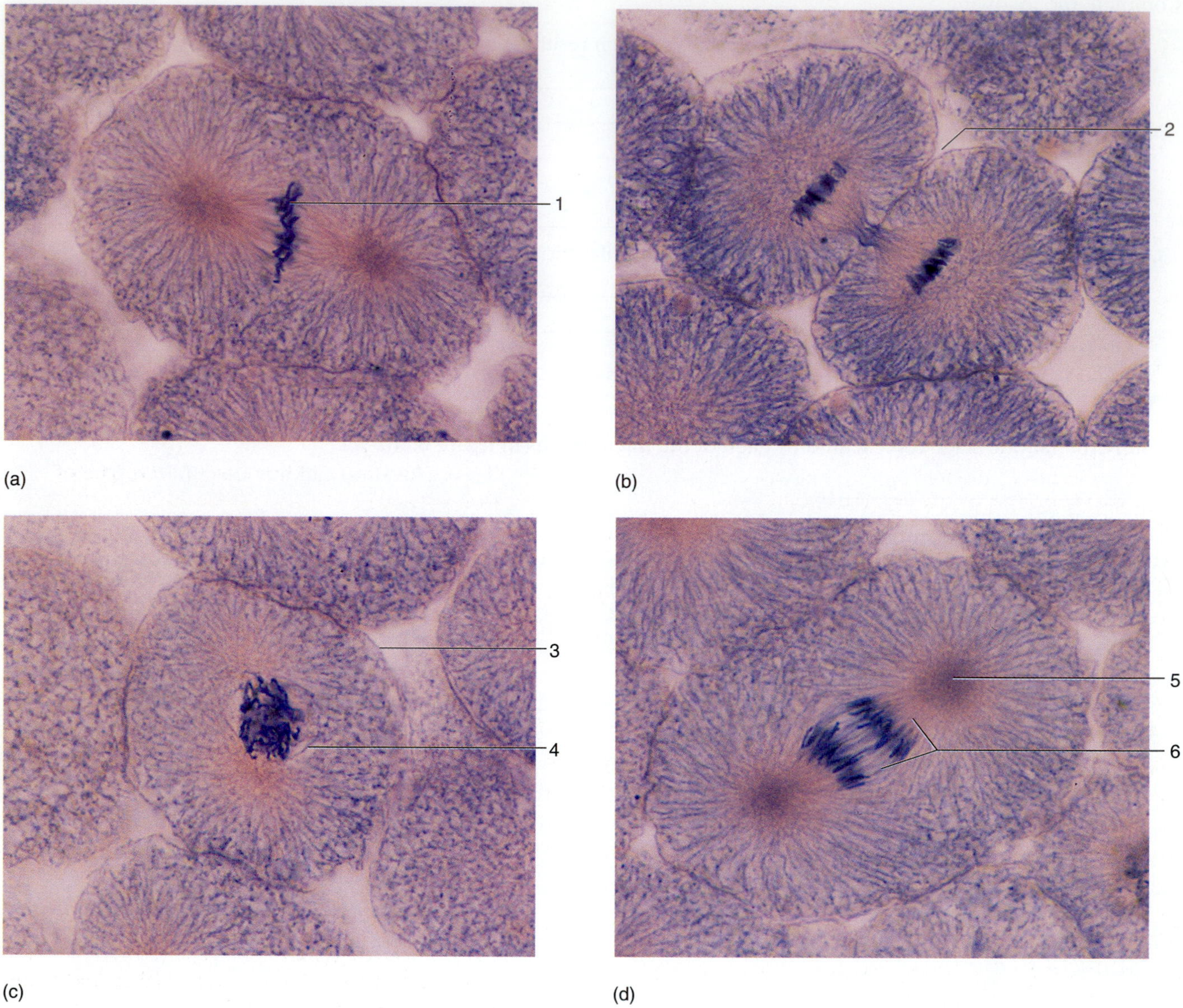

PART E

Identify the structures indicated in figure 6.5.

1. ______________________
2. ______________________
3. ______________________
4. ______________________
5. ______________________
6. ______________________

LABORATORY EXERCISE 7

EPITHELIAL TISSUES

MATERIALS NEEDED

Textbook
Compound microscope
Prepared slides of the following epithelial tissues:
- simple squamous epithelium (lung)
- simple cuboidal epithelium (kidney)
- simple columnar epithelium (small intestine)
- pseudostratified (ciliated) columnar epithelium (trachea)
- stratified squamous epithelium (esophagus)
- transitional epithelium (urinary bladder)

For Optional Activity:

Colored pencils

A tissue is composed of a layer or group of cells—cells that are similar in size, shape, and function. Within the human body, there are four major types of tissues: (1) epithelial, which cover the body's external and internal surfaces; (2) connective, which bind and support parts; (3) muscle, which make movement possible; and (4) nervous, which conduct impulses from one part of the body to another and help to control and coordinate body activities.

Epithelial tissues are tightly packed single (simple) to multiple (stratified) layers of cells that provide protective barriers. The underside of this tissue layer contains a basement membrane layer to which the epithelial cells anchor. Epithelial cells always have a free surface that is exposed to the outside or to an open space internally. Numerous shapes of the cells exist that are used to name and identify the variations. Many of the prepared slides contain more than the tissue to be studied, so care should be taken to be certain that your view matches the correct tissue. Also be aware that stained colors of all tissues might vary.

PURPOSE OF THE EXERCISE

To review the characteristics of epithelial tissues and to observe examples.

LEARNING OBJECTIVES

After completing this exercise, you should be able to

1. describe the general characteristics of epithelial tissues;
2. list six types of epithelial tissues;
3. describe the special characteristics of each type of epithelial tissue;
4. indicate a location and function of each type of epithelial tissue;
5. identify examples of epithelial tissues.

PROCEDURE

1. Review the section entitled "Epithelial Tissues" in chapter 5 of the textbook.
2. Complete Part A of Laboratory Report 7.
3. Use the microscope to observe the prepared slides of types of epithelial tissues. As you observe each tissue, look for its special distinguishing features as described in the textbook, such as cell size, shape, and arrangement. Compare your prepared slides of epithelial tissues to the micrographs in figure 7.1.
4. Complete Part B of the laboratory report.
5. Test your ability to recognize each type of epithelial tissue. To do this, have a laboratory partner select one of the prepared slides, cover its label, and focus the microscope on the tissue. Then see if you can correctly identify the tissue.

Web Quest

Identify tissues from micrographs and examine the structural components of tissues. Search these at www.mhhe.com/shieress8

Figure 7.1 Micrographs of epithelial tissues. *Note:* The brackets indicate the tissue.

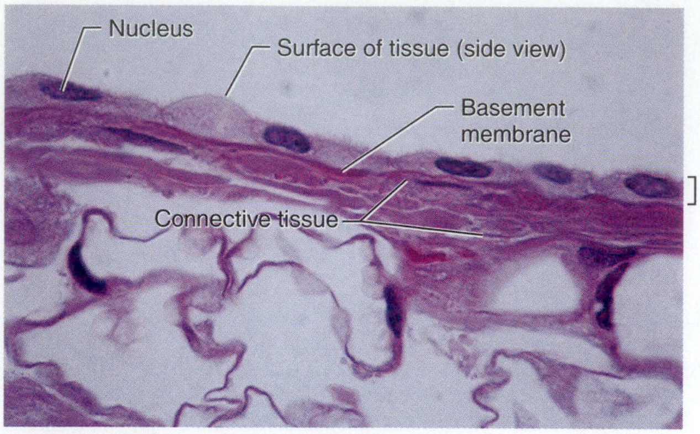

(a) Simple squamous epithelium (side view) (1,100×)

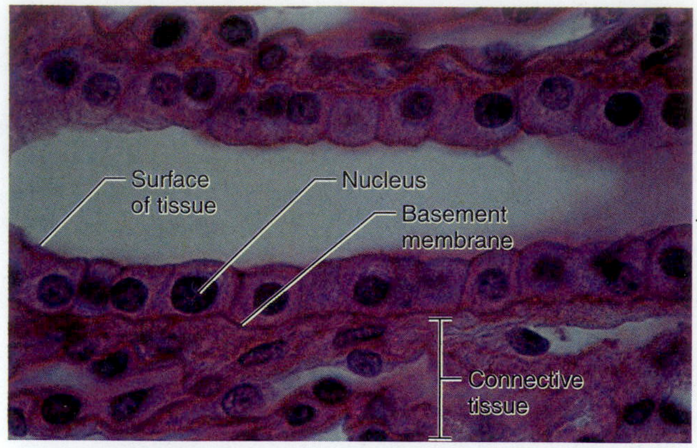

(b) Simple cuboidal epithelium (800×)

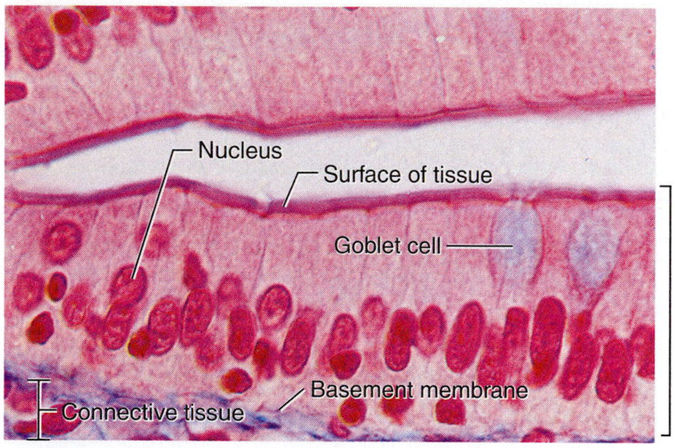

(c) Simple columnar epithelium (650×)

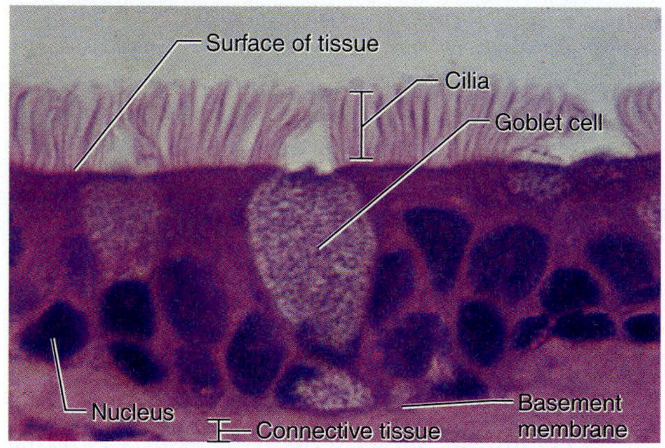

(d) Pseudostratified columnar epithelium with cilia (1,500×)

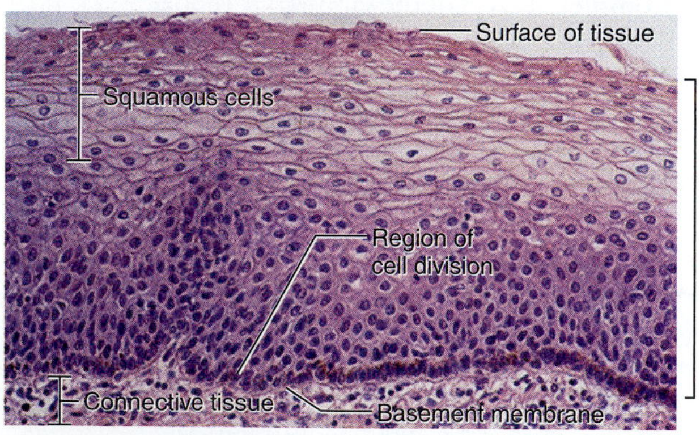

(e) Stratified squamous epithelium (370×)

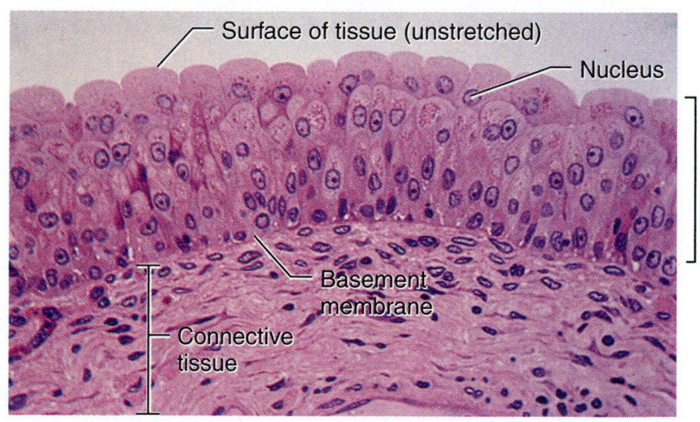

(f) Transitional epithelium (190×)

Laboratory Report **7**

Name ____________________

Date ____________________

Section ____________________

EPITHELIAL TISSUES

PART A

Match the tissues in column A with the characteristics in column B. Place the letter of your choice in the space provided. (Some answers may be used more than once.)

Column A

a. Simple columnar epithelium
b. Simple cuboidal epithelium
c. Simple squamous epithelium
d. Pseudostratified columnar epithelium
e. Stratified squamous epithelium
f. Transitional epithelium

Column B

_______ 1. Consists of several layers of cube-shaped and elongated cells

_______ 2. Commonly possesses cilia that move sex cells and mucus

_______ 3. Single layer of flattened cells

_______ 4. Nuclei located at different levels within cells

_______ 5. Forms walls of capillaries and air sacs of lungs

_______ 6. Forms linings of respiratory passages

_______ 7. Younger cells cuboidal, older cells flattened

_______ 8. Forms inner lining of urinary bladder

_______ 9. Lines kidney tubules and ducts of salivary glands

_______ 10. Forms lining of stomach and intestines

_______ 11. Nuclei located near basement membrane

_______ 12. Forms lining of oral cavity, anal canal, and vagina

PART B

In the space that follows, sketch a few cells of each type of epithelium you observed. For each sketch, label the major characteristics, indicate the magnification used, write an example of a location, and provide a function.

Simple squamous epithelium (____×) Location example ______________ Function ______________	Simple cuboidal epithelium (____×) Location example ______________ Function ______________
Simple columnar epithelium (____×) Location example ______________ Function ______________	Pseudostratified columnar epithelium with cilia (____×) Location example ______________ Function ______________
Stratified squamous epithelium (____×) Location example ______________ Function ______________	Transitional epithelium (____×) Location example ______________ Function ______________

OPTIONAL ACTIVITY

Use colored pencils to differentiate various cellular structures in Part B. Select a different color for a nucleus, cytoplasm, cell membrane, basement membrane, goblet cell, and cilia whenever visible.

Critical Thinking Application

As a result of your observations of epithelial tissues, which one(s) provide(s) the best protection? Explain your answer.

____

____

____

____

CONNECTIVE TISSUES

MATERIALS NEEDED

Textbook
Compound microscope
Prepared slides of the following:
- loose (areolar) connective tissue
- adipose tissue
- dense connective tissue (regular type)
- hyaline cartilage
- elastic cartilage
- fibrocartilage
- bone (compact, ground, cross section)
- blood (human smear)

For Optional Activity:

Colored pencils

Connective tissues contain a variety of cell types and occur in all regions of the body. They bind structures together, provide support and protection, fill spaces, store fat, and produce blood cells.

Connective tissue cells are often widely scattered in an abundance of intercellular matrix. The matrix consists of fibers and a ground substance of various densities. Many of the prepared slides contain more than the tissue to be studied, so care should be taken to be certain that your view matches the correct tissue. Additional study of bone and blood will be found in Laboratory Exercises 11 and 32.

PURPOSE OF THE EXERCISE

To review the characteristics of connective tissues and to observe examples of the major types.

LEARNING OBJECTIVES

After completing this exercise, you should be able to

1. describe the general characteristics of connective tissues;
2. list the major types of connective tissues;
3. describe the special characteristics of each of the major types of connective tissue;
4. indicate a location and function of each type of connective tissue;
5. identify the major types of connective tissues on microscope slides.

PROCEDURE

1. Review the section entitled "Connective Tissues" in chapter 5 of the textbook.
2. Complete Part A of Laboratory Report 8.
3. Use a microscope to observe the prepared slides of various connective tissues. As you observe each tissue, look for its special distinguishing features as described in the textbook. Compare your prepared slides of connective tissues to the micrographs in figure 8.1.
4. Complete Part B of the laboratory report.
5. Test your ability to recognize each of these connective tissues by having a laboratory partner select a slide, cover its label, and focus the microscope on this tissue. Then see if you correctly identify the tissue.

Web Quest

Identify tissues from micrographs and examine the structural components of tissues. Search these at www.mhhe.com/shieress8

Figure 8.1 Micrographs of connective tissues.

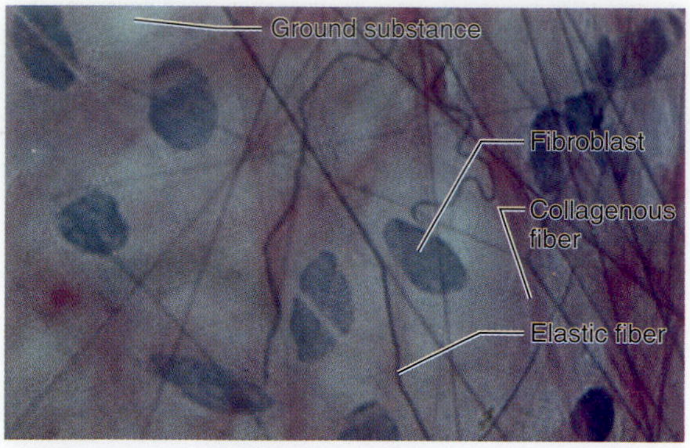

(a) Loose (areolar) connective (400×)

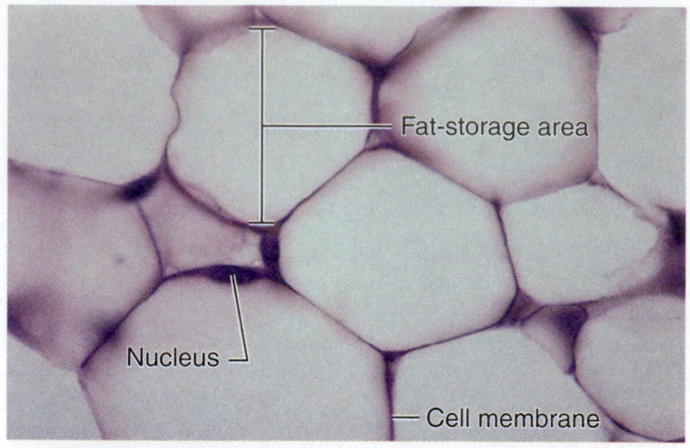

(b) Adipose (265×)

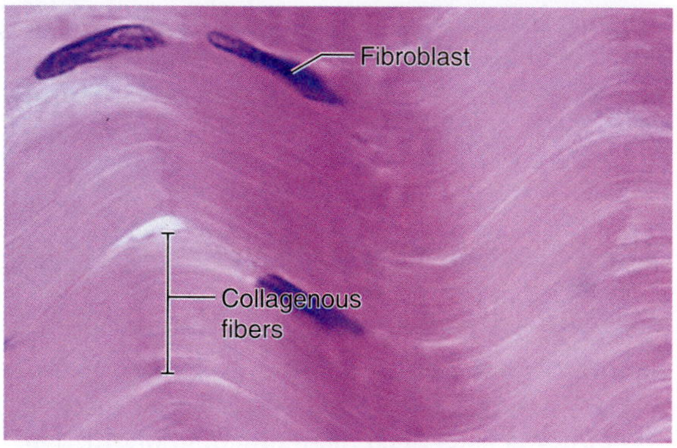

(c) Dense connective (regular) (1,000×)

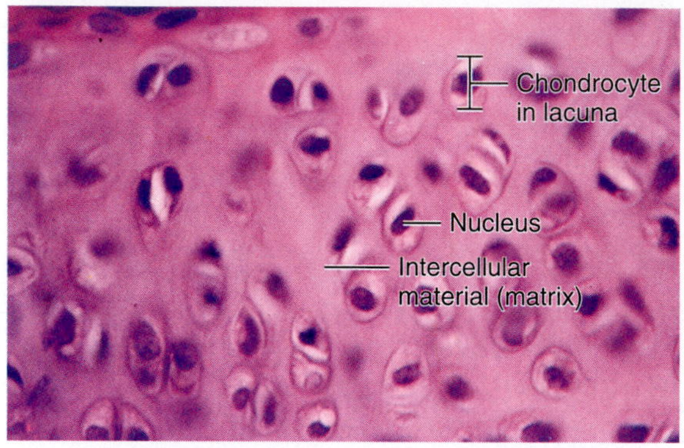

(d) Hyaline cartilage (500×)

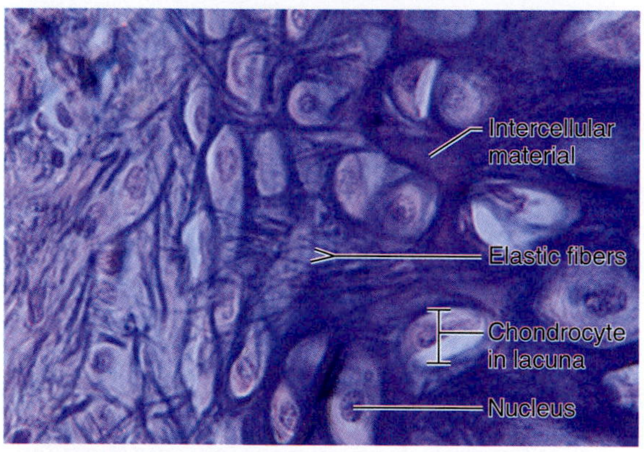

(e) Elastic cartilage (1,200×)

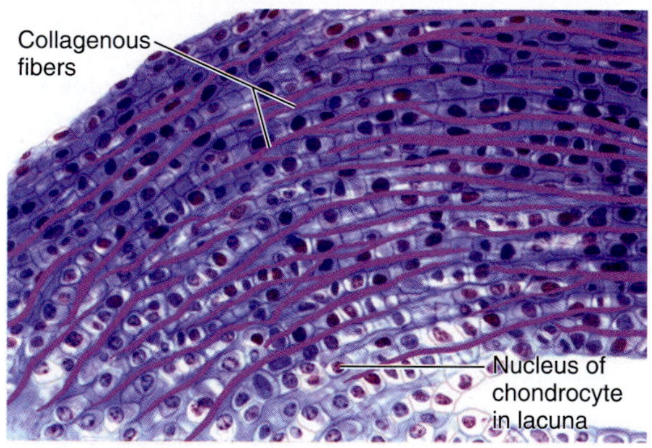

(f) Fibrocartilage (1,800×)

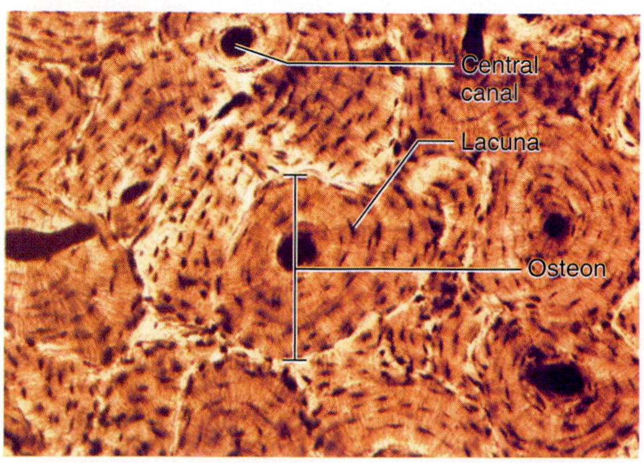

(g) Bone (100×)

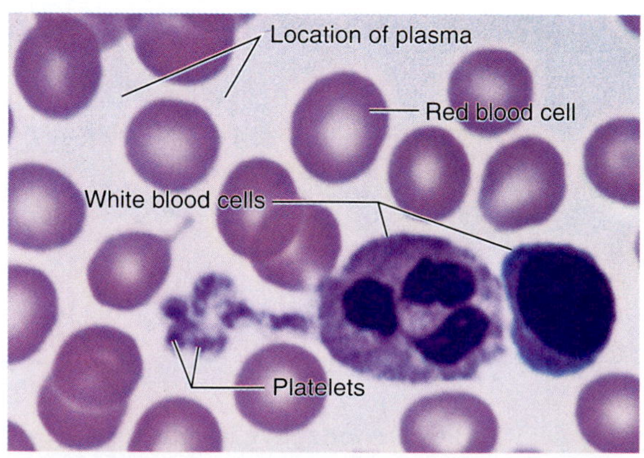

(h) Blood (1,200×)

Laboratory Report **8**

Name ____________________

Date ____________________

Section ____________________

CONNECTIVE TISSUES

PART A

Match the tissues in column A with the characteristics in column B. Place the letter of your choice in the space provided. (Some answers may be used more than once.)

Column A

a. Adipose tissue
b. Blood
c. Bone
d. Dense connective tissue
e. Elastic cartilage
f. Fibrocartilage
g. Hyaline cartilage
h. Loose (areolar) connective tissue

Column B

_______ 1. Forms framework of outer ear

_______ 2. Functions as heat insulator beneath skin

_______ 3. Contains large amounts of fluid and lacks fibers

_______ 4. Cells arranged around central canal

_______ 5. Binds skin to underlying organs

_______ 6. Main tissue of tendons and ligaments

_______ 7. Provides stored energy supply in fat vacuoles

_______ 8. Forms the flexible soft part of the nasal septum

_______ 9. Pads between vertebrae that are shock absorbers

_______ 10. Forms supporting rings of respiratory passages

_______ 11. Cells greatly enlarged with nuclei pushed to sides

_______ 12. Forms delicate, thin layers between muscles

_______ 13. Matrix contains collagen fibers and mineral salts

PART B

In the space provided, sketch a small section of each of the types of connective tissues you observed. For each sketch, label the major characteristics, indicate the magnification used, write an example of a location, and provide a function.

Loose (areolar) connective tissue (____×) Location example ______________ Function ______________	Adipose tissue (____×) Location example ______________ Function ______________
Dense connective tissue (____×) Location example ______________ Function ______________	Hyaline cartilage (____×) Location example ______________ Function ______________
Elastic cartilage (____×) Location example ______________ Function ______________	Fibrocartilage (____×) Location example ______________ Function ______________

Bone (____×) Location example ______________ Function ______________	Blood (____×) Location example ______________ Function ______________

OPTIONAL ACTIVITY

Use colored pencils to differentiate various cellular structures in Part B. Select a different color for the cells, fibers, and ground substance whenever visible.

MUSCLE AND NERVOUS TISSUES

MATERIALS NEEDED

Textbook
Compound microscope
Prepared slides of the following:
- skeletal muscle tissue
- smooth muscle tissue
- cardiac muscle tissue
- nervous tissue (spinal cord smear and/or cerebellum)

For Optional Activity:

Colored pencils

Muscle tissues are characterized by the presence of elongated cells or muscle fibers that can contract. As they shorten, these fibers pull at their attached ends and cause body parts to move. The three types of muscle tissues are skeletal, smooth, and cardiac.

Nervous tissues occur in the brain, spinal cord, and nerves. They consist of neurons (nerve cells), which are the impulse-conducting cells of the nervous system, and neuroglial cells, which perform supportive and protective functions.

PURPOSE OF THE EXERCISE

To review the characteristics of muscle and nervous tissues and to observe examples of these tissues.

LEARNING OBJECTIVES

After completing this exercise, you should be able to

1. list the three types of muscle tissues;
2. describe the general characteristics of muscle tissues;
3. describe the special characteristics of each type of muscle tissue;
4. indicate a location and function of each type of muscle tissue;
5. identify examples of muscle tissues;
6. describe the general characteristics of nervous tissues;
7. identify nervous tissue.

PROCEDURE

1. Review the sections entitled "Muscle Tissues" and "Nervous Tissues" in chapter 5 of the textbook.
2. Complete Part A of Laboratory Report 9.
3. Using the microscope, observe each of the types of muscle tissues on the prepared slides. Look for the special features of each type, as described in the textbook. Compare your prepared slides of muscle tissues to the micrographs in figure 9.1 and the muscle tissue characteristics in table 9.1.
4. Observe the prepared slide of nervous tissue and identify neurons (nerve cells), neuron processes, and neuroglial cells. Compare your prepared slide of nervous tissue to the micrograph in figure 9.1.
5. Complete Part B of the laboratory report.
6. Test your ability to recognize each of these muscle and nervous tissues by having a laboratory partner select a slide, cover its label, and focus the microscope on this tissue. Then see if you correctly identify the tissue.

Web Quest

Identify tissues from micrographs, and examine the structural components of tissues. Search these at www.mhhe.com/shieress8

Figure 9.1 Micrographs of muscle and nervous tissues.

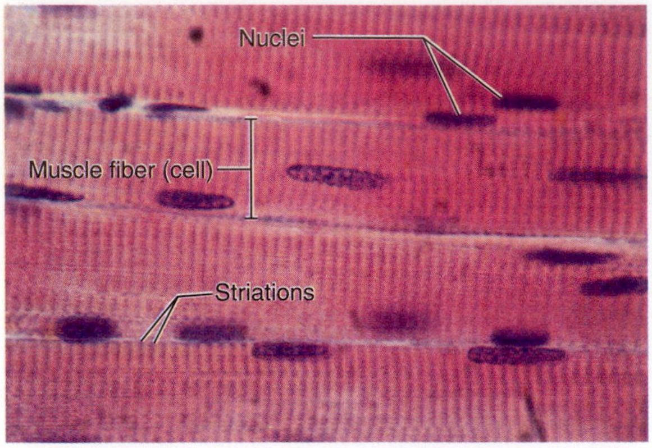

(a) Skeletal muscle (700×)

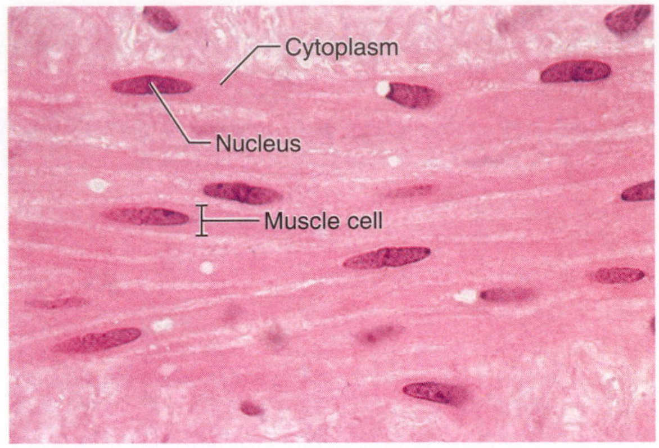

(b) Smooth muscle (900×)

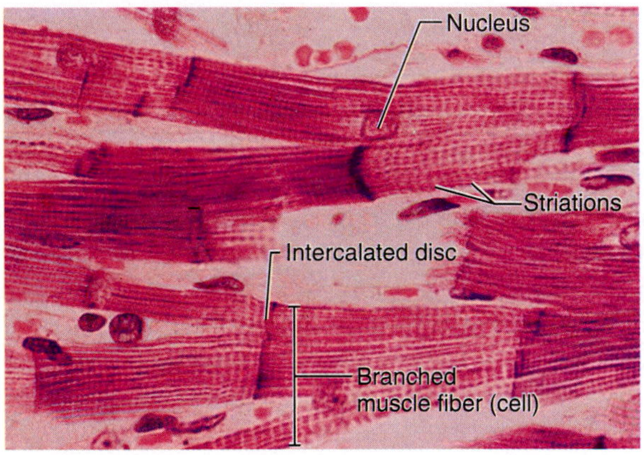

(c) Cardiac muscle (400×)

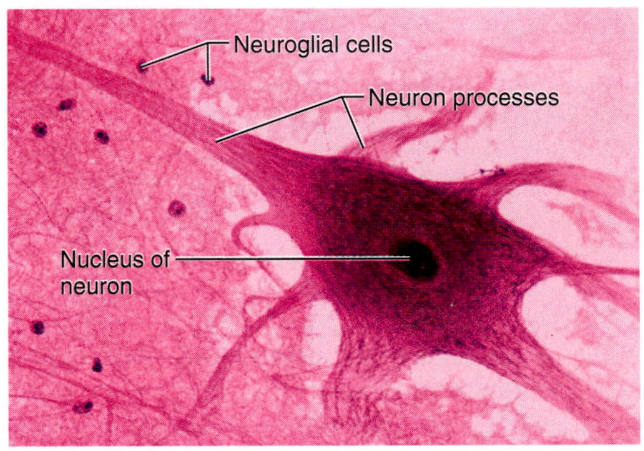

(d) Nervous tissue (300×)

Table 9.1 Muscle Tissue Characteristics

Characteristic	Skeletal Muscle	Cardiac Muscle	Smooth Muscle
Appearance of cells	Unbranched and relatively parallel	Branched and connected in complex networks	Spindle-shaped
Striations	Present and obvious	Present	Absent
Nucleus	Multinucleated	Uninucleated (usually)	Uninucleated
Intercalated discs	Absent	Present	Absent
Control	Voluntary	Involuntary	Involuntary

Laboratory Report **9**

Name ______________________

Date ______________________

Section ______________________

MUSCLE AND NERVOUS TISSUES

PART A

Match the tissues in column A with the characteristics in column B. Place the letter of your choice in the space provided. (Some answers may be used more than once.)

Column A

a. Cardiac muscle
b. Nervous tissue
c. Skeletal muscle
d. Smooth muscle

Column B

_______ 1. Coordinates, regulates, and integrates body functions

_______ 2. Contains intercalated discs

_______ 3. Muscle that lacks striations

_______ 4. Striated and involuntary

_______ 5. Striated and voluntary

_______ 6. Contains neurons and neuroglial cells

_______ 7. Muscle attached to bones

_______ 8. Muscle that composes heart

_______ 9. Moves food through the digestive tract

_______ 10. Transmits impulses along cytoplasmic extensions

PART B

In the space that follows, sketch a few cells or fibers of each of the three types of muscle tissues and of nervous tissue as they appear through the microscope. For each sketch, label the major structures of the cells or fibers, indicate the magnification used, write an example of a location, and provide a function.

Skeletal muscle tissue (____×) Location example ______________ Function ______________	Smooth muscle tissue (____×) Location example ______________ Function ______________
Cardiac muscle tissue (____×) Location example ______________ Function ______________	Nervous tissue (____×) Location example ______________ Function ______________

OPTIONAL ACTIVITY

Use colored pencils to differentiate various cellular structures in Part B.

INTEGUMENTARY SYSTEM

MATERIALS NEEDED

Textbook
Hand magnifier or dissecting microscope
Compound microscope
Prepared microscope slide of human scalp or axilla
Forceps
Microscope slide and coverslip

For Optional Activity:

Prepared slide of heavily pigmented human skin

The integumentary system includes the skin, hair, nails, sebaceous glands, and sweat glands. These organs provide a protective covering for deeper tissues, aid in regulating body temperature, retard water loss, house sensory receptors, synthesize various chemicals, and excrete small quantities of wastes.

PURPOSE OF THE EXERCISE

To observe the organs and tissues of the integumentary system and to review the functions of these parts.

LEARNING OBJECTIVES

After completing this exercise, you should be able to

1. name the organs of the integumentary system;
2. describe the major functions of these organs;
3. distinguish among epidermis, dermis, and the subcutaneous layer;
4. identify the layers of the skin, a hair follicle, an arrector pili muscle, a sebaceous gland, and a sweat gland on a microscope slide, diagram, or model.

PROCEDURE

1. Review the sections entitled "Skin and Its Tissues" and "Accessory Organs of the Skin" in chapter 6 of your textbook.
2. As a review activity, label figures 10.1 and 10.2.
3. Complete Part A of Laboratory Report 10.
4. Use the hand magnifier or dissecting microscope to do the following:
   a. Observe the skin, hair, and nails of your hand.
   b. Compare the type and distribution of hairs on the front and back of your forearm.
5. Pull out a single hair with the forceps and mount it on a microscope slide under a coverslip. Use the hand magnifier or dissecting microscope to observe the root and shaft of the hair. Note the scalelike parts that make up the shaft.
6. Complete Part B of the laboratory report.
7. As vertical sections of human skin are observed, remember that the lenses of the microscope invert and reverse images. It is important to orient the position of the epidermis, dermis, and subcutaneous (hypodermis) layers using scan magnification before continuing with additional observations. Compare all of your skin observations to figure 10.3. Use low-power magnification of the compound microscope and proceed as follows:
   a. Observe the prepared slide of human scalp or axilla.
   b. Locate the epidermis, dermis, and subcutaneous layer, a hair follicle, an arrector pili muscle, a sebaceous gland, and a sweat gland.
   c. Focus on the epidermis with high power and locate the stratum corneum and stratum basale (stratum germinativum). Note how the shapes of the cells in these two layers differ.
   d. Observe the dense connective tissue (irregular type) that makes up the bulk of the dermis.
   e. Observe the adipose tissue that composes most of the subcutaneous layer.
8. Complete Part C of the laboratory report.
9. Using low-power magnification, locate a hair follicle that has been sectioned longitudinally through its bulblike base, a sebaceous gland close to the follicle, and a sweat gland. Observe the detailed structure of these parts with high-power magnification.
10. Complete Parts D and E of the laboratory report.

Figure 10.1 Label this vertical section of the skin.

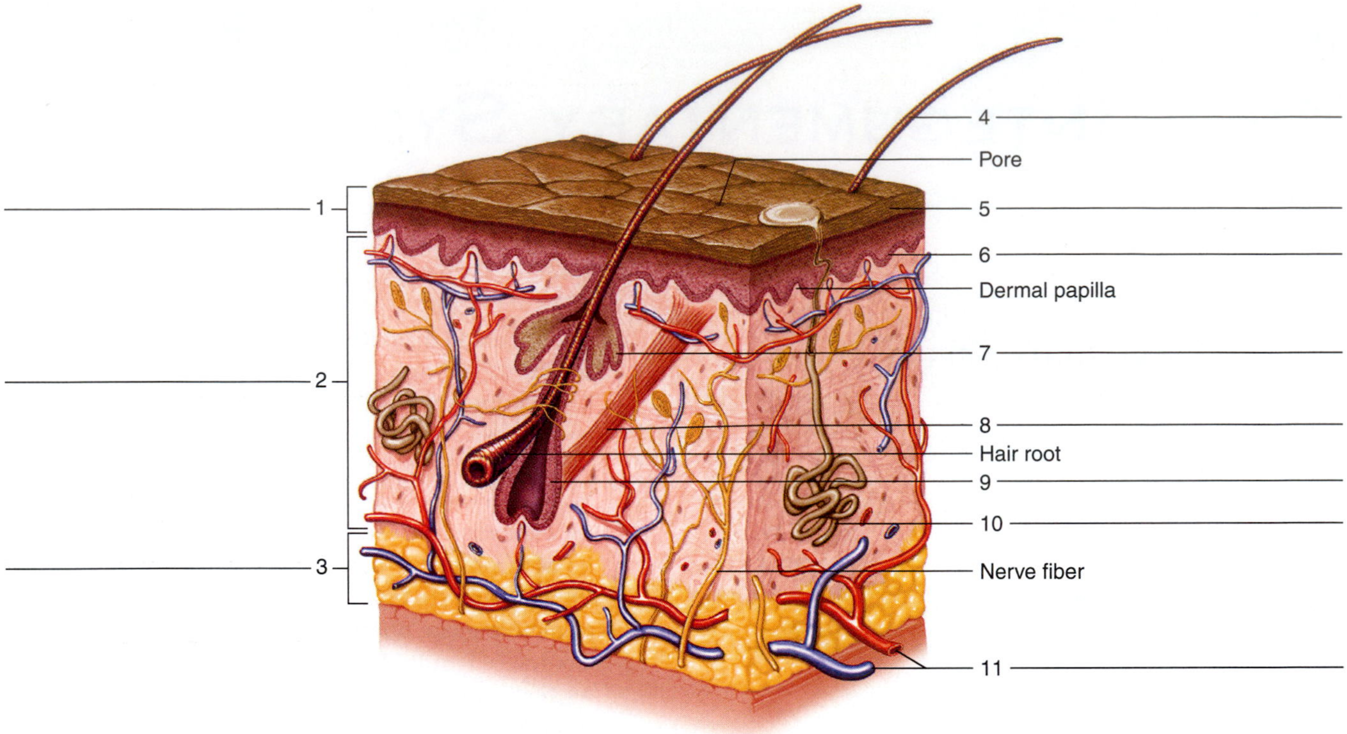

OPTIONAL ACTIVITY

Observe the prepared slide of heavily pigmented skin with low-power magnification. Note that the pigment is most abundant in the epidermis. Focus on this region with the high-power objective. The pigment-producing cells, or melanocytes, are located among the deeper layers of epidermal cells and the underlying connective tissue cells of the dermis. Differences in skin color are primarily due to the amount of pigment (melanin) produced by these cells. The number of melanocytes in the skin is about the same for all humans.

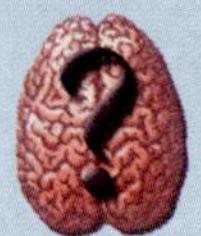

Critical Thinking Application

Explain the advantage for melanin granules being located in the deep layer of the epidermis.

____

____

____

____

Web Quest

Identify skin layers from micrographs and review the functions of the skin structures. Search these at www.mhhe.com/shieress8

Figure 10.2 Label the features associated with this hair follicle.

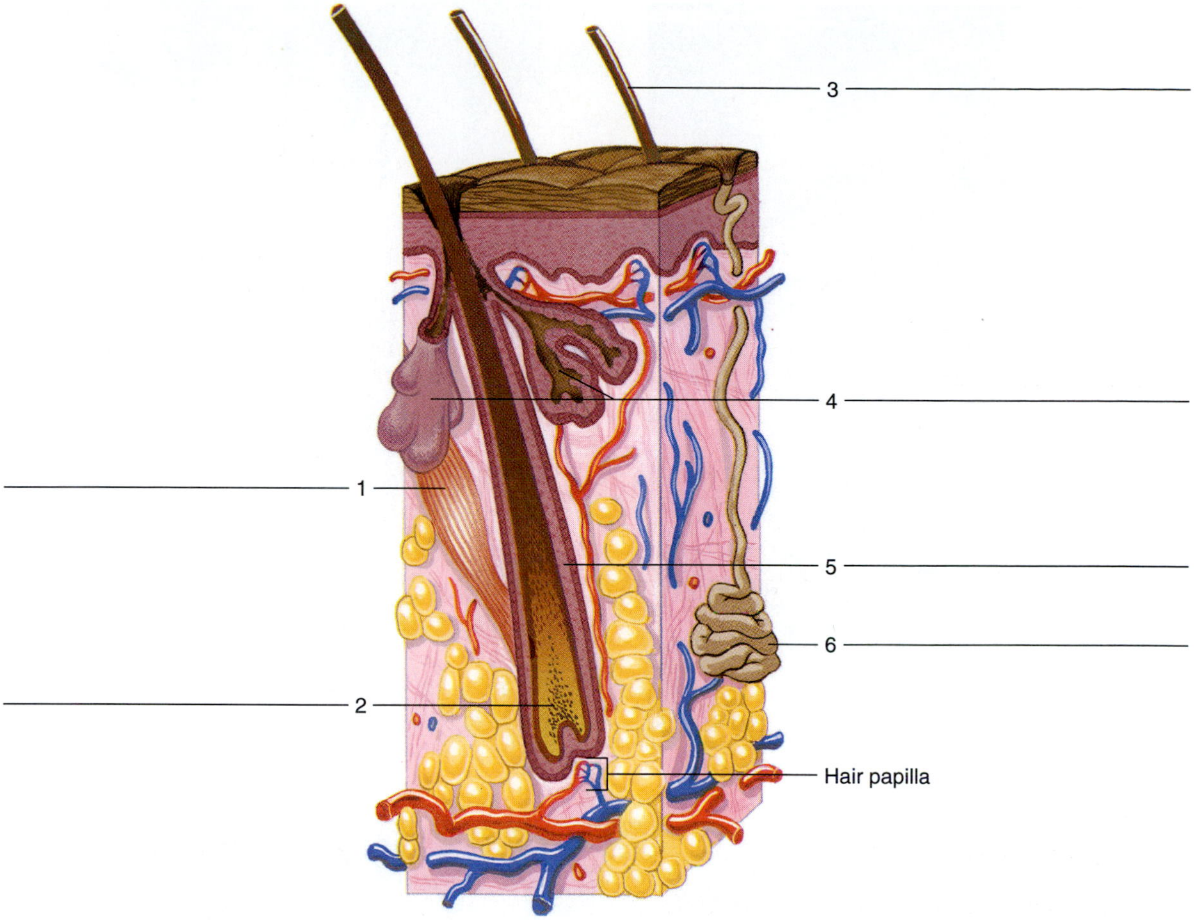

Figure 10.3 Features of human skin are indicated in these micrographs. Magnifications: (*a*) 290×; (*b*) 30× micrograph enlarged to 280×; (*c*) 45×; (*d*) 110×.

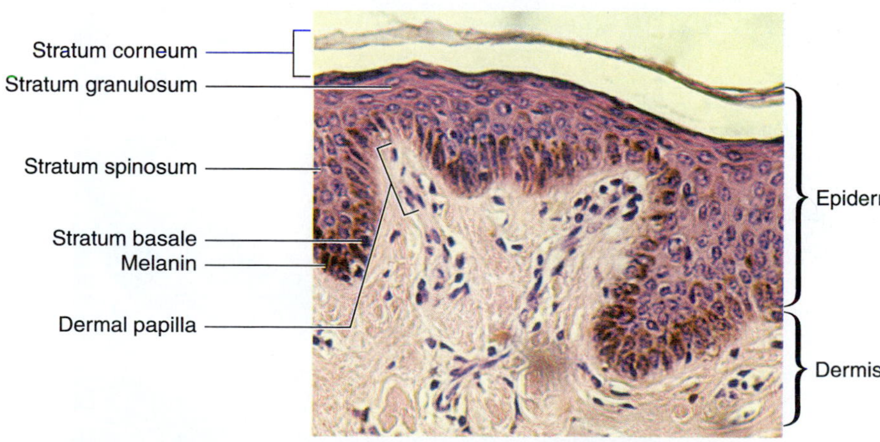

(a)

Figure 10.3 *Continued.*

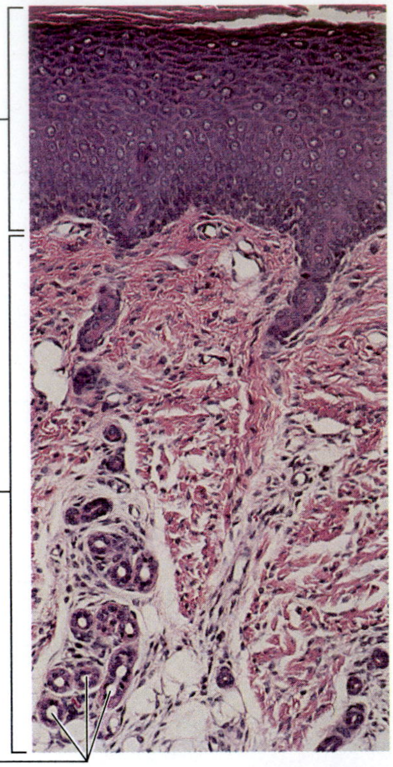

(b)

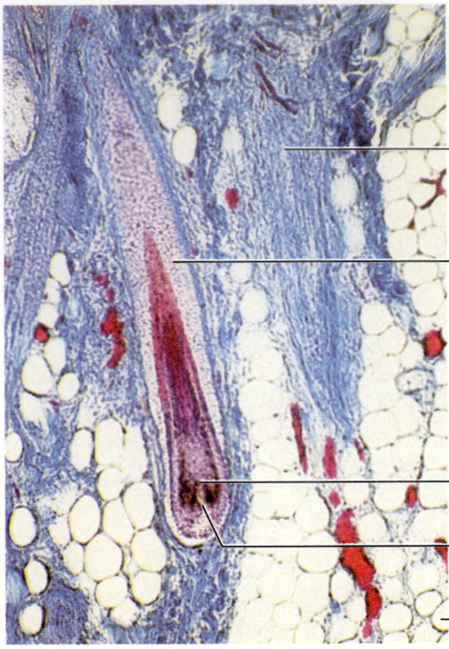

(c)

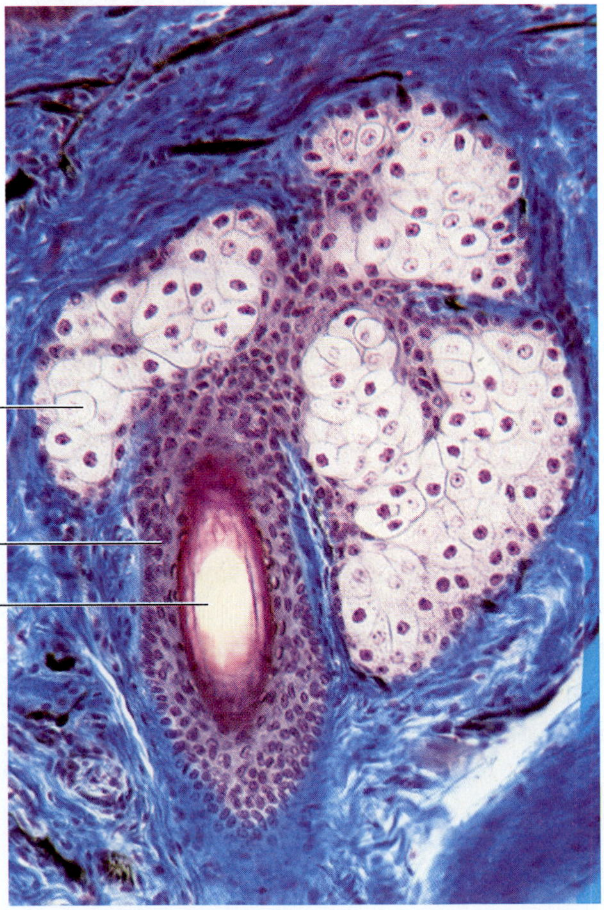

(d)

Laboratory Report **10**

Name ______________________

Date ______________________

Section ______________________

INTEGUMENTARY SYSTEM

PART A

Match the structures in column A with the description and functions in column B. Place the letter of your choice in the space provided.

Column A

a. Apocrine sweat gland
b. Arrector pili muscle
c. Dermis
d. Eccrine sweat gland
e. Epidermis
f. Hair follicle
g. Keratin
h. Melanin
i. Melanocyte
j. Sebaceous gland
k. Sebum
l. Stratum basale
m. Stratum corneum
n. Subcutaneous layer

Column B

_______ 1. An oily secretion that helps to waterproof body surface
_______ 2. Outermost layer of epidermis
_______ 3. Become active at puberty
_______ 4. Epidermal pigment
_______ 5. Inner layer of skin
_______ 6. Responds to elevated body temperature
_______ 7. Pigment-producing cell
_______ 8. General name of entire superficial layer of the skin
_______ 9. Gland that secretes an oily substance
_______ 10. Hard protein of nails and hair
_______ 11. Binds skin to underlying organs
_______ 12. Cell division and deepest layer of epidermis
_______ 13. Tubelike part that contains the root of the hair
_______ 14. Causes hair to stand erect and goose bumps to appear

PART B

Complete the following:

1. How does the skin on your palm differ from that on the back (posterior) of your hand? ______________________
____
____

2. Describe the differences you observed in the type and distribution of hair on the front (anterior) and back (posterior) of your forearm. ______________________
____
____

3. Explain how the hair is formed. ____

____

4. What cells produce the pigment in hair? ____

____

PART C

Complete the following:

1. Distinguish among the epidermis, dermis, and subcutaneous layers. ____

____

2. How do the cells of stratum corneum and stratum basale differ?____

____

3. What special qualities does the connective tissue of the dermis have? ____

____

PART D

Complete the following:

1. In which layer of skin are follicles usually found? ____
2. How are sebaceous glands associated with hair follicles? ____

____

3. In which layer of skin are sweat glands usually located? ____

PART E

Sketch and label a vertical section of human skin, using the scanning objective.

Laboratory Exercise 11

Structure of Bone

MATERIALS NEEDED

Textbook
Human long bone, sectioned longitudinally
Fresh animal bones, sectioned longitudinally and transversely
Dissecting microscope
Prepared microscope slide of ground compact bone

For Demonstration:

Fresh chicken bones (radius and ulna from wings)
Vinegar or dilute hydrochloric acid

SAFETY

- Wear disposable gloves for handling fresh bones and for the demonstration of a bone soaked in vinegar or dilute hydrochloric acid.
- Wash your hands before leaving the laboratory.

A bone represents an organ of the skeleton system. As such, it is composed of a variety of tissues including bone tissue, cartilage, dense connective tissue, blood, and nervous tissue.

Although various bones of the skeleton vary greatly in size and shape, they have much in common structurally and functionally.

PURPOSE OF THE EXERCISE

To examine the structure of a long bone.

LEARNING OBJECTIVES

After completing this exercise, you should be able to

1. locate and name the major structures of a long bone;
2. describe the functions of various structures of a bone;
3. distinguish between compact and spongy bone.

PROCEDURE

1. Review the section entitled "Bone Structure" in chapter 7 of the textbook.
2. As a review activity, label figures 11.1 and 11.2.
3. Examine the sectioned bones and locate the following:

   epiphysis (proximal and distal)
   epiphyseal plate
   articular cartilage
   diaphysis
   periosteum
   compact bone
   spongy bone
   medullary cavity
   endosteum
   red marrow
   yellow marrow

4. Use the dissecting microscope to observe the compact bone and spongy bone of the sectioned specimens. Also examine the marrow in the medullary cavity and the spaces within the spongy bone of the fresh specimen.
5. Reexamine the microscopic structure of bone tissue by observing a prepared microscope slide of ground compact bone. Use the figures of bone tissue in chapter 5 of the textbook to locate the following features:

   osteon (Haversian system)
   central canal (Haversian canal)
   lacuna (small chamber for an osteocyte)
   lamella
   canaliculus

Figure 11.1 Label the major structures of this long bone (femur).

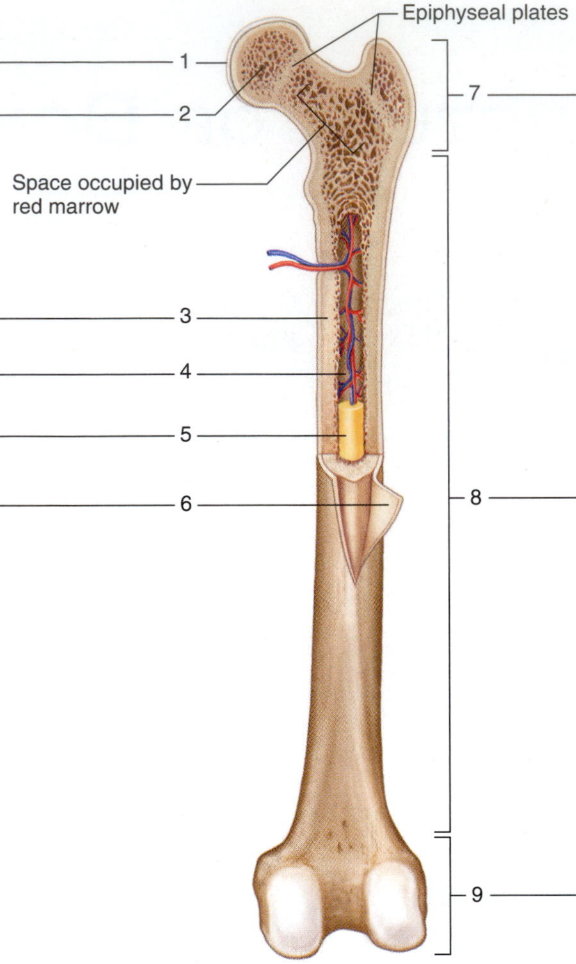

Critical Thinking Application

Explain how bone cells embedded in a solid ground substance obtain nutrients and eliminate wastes.

____

____

____

____

6. Complete Parts A and B of Laboratory Report 11.

DEMONSTRATION

Examine a fresh chicken bone and a chicken bone that has been soaked for several days in vinegar or exposed overnight in dilute hydrochloric acid. Wear disposable gloves for handling these bones. This acid treatment removes the inorganic salts from the bone matrix. Rinse the bones in water and note the texture and flexibility of each. Based on your observations, what quality of the fresh bone seems to be due to the inorganic salts that were removed by the acid treatment?

____

____

Examine the specimen of chicken bone that has been exposed to high temperature (baked at 121°C [250°F] for 2 hours). This treatment removes the protein and other organic substances from the bone matrix. What quality of the fresh bone seems to be due to these organic materials?

____

____

Figure 11.2 Label the features associated with the microscopic structure of bone.

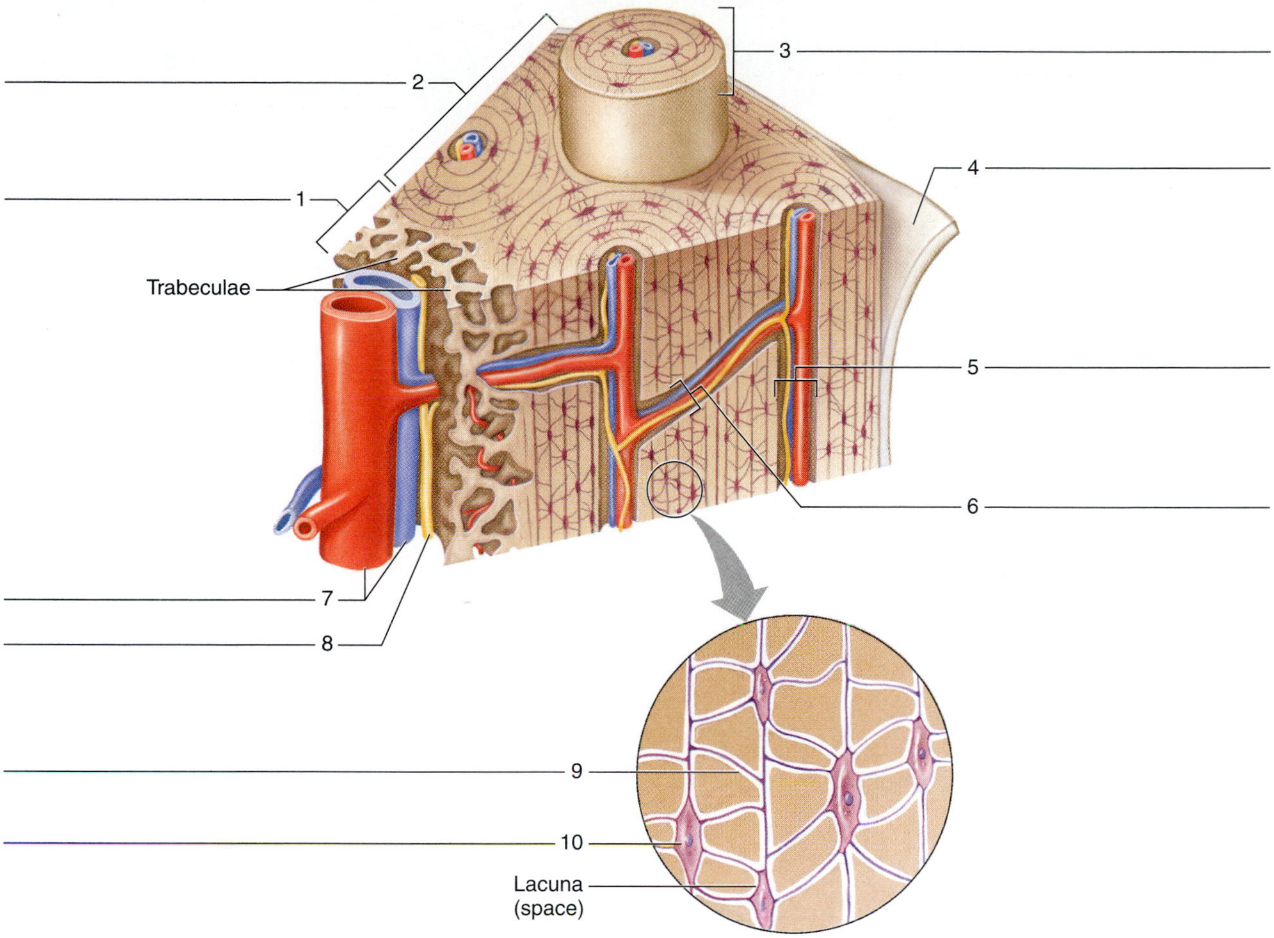

Laboratory Report **11**

Name ______________________

Date ______________________

Section ______________________

Structure of Bone

PART A

Complete the following:

1. Where in the human skeleton are long bones found?

2. Distinguish between the epiphysis and the diaphysis of a long bone.

3. Where is cartilage found on a long bone?

4. Where is dense connective tissue found on a long bone?

5. In general, what is the function of bony processes?

6. Distinguish between the periosteum and the endosteum.

7. What differences did you note between the structure of the compact bone and the spongy bone?

8. How are these structural differences related to the locations and functions of these two types of bone?

9. From your observations, how does the marrow in the medullary cavity compare with the marrow in the spaces of the spongy bone?

PART B

Identify the structures indicated in figure 11.3.

Figure 11.3 Identify the structures indicated in (*a*) the unsectioned long bone (fifth metatarsal) and (*b*) the partially sectioned long bone, using the terms provided.

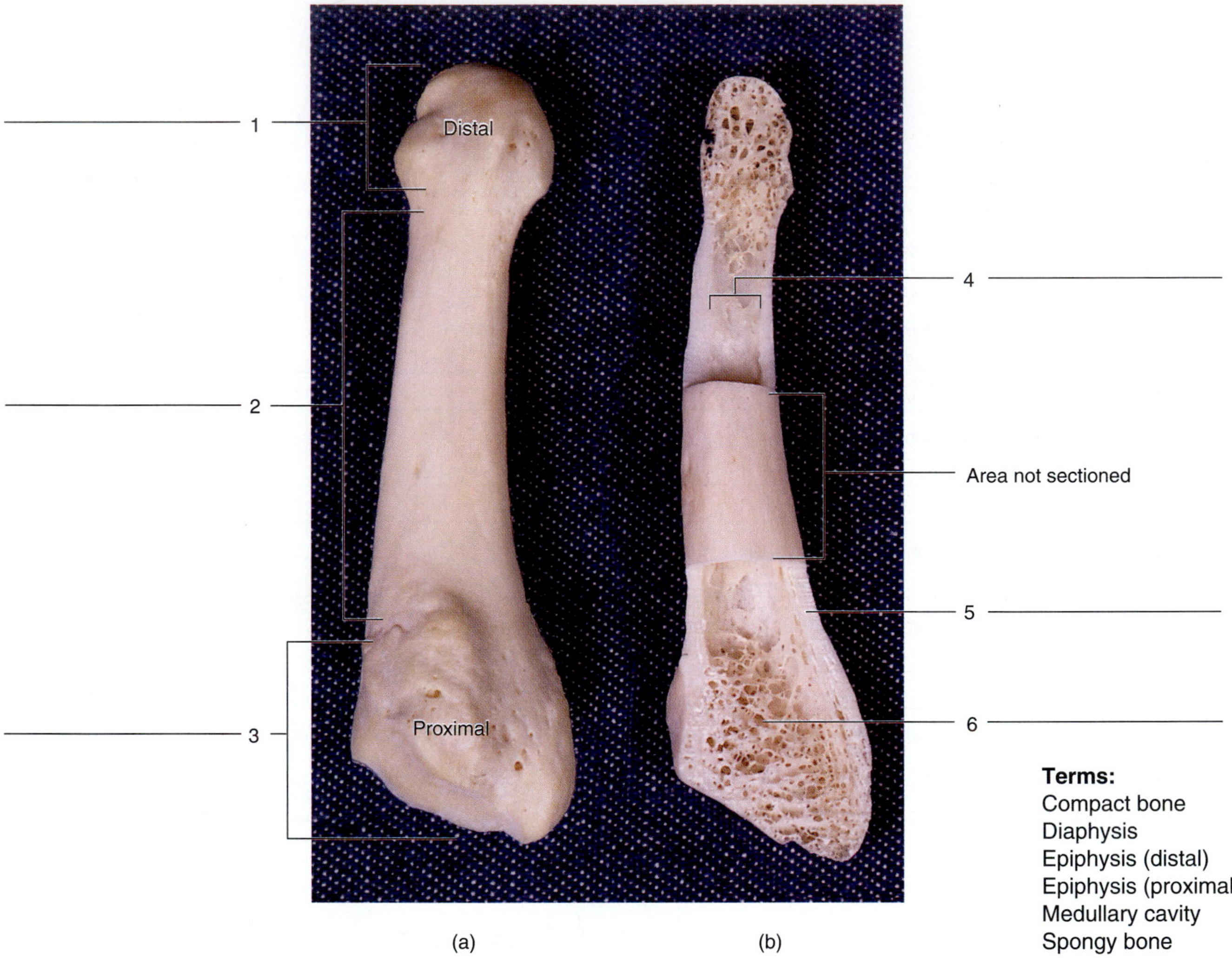

Terms:
Compact bone
Diaphysis
Epiphysis (distal)
Epiphysis (proximal)
Medullary cavity
Spongy bone

ORGANIZATION OF THE SKELETON

MATERIALS NEEDED

Textbook
Human skeleton, articulated

For Optional Activity:

Colored pencils

For Demonstration:

Radiographs (X rays) of skeletal structures

The skeleton can be divided into two major portions: (1) the axial skeleton, which consists of the bones and cartilages of the head, neck, and trunk, and (2) the appendicular skeleton, which consists of the bones of the limbs and those that anchor the limbs to the axial skeleton.

PURPOSE OF THE EXERCISE

To review the organization of the skeleton, the major bones of the skeleton, and the terms used to describe skeletal structures.

LEARNING OBJECTIVES

After completing this exercise, you should be able to

1. distinguish between the axial skeleton and the appendicular skeleton;
2. locate and name the major bones of the human skeleton;
3. define the terms used to describe skeletal structures and locate examples of such structures on the human skeleton.

PROCEDURE

1. Review the section entitled "Skeletal Organization" in chapter 7 of the textbook. (Note that pronunciations of the names for major skeletal structures are included within the narrative of chapter 7.)
2. As a review activity, label figure 12.1.
3. Examine the human skeleton and locate the following parts. As you locate the following bones, note the number of each in the skeleton. Palpate as many of the corresponding bones in your own skeleton as possible.

axial skeleton

skull

- cranium 8
- face 14

middle ear bone 6

hyoid bone 1

vertebral column

- vertebra 24
- sacrum 1
- coccyx 1

thoracic cage

- rib 24
- sternum 1

appendicular skeleton

pectoral girdle

- scapula 2
- clavicle 2

upper limbs

- humerus 2
- radius 2
- ulna 2
- carpal 16
- metacarpal 10
- phalanx 28

Figure 12.1 Label the major bones of the skeleton: (*a*) anterior view; (*b*) posterior view.

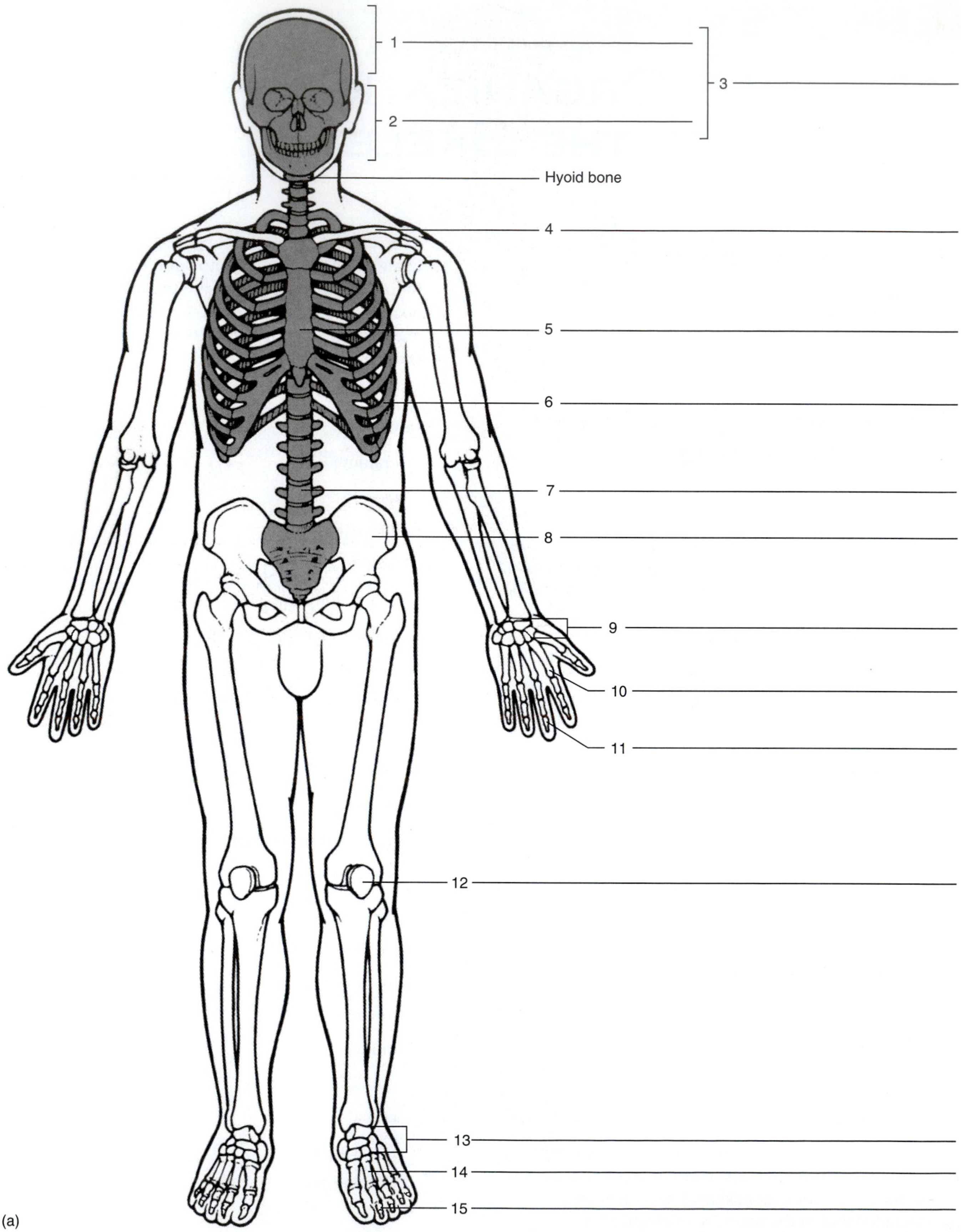

Figure 12.1 *Continued.*

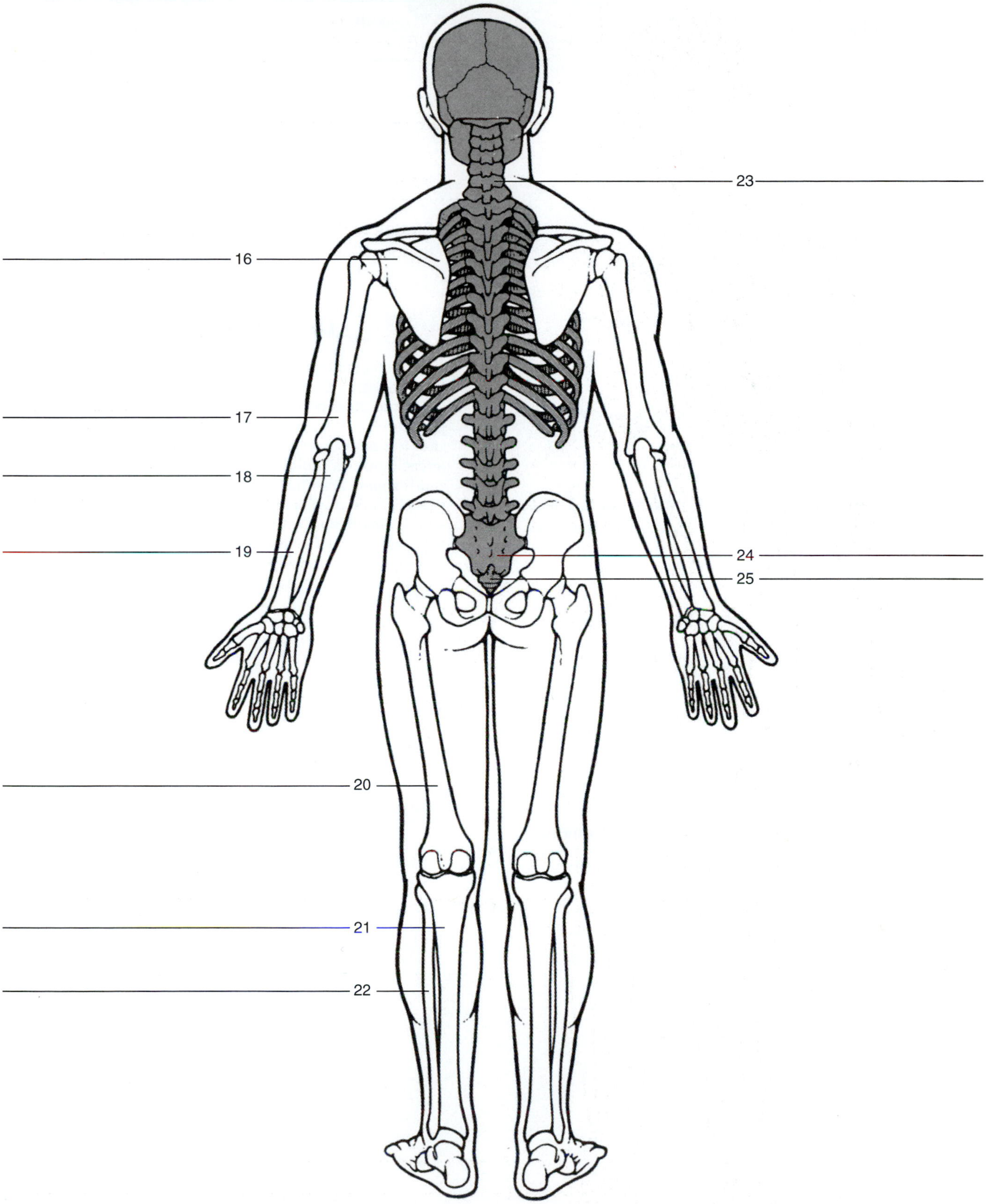

(b)

pelvic girdle

coxa (hipbone; innominate) 2

lower limbs

femur 2

tibia 2

fibula 2

patella 2

tarsal 14

metatarsal 10

phalanx 28

Total 206 bones

OPTIONAL ACTIVITY

Use different colored pencils to distinguish the individual bones in figure 12.1.

4. Study table 7.2 in chapter 7 of the textbook. Locate each of the following features (bone markings) on the bone listed, noting the size, shape, and location in the human skeleton:

condyle—occipital	meatus—temporal
crest—coxa (hipbone)	process—temporal
epicondyle—femur	sinus—frontal
facet—vertebra	spine—scapula
fontanel—skull	suture—skull
foramen—vertebra	trochanter—femur
fossa—humerus	tubercle—humerus
fovea—femur	tuberosity—tibia
head—humerus	

Critical Thinking Application

Locate and name the largest foramen in the skull.

____

____

Locate and name the largest foramen in the skeleton.

____

____

5. Complete Parts A, B, and C of Laboratory Report 12.

DEMONSTRATION

Images on radiographs (X rays) are produced by allowing X rays from an X-ray tube to pass through a body part and to expose photographic film that is positioned on the opposite side of the part. The image that appears on the film after it is developed reveals the presence of parts with different densities. Bone, for example, is very dense tissue and is a good absorber of X rays. Thus, bone generally appears light on the film. Air-filled spaces, on the other hand, absorb almost no X rays and appear as dark areas on the film. Liquids and soft tissues absorb intermediate quantities of X rays, so they usually appear in various shades of gray.

Examine the available radiographs (X rays) of skeletal structures by holding each film in front of a light source. Identify as many of the bones and features as you can.

Laboratory Report **12**

Name ______________________

Date ______________________

Section ______________________

ORGANIZATION OF THE SKELETON

PART A

Complete the following statements:

1. The two divisions of the skeleton are the ______________________ skeleton and the appendicular skeleton.
2. The ______________________ bone supports the tongue.
3. The ______________________ at the inferior end of the sacrum is composed of several fused vertebrae.
4. The ribs are attached posteriorly to the ______________________.
5. The thoracic cage is composed of ______________________ pairs of ribs.
6. The scapulae and clavicles together form the ______________________.
7. The humerus, radius, and ______________________ articulate to form the elbow joint.
8. The wrist is composed of eight bones called ______________________.
9. The coxae (hipbones) are attached posteriorly to the ______________________.
10. The pelvic girdle (coxae), sacrum, and coccyx together form the ______________________.
11. The ______________________ covers the anterior surface of the knee.
12. The bones that articulate with the distal ends of the tibia and fibula are called ______________________.
13. All finger and toe bones are called ______________________.

PART B

Match the terms in column A with the definitions in column B. Place the letter of your choice in the space provided.

Column A		Column B
a. Condyle	_______	1. Small, nearly flat articular surface
b. Crest	_______	2. Deep depression
c. Facet	_______	3. Rounded process
d. Fontanel	_______	4. Opening or passageway
e. Foramen	_______	5. Interlocking line of union
f. Fossa	_______	6. Narrow, ridgelike projection
g. Suture	_______	7. Soft region between bones of skull

PART C

Match the terms in column A with the definitions in column B. Place the letter of your choice in the space provided.

Column A		Column B
a. Fovea	_______	1. Tubelike passageway
b. Head	_______	2. Tiny pit or depression
c. Meatus	_______	3. Small, knoblike process
d. Sinus	_______	4. Rounded enlargement at end of bone
e. Spine	_______	5. Air-filled cavity within bone
f. Trochanter	_______	6. Relatively large process
g. Tubercle	_______	7. Thornlike projection

Laboratory Exercise 13

The Skull

MATERIALS NEEDED

Textbook
Human skull, articulated
Human skull, disarticulated (Beauchene)
Human skull, sagittal section

For Optional Activity:

Colored pencils

For Demonstration:

Fetal skull

A human skull consists of twenty-two bones that, except for the lower jaw, are firmly interlocked along sutures. Eight of these immovable bones make up the braincase, or cranium, and thirteen more immovable bones and mandible form the facial skeleton.

PURPOSE OF THE EXERCISE

To examine the structure of the human skull and to identify the bones and major features of the skull.

LEARNING OBJECTIVES

After completing this exercise, you should be able to

1. distinguish between the cranium and the facial skeleton;
2. locate and name the bones of the skull and their major features;
3. locate and name the major sutures of the cranium;
4. locate and name the sinuses of the skull.

PROCEDURE

1. Review the section entitled "Skull" in chapter 7 of the textbook.
2. As a review activity, label figures 13.1, 13.2, 13.3, 13.4, and 13.5.
3. Examine the **cranial bones** of the articulated human skull and the sectioned skull. Also observe the corresponding disarticulated bones. Locate the following bones and features in the laboratory specimens and, at the same time, palpate as many of these bones and features in your own skull as possible.

frontal bone 1

supraorbital foramen

frontal sinuses

parietal bones 2

sagittal suture

coronal suture

occipital bone 1

lambdoidal suture

foramen magnum

occipital condyles

temporal bones 2

squamosal suture

external auditory meatus

mandibular fossae

mastoid process

styloid process

zygomatic process

sphenoid bone 1

sella turcica

sphenoidal sinuses

ethmoid bone 1

cribriform plates

perpendicular plate

superior nasal concha

middle nasal concha

ethmoidal sinuses

crista galli

Figure 13.1 Label the anterior bones and features of the skull. (If the line lacks the word *bone,* label the particular feature of that bone.)

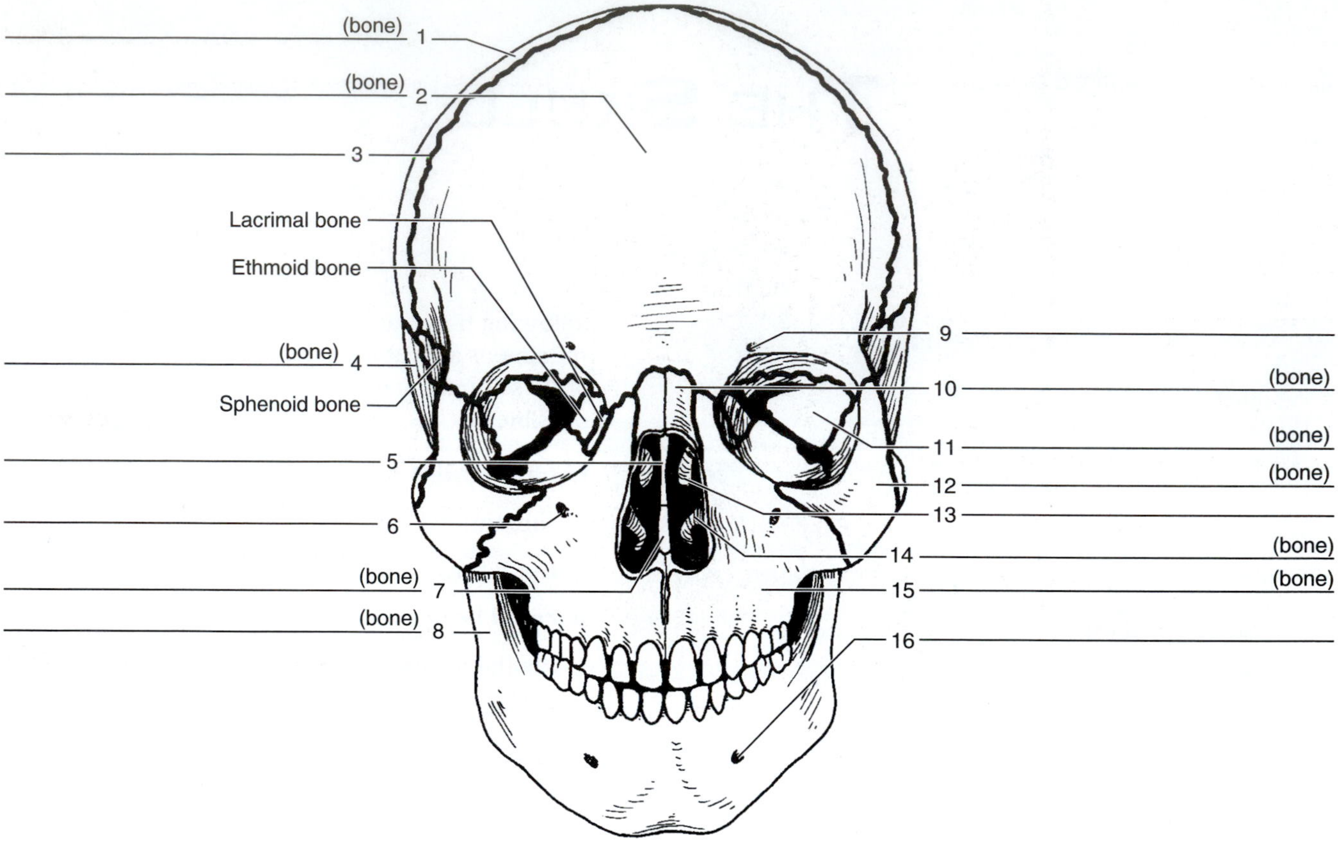

Figure 13.2 Label the lateral bones and features of the skull.

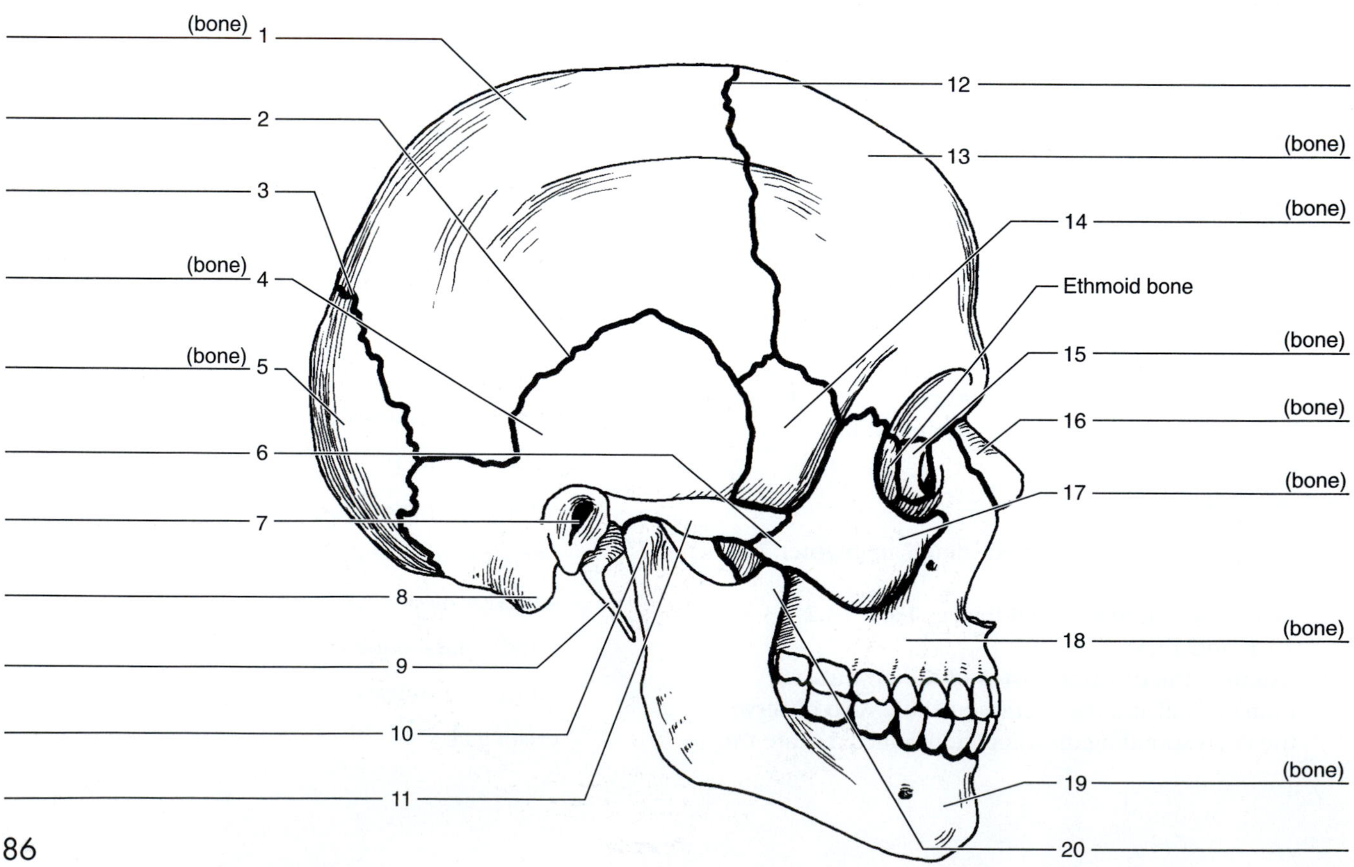

Figure 13.3 Label the inferior bones and features of the skull.

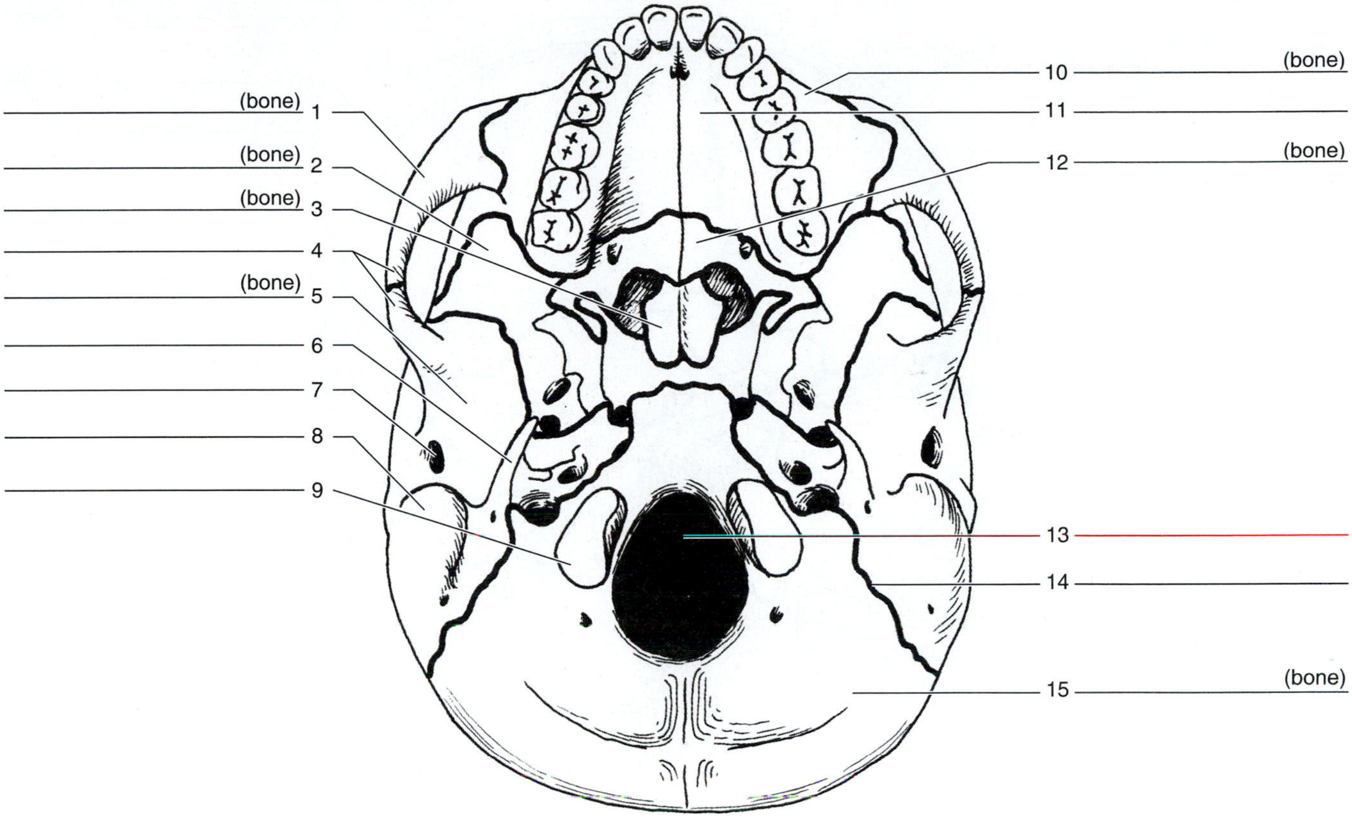

Figure 13.4 Label the bones and features of the floor of the cranial cavity as viewed from above.

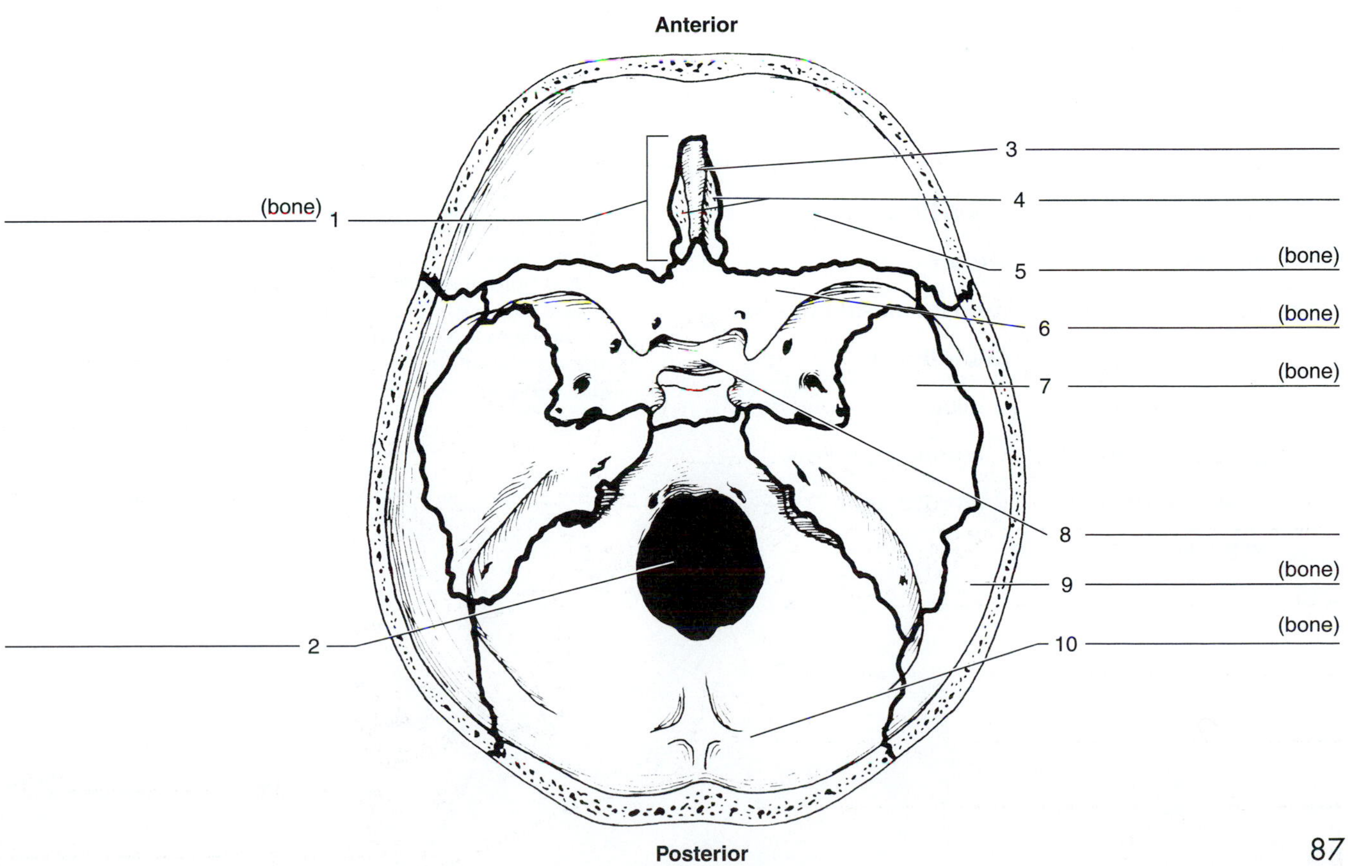

Figure 13.5 Label the bones and features of the sagittal section of the skull.

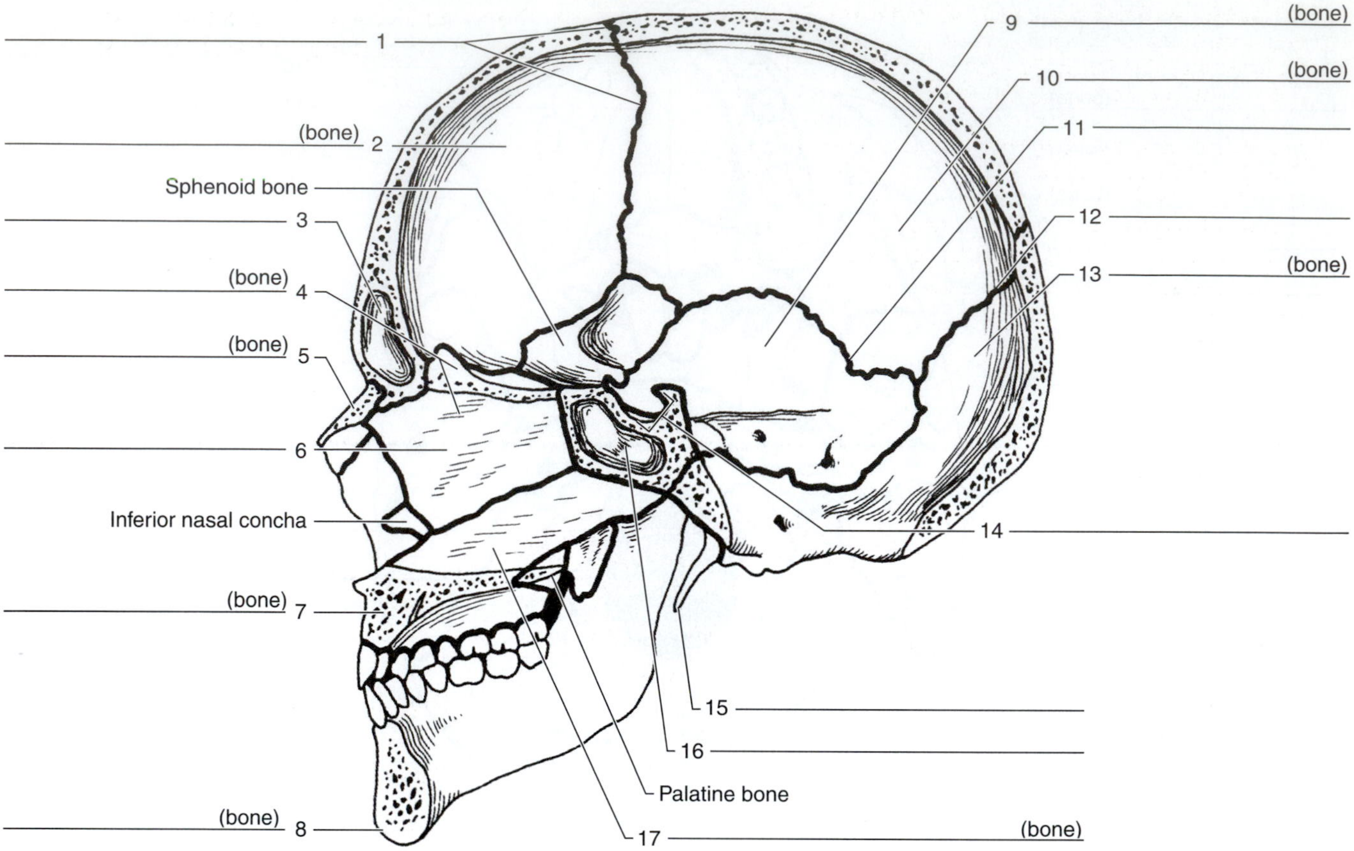

4. Complete Parts A and B of Laboratory Report 13.
5. Examine the **facial bones** of the articulated and sectioned skulls and the corresponding disarticulated bones. Locate the following:

maxillary bones 2
- maxillary sinuses
- palatine processes
- alveolar processes
- alveolar arch

palatine bones 2

zygomatic bones 2
- temporal process
- zygomatic arch

lacrimal bones 2

nasal bones 2

vomer bone 1

inferior nasal conchae 2

mandible 1
- mandibular condyle
- coronoid process
- alveolar arch

6. Complete Parts C and D of the laboratory report.

OPTIONAL ACTIVITY

Use colored pencils to color the bones illustrated in figures 13.1–13.5. Select a different color for each individual bone in the series. This activity should help you locate various bones that are shown in different views in the figures. You can check your work by referring to the corresponding figures in the textbook, which are presented in full color.

DEMONSTRATION

Examine the fetal skull (fig. 13.6). Note that the skull is incompletely developed and that the cranial bones are separated by fibrous membranes. These membranous areas are called *fontanels,* or "soft spots." The fontanels close as the cranial bones grow together. The posterior fontanel usually closes within a few months after birth, whereas the anterior fontanel may not close until the middle or end of the second year. What other features characterize the fetal skull? ______________

____

____

____

Web Quest

What are the functions of individual bones and features? Search these and review the anatomy of the skeleton at www.mhhe.com/shieress8

Figure 13.6 Human fetal skeleton.

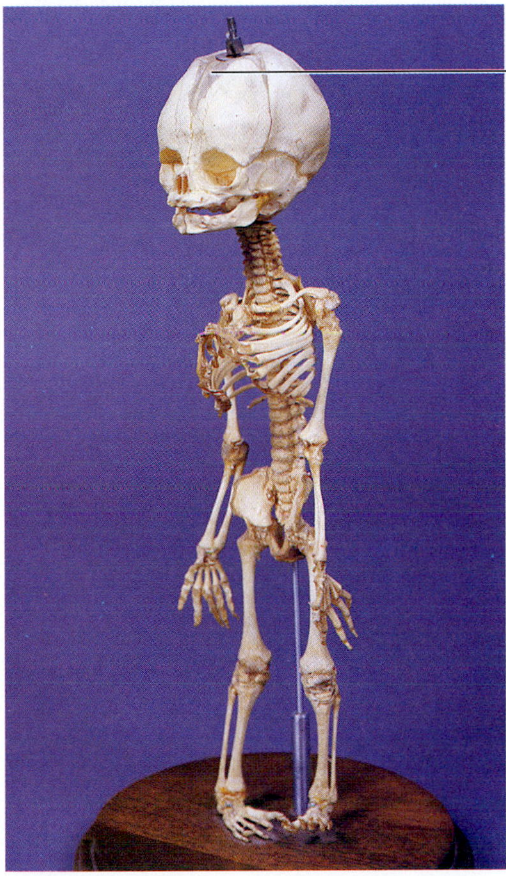

Laboratory Report 13

Name ______________________________

Date ______________________________

Section ______________________________

The Skull

PART A

Match the bones in column A with the features in column B. Place the letter of your choice in the space provided.

Column A

a. Ethmoid bone
b. Frontal bone
c. Occipital bone
d. Parietal bone
e. Sphenoid bone
f. Temporal bone

Column B

_______ 1. Forms sagittal, coronal, squamosal, and lambdoidal sutures
_______ 2. Cribriform plate
_______ 3. Crista galli
_______ 4. External auditory meatus
_______ 5. Foramen magnum
_______ 6. Mandibular fossa
_______ 7. Mastoid process
_______ 8. Middle nasal concha
_______ 9. Occipital condyle
_______ 10. Sella turcica
_______ 11. Styloid process
_______ 12. Supraorbital foramen

PART B

Complete the following statements:

1. The ______________________ suture joins the frontal bone to the parietal bones.
2. The parietal bones are fused along the midline by the ______________________ suture.
3. The ______________________ suture joins the parietal bones to the occipital bone.
4. The temporal bones are joined to the parietal bones along the ______________________ sutures.
5. Name the three cranial bones that contain sinuses. ______________________________________
____
6. Name a facial bone that contains a sinus. ______________________________________

PART C

Match the bones in column A with the characteristics in column B. Place the letter of your choice in the space provided.

Column A

a. Inferior nasal concha
b. Lacrimal bone
c. Mandible
d. Maxillary bone
e. Nasal bone
f. Palatine bone
g. Vomer bone
h. Zygomatic bone

Column B

_______ 1. Forms bridge of nose
_______ 2. Only movable bone in facial skeleton
_______ 3. Contains coronoid process
_______ 4. Creates prominence of cheek inferior and lateral to the eye
_______ 5. Contains sockets of upper teeth
_______ 6. Forms inferior portion of nasal septum in back
_______ 7. Forms anterior portion of zygomatic arch
_______ 8. Scroll-shaped bone
_______ 9. Forms anterior roof of mouth
_______ 10. Forms posterior roof of mouth
_______ 11. Scalelike part in medial wall of orbit

PART D

Identify the numbered bones and features of the skulls in figures 13.7, 13.8, 13.9, and 13.10.

Figure 13.7 Identify the bones and features indicated on this anterior view of the skull, using the terms provided.

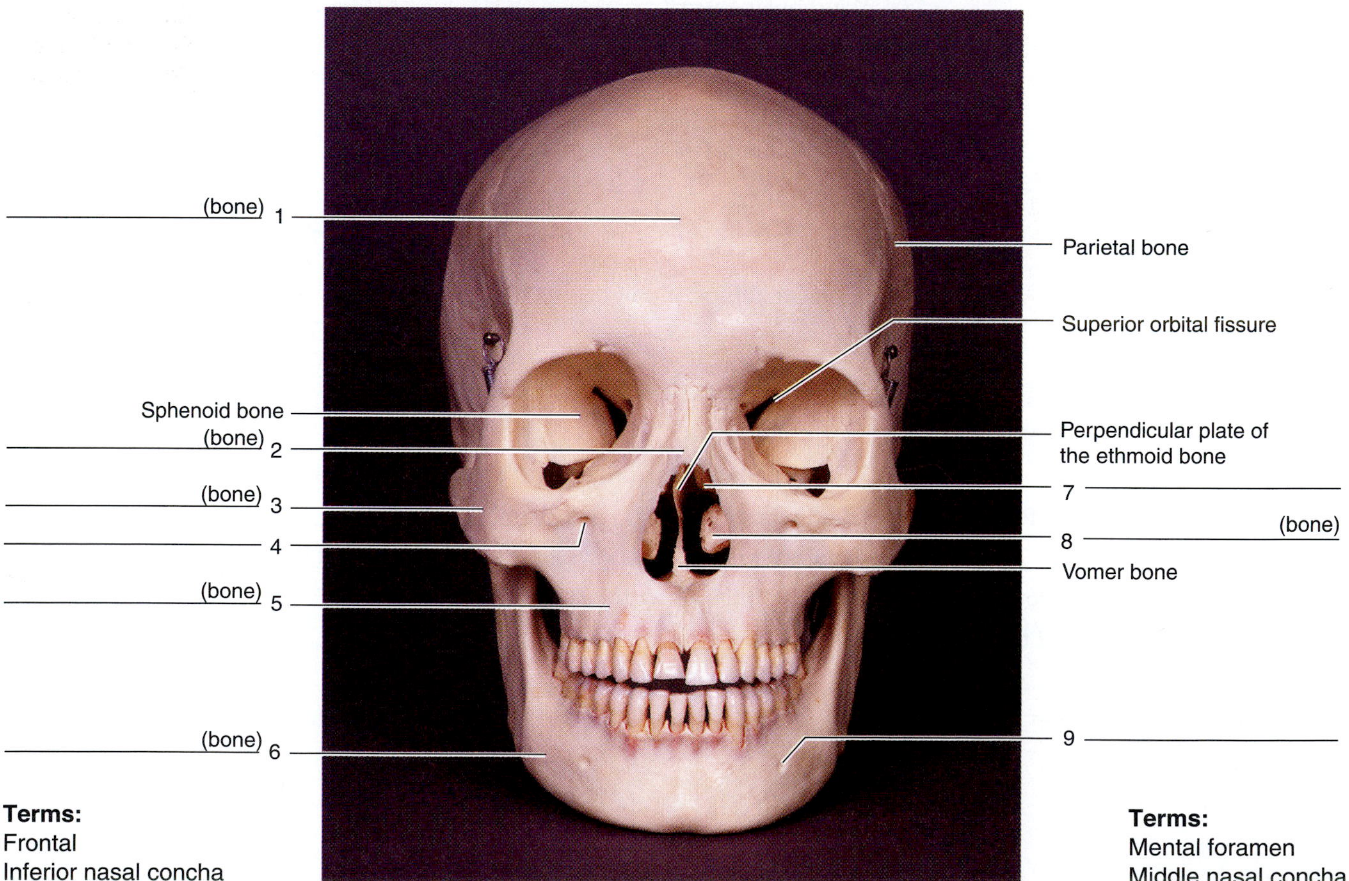

Terms:
Frontal
Inferior nasal concha
Infraorbital foramen
Mandible
Maxilla

Terms:
Mental foramen
Middle nasal concha (of ethmoid bone)
Nasal
Zygomatic

Figure 13.8 Identify the bones and features on this floor of the cranial cavity of a skull, using the terms provided.

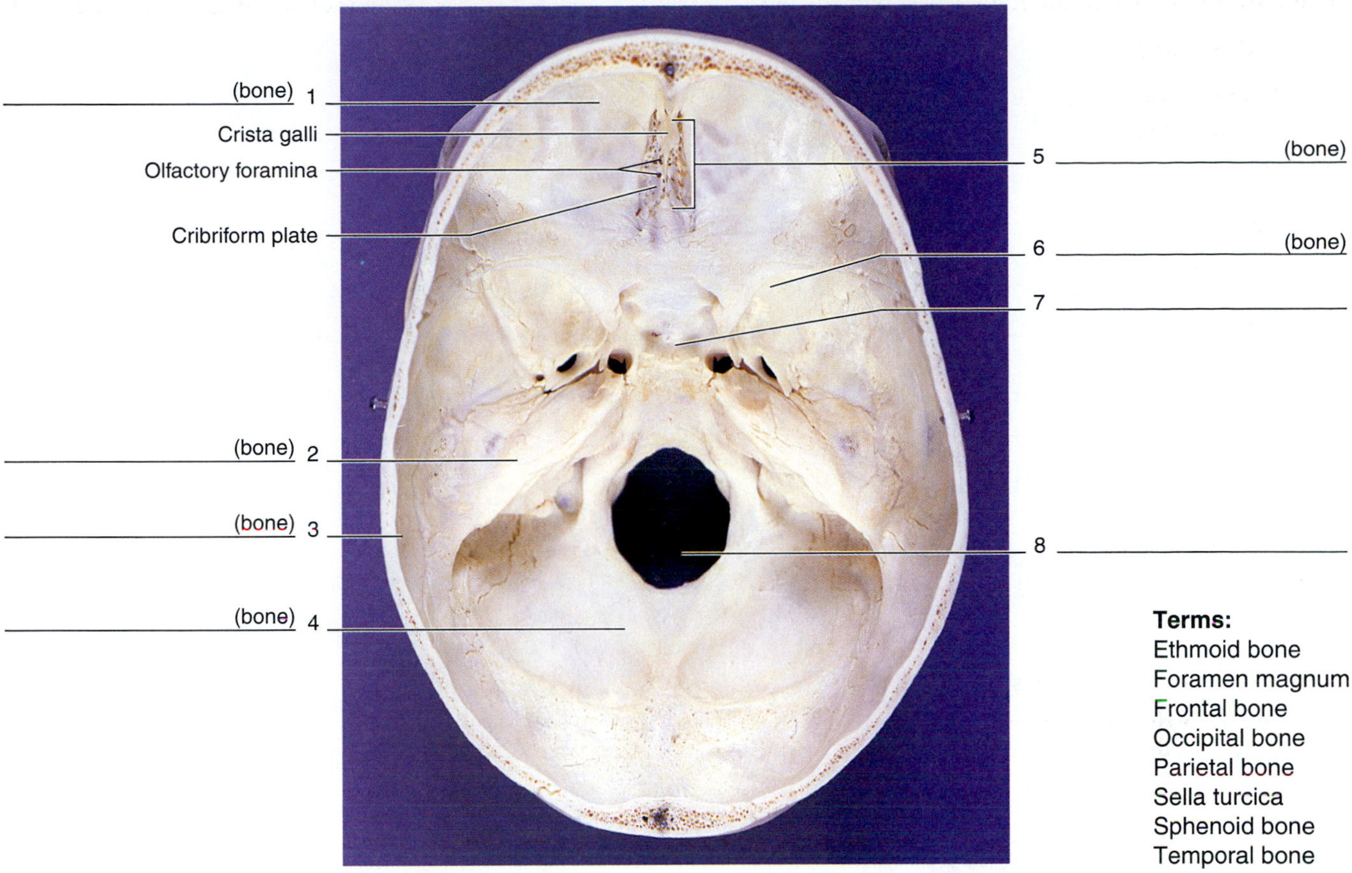

Terms:
Ethmoid bone
Foramen magnum
Frontal bone
Occipital bone
Parietal bone
Sella turcica
Sphenoid bone
Temporal bone

Figure 13.9 Identify the bones and features indicated on this inferior view of the skull, using the terms provided.

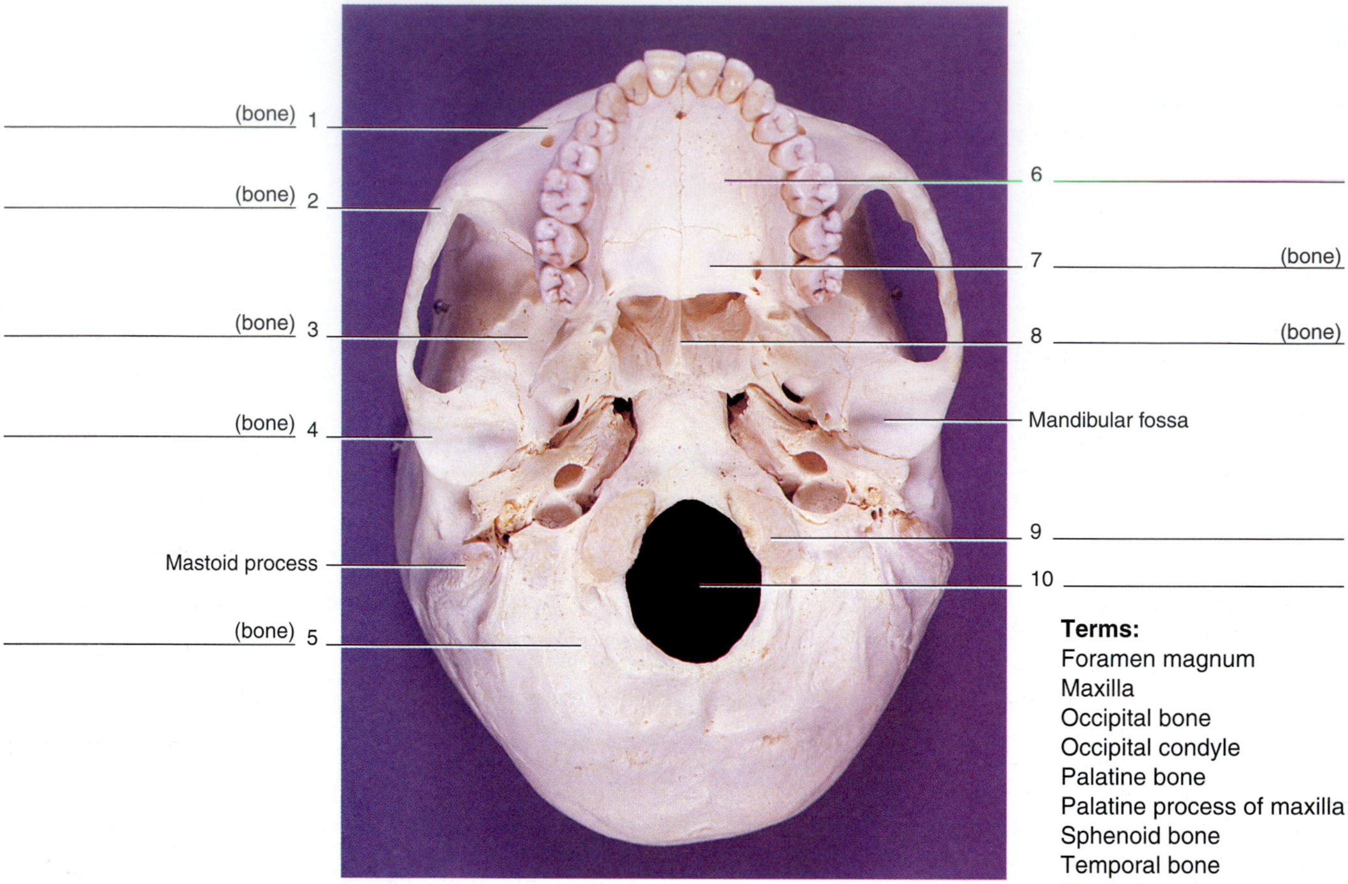

Terms:
Foramen magnum
Maxilla
Occipital bone
Occipital condyle
Palatine bone
Palatine process of maxilla
Sphenoid bone
Temporal bone
Vomer bone
Zygomatic bone

Figure 13.10 Identify the bones on this disarticulated skull, using the terms provided.

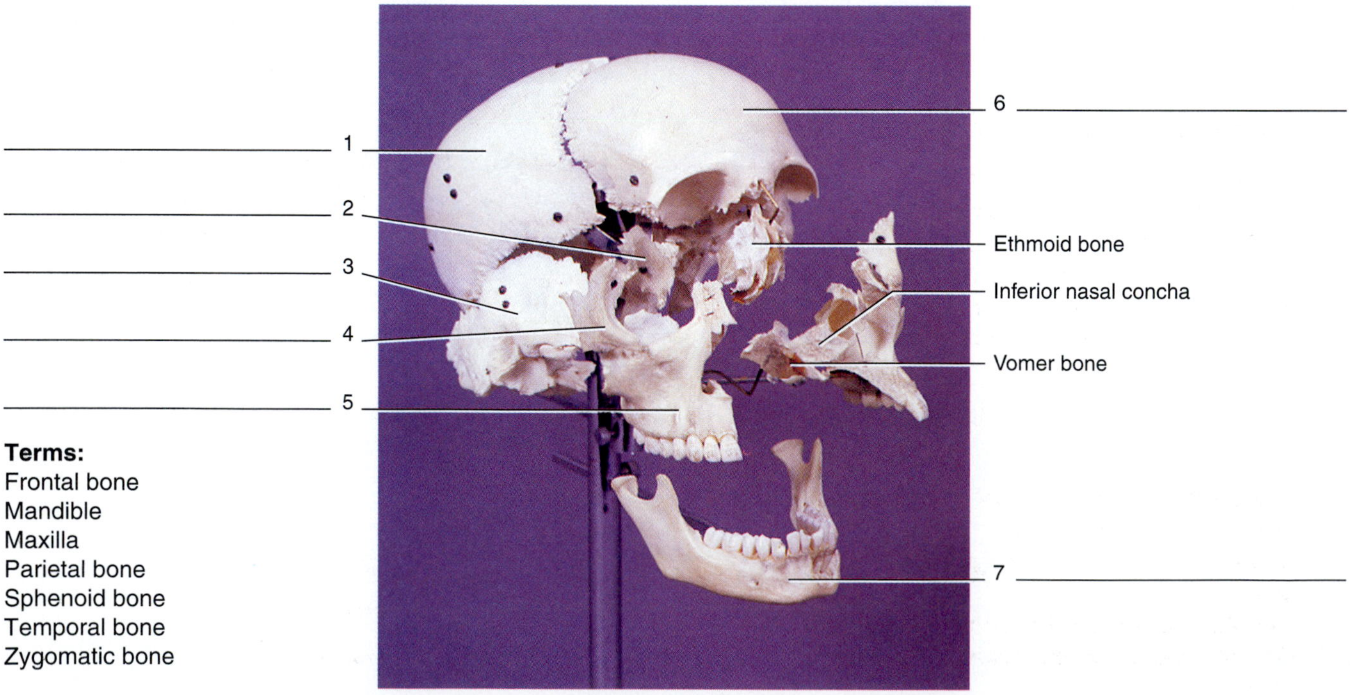

Terms:
Frontal bone
Mandible
Maxilla
Parietal bone
Sphenoid bone
Temporal bone
Zygomatic bone

VERTEBRAL COLUMN AND THORACIC CAGE

MATERIALS NEEDED

Textbook
Human skeleton, articulated
Samples of cervical, thoracic, and lumbar vertebrae
Human skeleton, disarticulated

The vertebral column, consisting of twenty-six bones, extends from the skull to the pelvis and forms the vertical axis of the human skeleton. The column is composed of many vertebrae, which are separated from one another by cartilaginous intervertebral discs and are held together by ligaments.

The thoracic cage surrounds the thoracic and upper abdominal cavities. It includes the ribs, the thoracic vertebrae, the sternum, and the costal cartilages.

PURPOSE OF THE EXERCISE

To examine the vertebral column and the thoracic cage of the human skeleton and to identify the bones and major features of these parts.

LEARNING OBJECTIVES

After completing this exercise, you should be able to

1. identify the major features of the vertebral column;
2. name the features of a typical vertebra;
3. distinguish among a cervical, thoracic, and lumbar vertebra and locate the sacrum and coccyx;
4. identify the structures of the thoracic cage;
5. distinguish between true and false ribs.

PROCEDURE A—THE VERTEBRAL COLUMN

1. Review the section entitled "Vertebral Column" in chapter 7 of the textbook.
2. As a review activity, label figures 14.1, 14.2, 14.3, and 14.4.
3. Examine the vertebral column of the human skeleton and locate the following bones and features. At the same time, locate as many of the corresponding bones and features in your own skeleton as possible.

atlas 1	sacrum 1
axis 1	coccyx 1
cervical vertebrae 7	cervical curvature
thoracic vertebrae 12	thoracic curvature
lumbar vertebrae 5	lumbar curvature
intervertebral discs	pelvic curvature
vertebral canal	intervertebral foramina

Critical Thinking Application

Note the four curvatures of the vertebral column. What functional advantages exist with curvatures for skeletal structure instead of a straight vertebral column?

4. Compare the available samples of cervical, thoracic, and lumbar vertebrae by noting differences in size, shape, and by locating the following features:

body	transverse foramina
pedicles	facets
vertebral foramen	dens of axis
laminae	
spinous process	
vertebral arch	
transverse processes	
superior articular processes	
inferior articular processes	

Figure 14.1 Label the bones and features of the vertebral column.

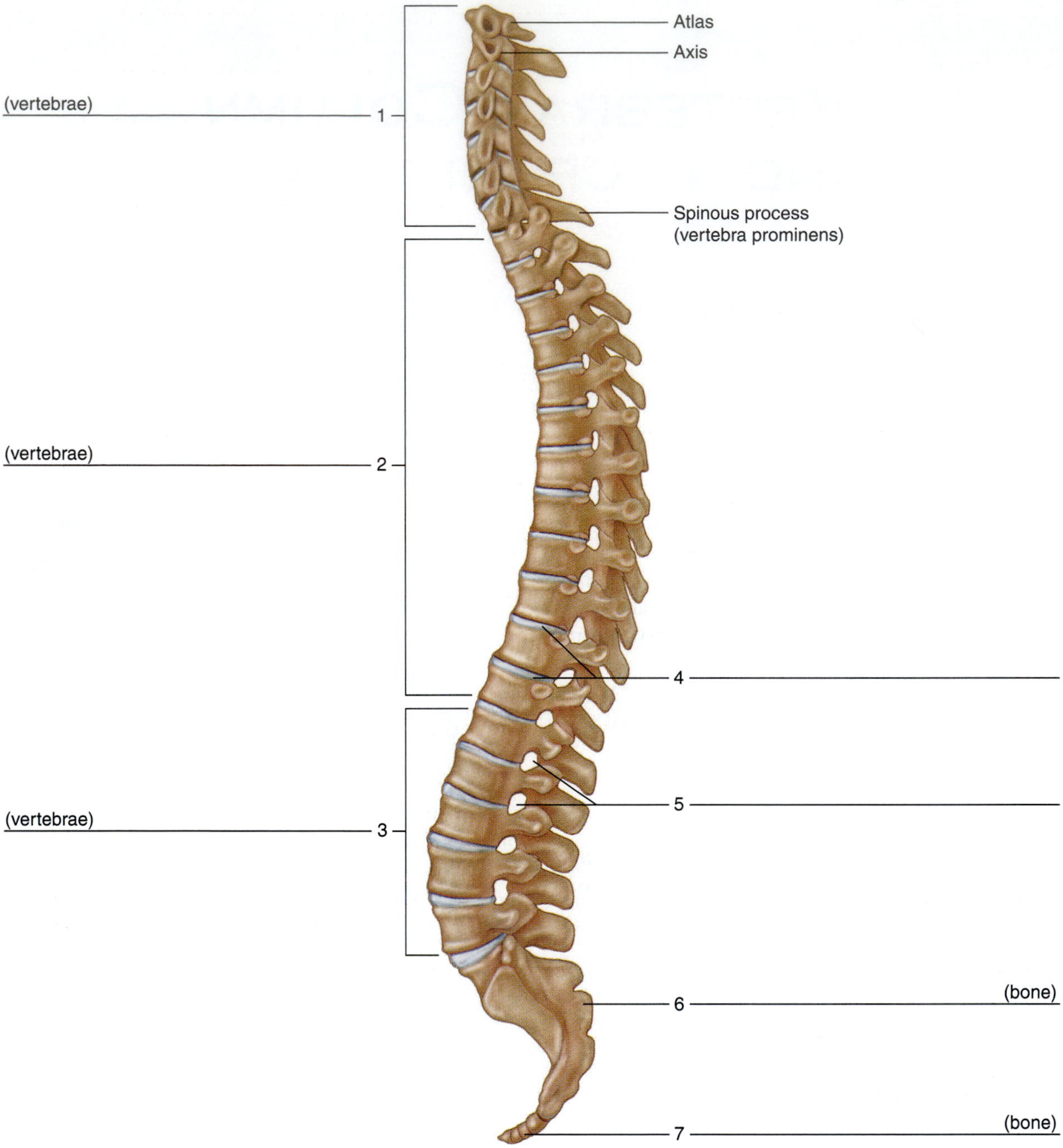

Figure 14.2 Label the superior features of (*a*) the atlas and (*b*) the axis.

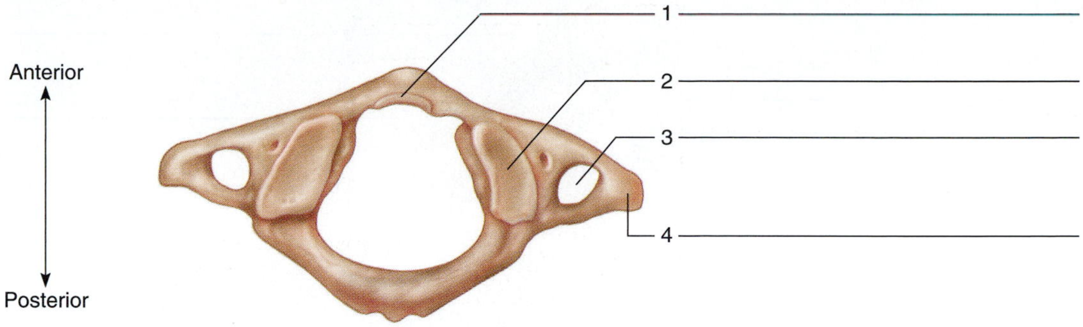

(a)

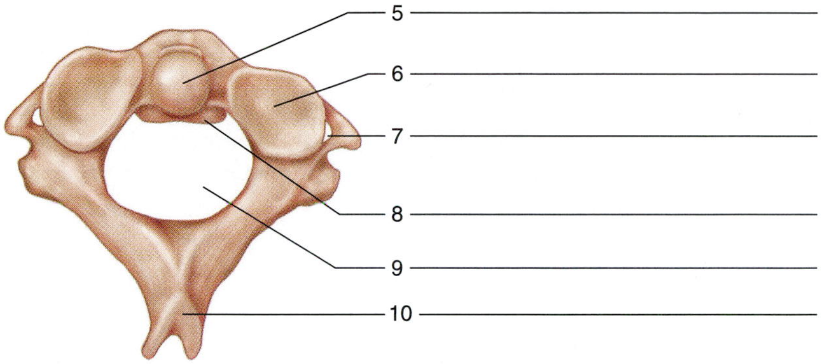

(b)

5. Examine the sacrum and coccyx. Locate the following features:

   sacrum

   superior articular process

   dorsal sacral foramen

   pelvic sacral foramen

   sacral canal

   tubercles

   sacral hiatus

   coccyx

6. Complete Parts A and B of Laboratory Report 14.

PROCEDURE B—THE THORACIC CAGE

1. Review the section entitled "Thoracic Cage" in chapter 7 of the textbook.
2. As a review activity, label figure 14.5.
3. Examine the thoracic cage of the human skeleton and locate the following bones and features:

   rib

   head

   tubercle

   facets

   true ribs

   false ribs

   floating ribs

   costal cartilages

   sternum

   manubrium

   body

   xiphoid process

4. Complete Parts C and D of the laboratory report.

Figure 14.3 Label the superior features of the (*a*) cervical, (*b*) thoracic, and (*c*) lumbar vertebrae.

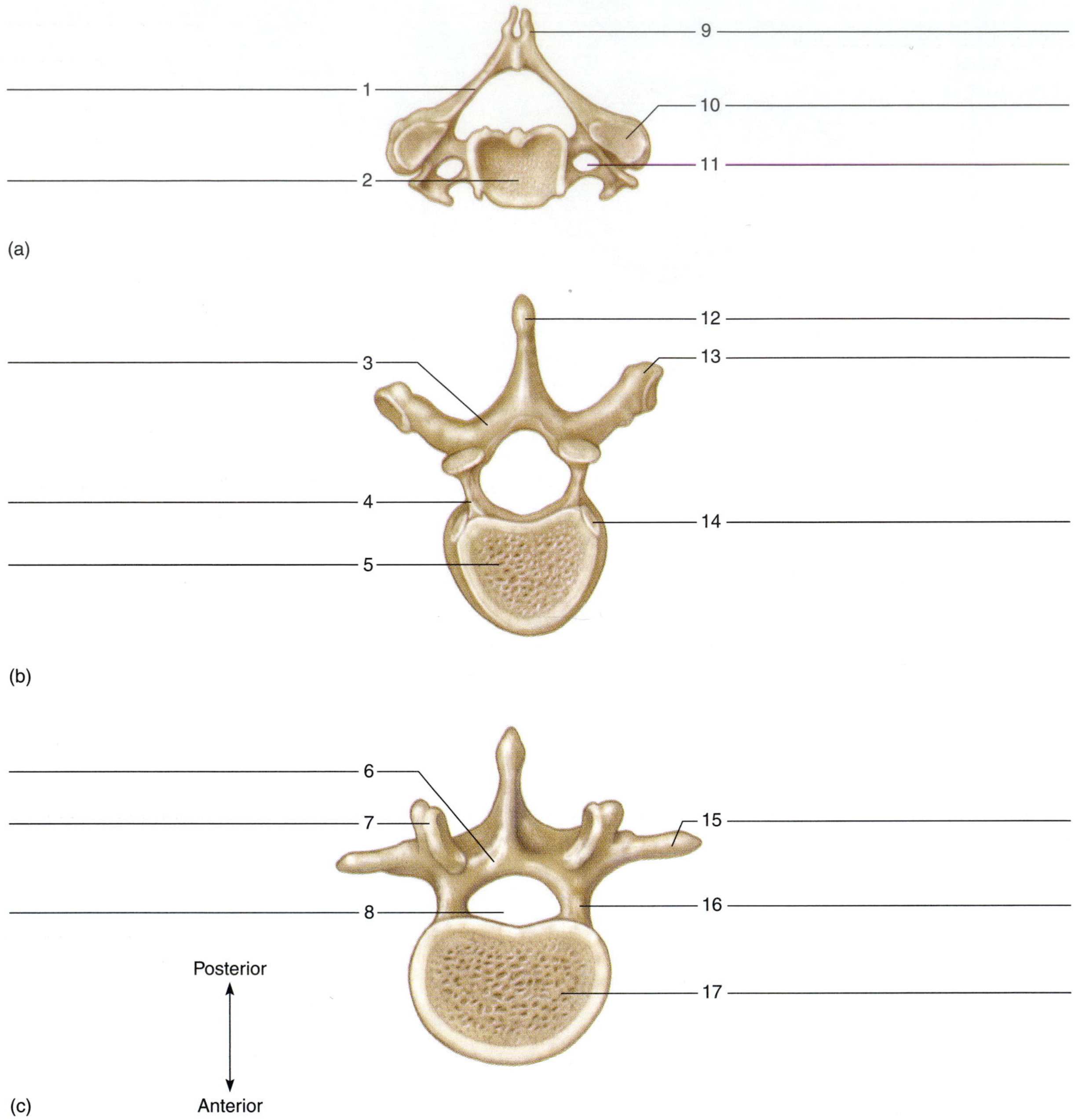

Web Quest

What are the functions of individual bones and features? Search these and review the anatomy of the skeleton at www.mhhe.com/shieress8

Figure 14.4 Label the coccyx and the features of the sacrum: (*a*) anterior view; (*b*) posterior view.

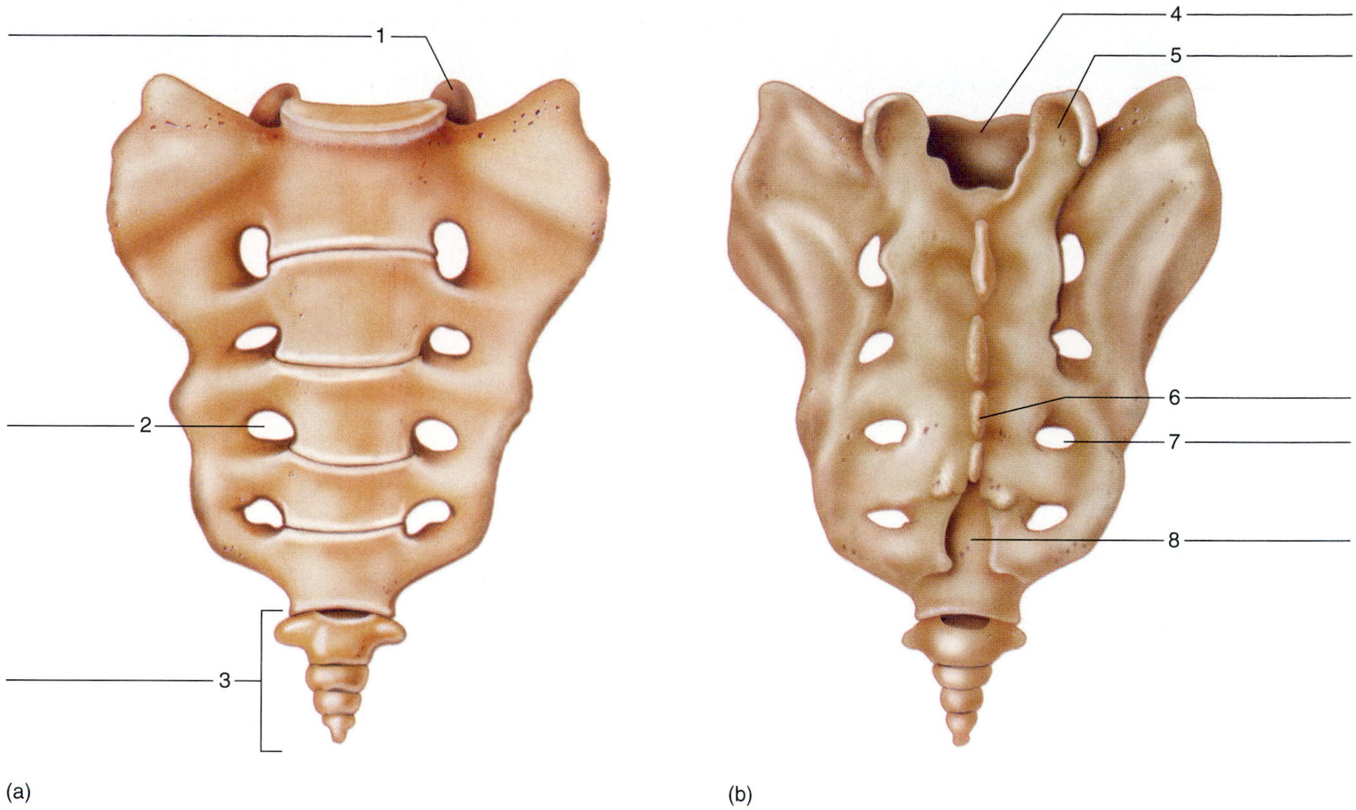

Figure 14.5 Label the bones and features of the thoracic cage.

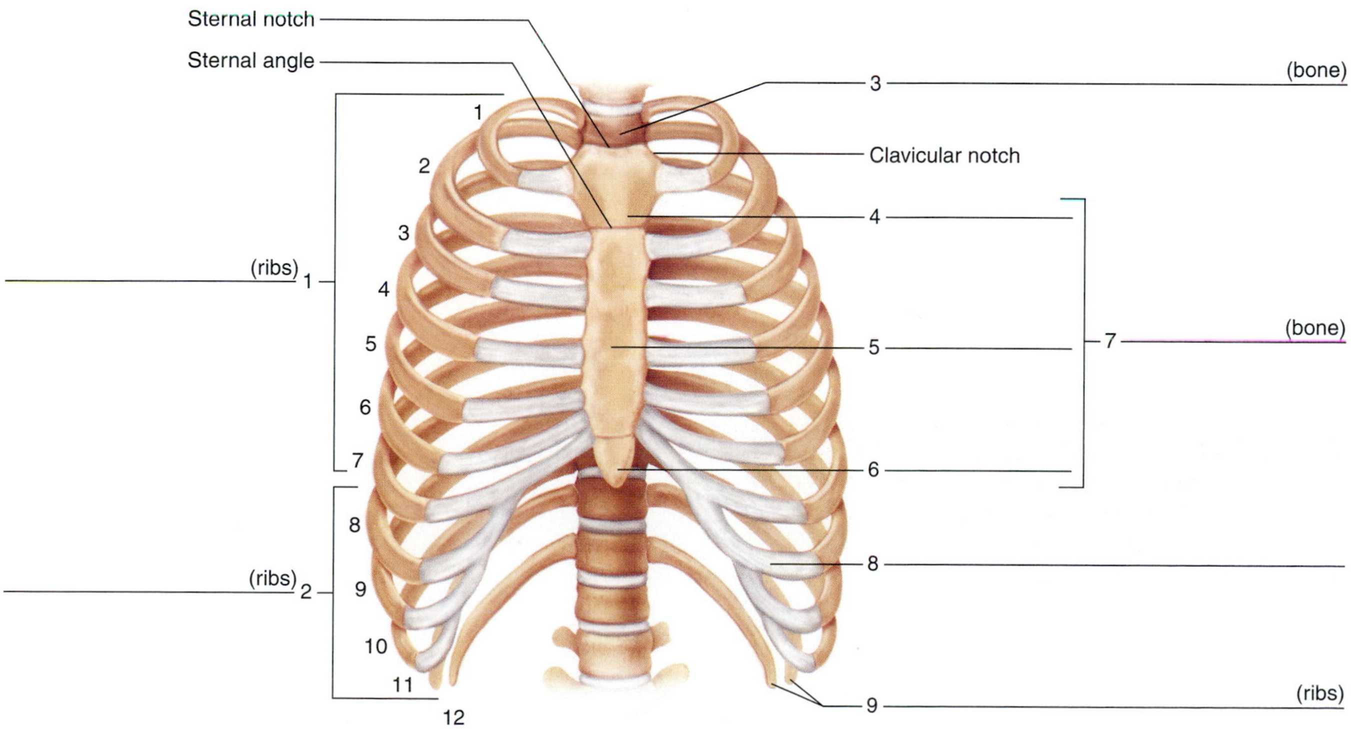

Laboratory Report **14**

Name ______________________________

Date ______________________________

Section ______________________________

Vertebral Column and Thoracic Cage

PART A

Complete the following statements:

1. The vertebral column encloses and protects the ______________________ .
2. The ______________________ of the vertebrae support the weight of the head and trunk.
3. The ______________________ separate adjacent vertebrae.
4. The pedicles, laminae, and spinous process of a vertebra form the ______________________ .
5. The intervertebral foramina provide passageways for ______________________ .
6. Transverse foramina of cervical vertebrae serve as passageways for ______________________ leading to the brain.
7. The first vertebra is also called the ______________________ .
8. When the head is moved from side to side, the first vertebra pivots around the ______________________ of the second vertebra.
9. The ______________________ vertebrae have the largest and strongest bodies.
10. The number of vertebrae that fuse to form the sacrum is ______________________ .
11. An opening called the ______________________ exists at the tip of the sacral canal.

PART B

Based on your observations, compare typical cervical, thoracic, and lumbar vertebrae in relation to the characteristics indicated in the table.

Vertebra	Number	Size	Body	Spinous Process	Transverse Foramina
Cervical					
Thoracic					
Lumbar					

PART C

Complete the following statements:

1. The adult skeleton of most men and women contains a total number of ______________________ bones.
2. The last two pairs of ribs that have no cartilaginous attachments to the sternum are sometimes called ______________________ ribs.
3. The tubercles of the ribs articulate with facets on the ______________________ processes of the thoracic vertebrae.
4. The manubrium articulates with the ______________________ on its superior border.
5. List three general functions of the thoracic cage.

   a. ______________________

   b. ______________________

   c. ______________________

PART D

Identify the bones and features indicated in the radiograph of the neck in figure 14.6.

Figure 14.6 Identify the bones and features indicated in this radiograph (X ray) of the neck (lateral view), using the terms provided.

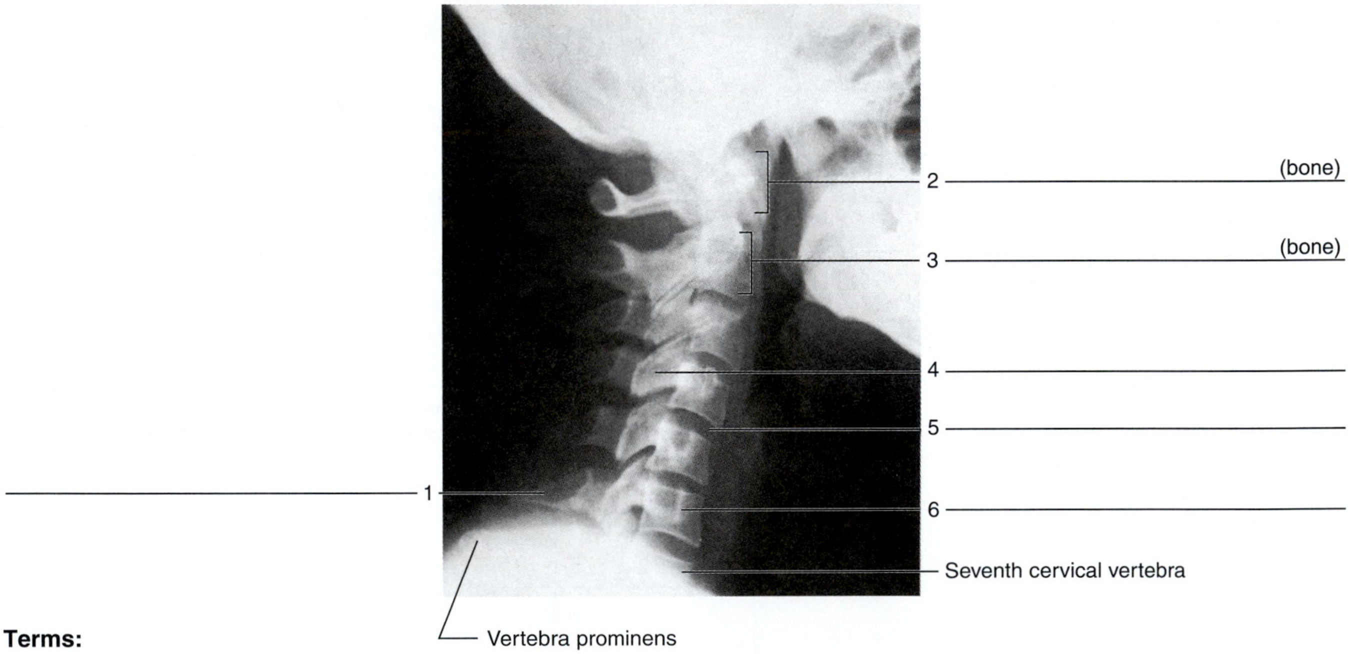

Terms:
Atlas
Axis
Body
Intervertebral disc
Spinous process
Transverse process

Laboratory Exercise 15

Pectoral Girdle and Upper Limb

MATERIALS NEEDED

Textbook
Human skeleton, articulated
Human skeleton, disarticulated

For Optional Activity:

Colored pencils

The pectoral girdle consists of two clavicles and two scapulae. These parts function to support the upper limbs and to serve as attachments for various muscles that move these limbs.

Each upper limb includes a humerus, a radius, an ulna, eight carpals, five metacarpals, and fourteen phalanges. These bones form the framework of the arm, forearm, and hand. They also function as parts of levers when the limbs are moved.

PURPOSE OF THE EXERCISE

To examine the bones of the pectoral girdle and upper limb and to identify the major features of these bones.

LEARNING OBJECTIVES

After completing this exercise, you should be able to

1. locate and identify the bones of the pectoral girdle and their major features;
2. locate and identify the bones of the upper limb and their major features.

PROCEDURE A—THE PECTORAL GIRDLE

1. Review the section entitled "Pectoral Girdle" in chapter 7 of the textbook.
2. As a review activity, label figures 15.1 and 15.2.
3. Examine the bones of the pectoral girdle and locate the following features. At the same time, locate as many of the corresponding surface bones and features of your own skeleton as possible.

   clavicle

   scapula

   spine

   acromion process

   coracoid process

   glenoid cavity (fossa)

Critical Thinking Application

Why is the clavicle a bone that can easily fracture?

___

___

___

___

4. Complete Part A of Laboratory Report 15.

Figure 15.1 Label the bones and features of the shoulder and upper limb (anterior view).

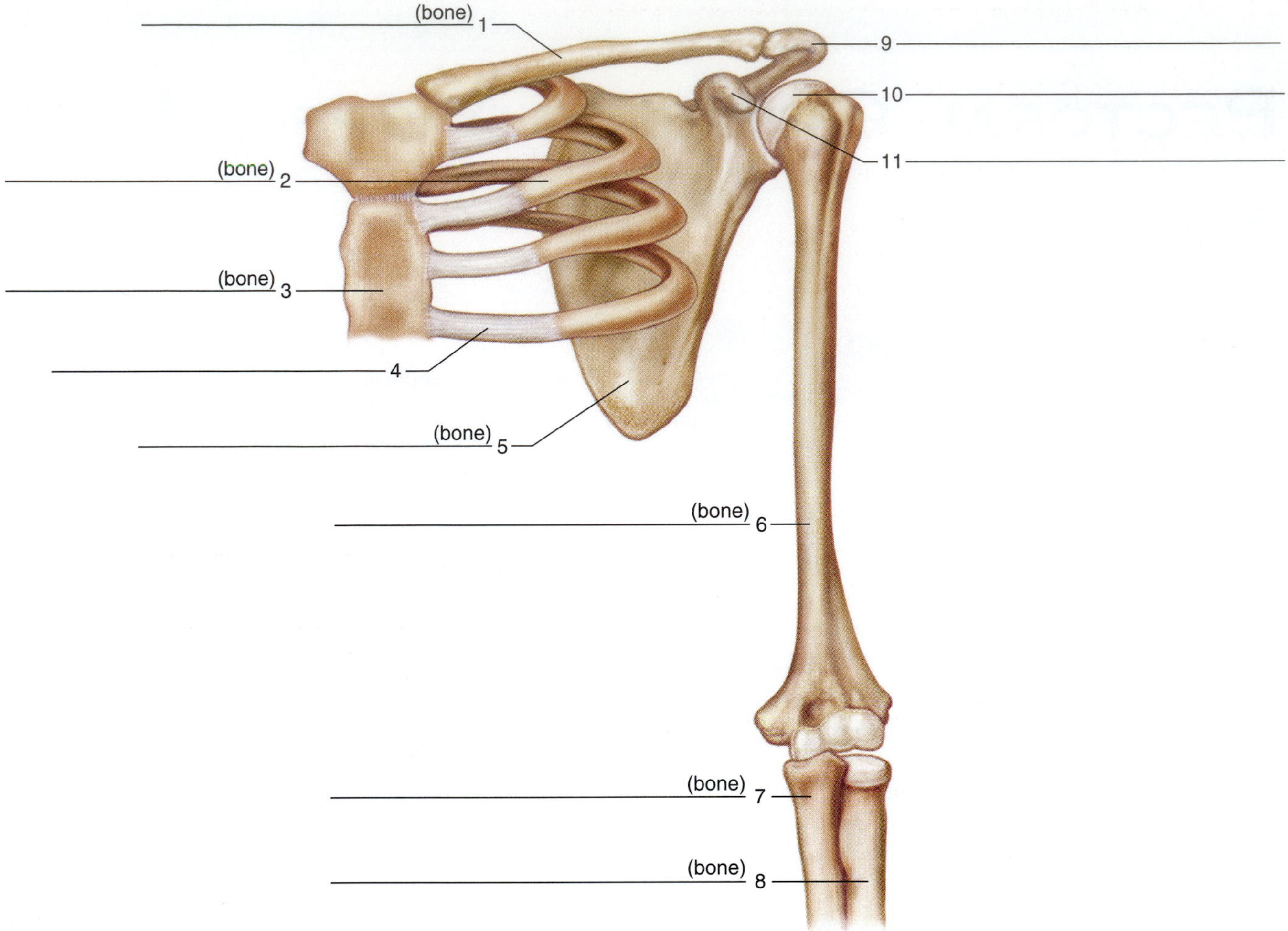

Figure 15.2 Label (*a*) the posterior surface of the right scapula and (*b*) the lateral aspect of the right scapula.

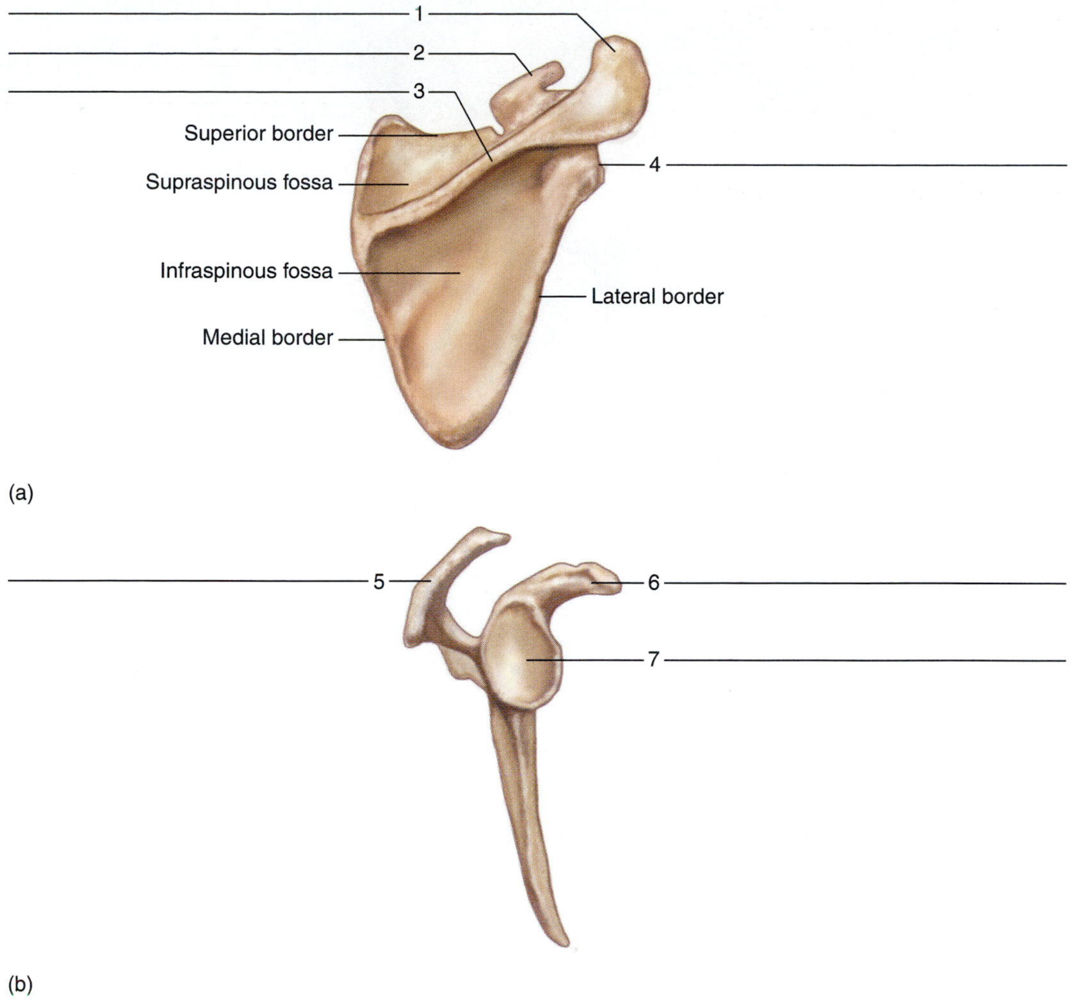

PROCEDURE B—THE UPPER LIMB

1. Review the section entitled "Upper Limb" in chapter 7 of the textbook.
2. As a review activity, label figures 15.3, 15.4, and 15.5.
3. Examine the following bones and features of the upper limb:

humerus

head

greater tubercle

lesser tubercle

intertubercular groove

anatomical neck

surgical neck

deltoid tuberosity

capitulum

trochlea

medial epicondyle

lateral epicondyle

coronoid fossa

olecranon fossa

radius

head

radial tuberosity

styloid process

ulna

trochlear notch (semilunar notch)

olecranon process

coronoid process

styloid process

head

Figure 15.3 Label the (*a*) posterior surface and (*b*) anterior surface of the left humerus.

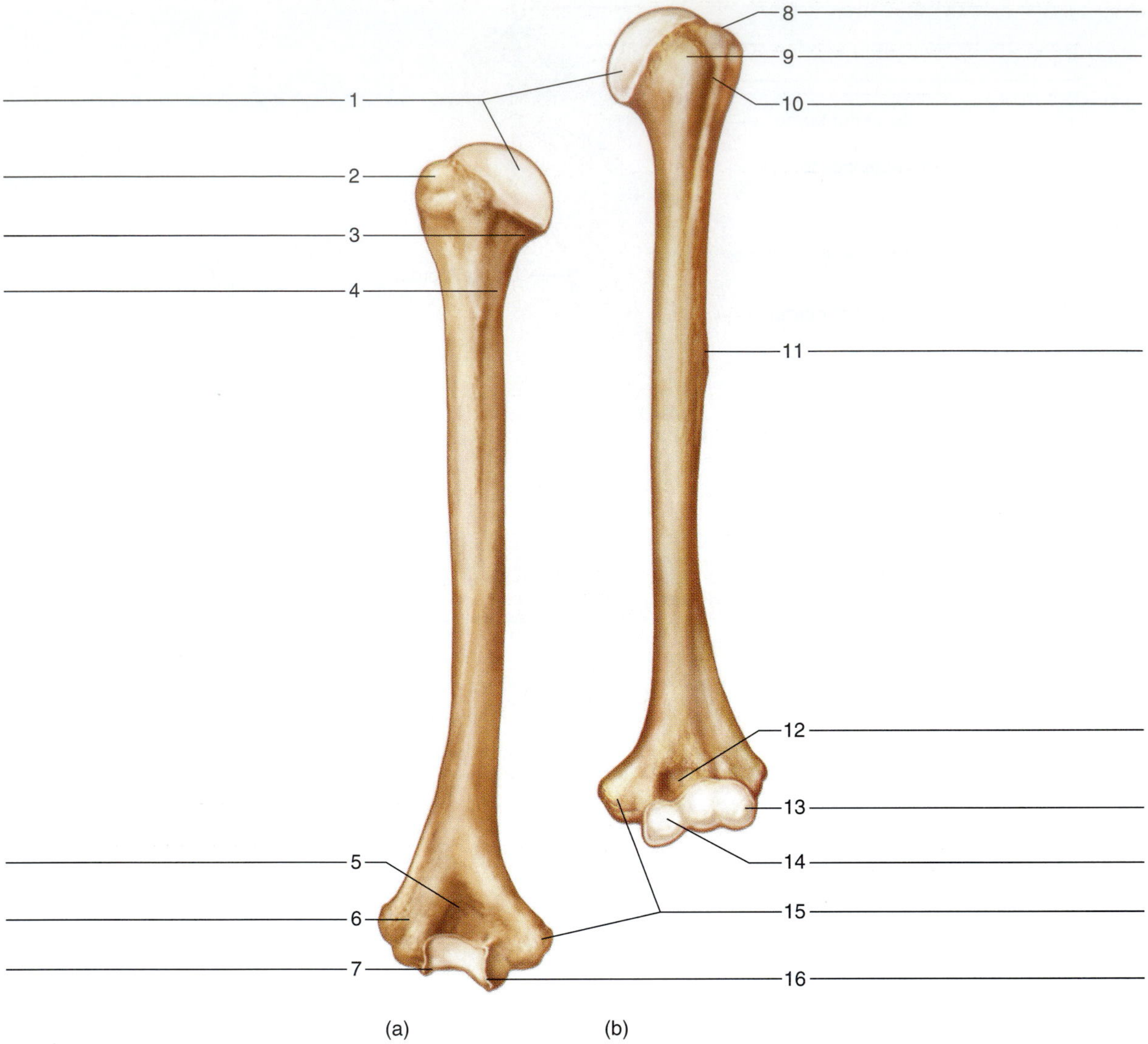

Figure 15.4 Label the major anterior features of the right radius and ulna.

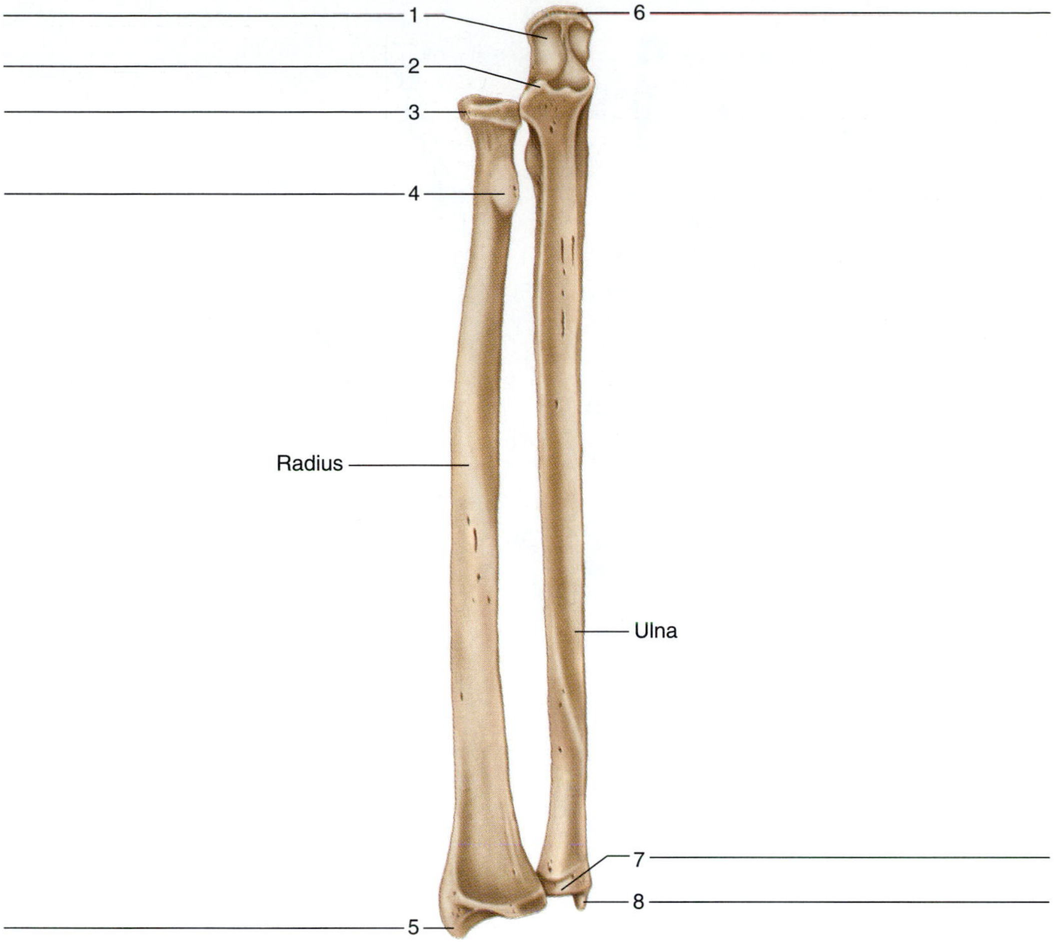

Figure 15.5 Label the bones and groups of bones in this posterior view of the left hand.

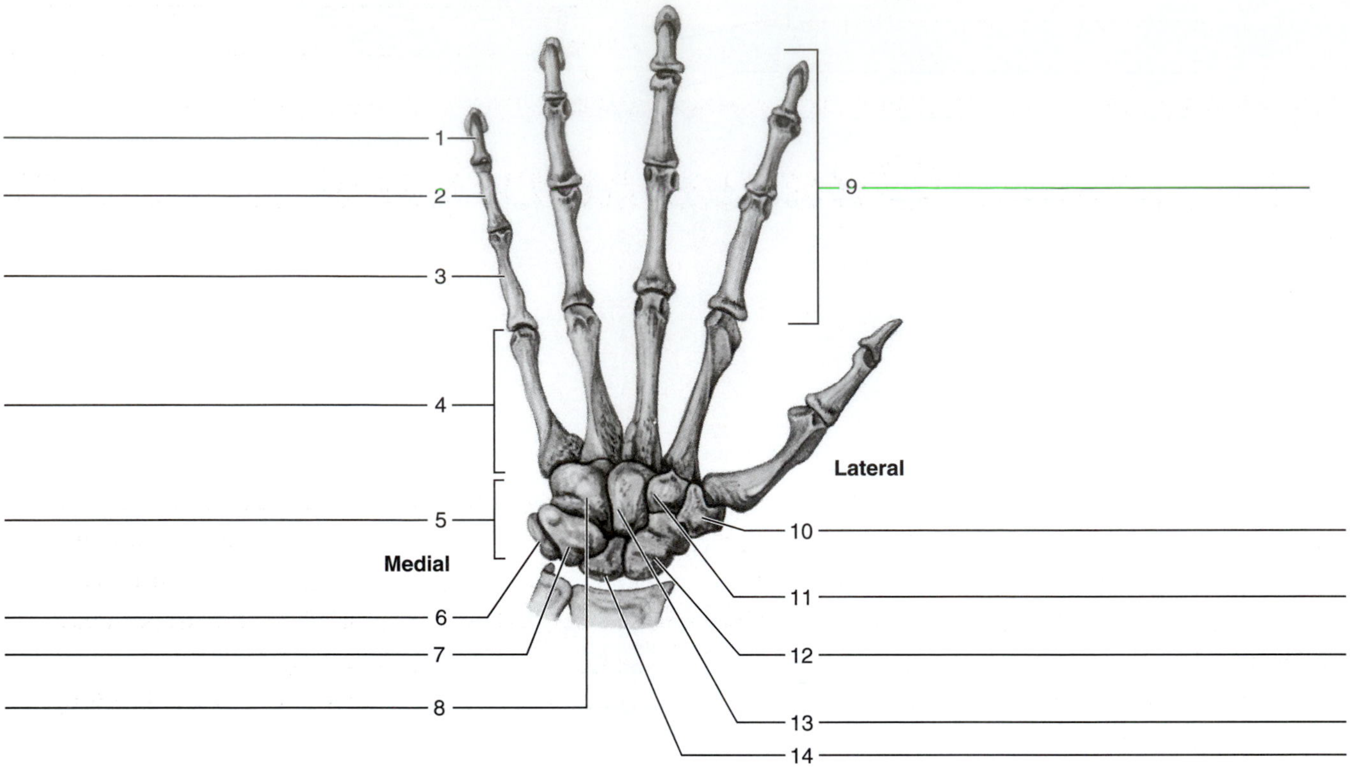

carpal bones

proximal row (listed lateral to medial)

scaphoid

lunate

triquetrum

pisiform

distal row (listed medial to lateral)

hamate

capitate

trapezoid

trapezium

metacarpal bones

phalanges

proximal phalanx

middle phalanx

distal phalanx

4. Complete Parts B and C of the laboratory report.

The following mnemonic device will help you learn the eight carpals:

So **L**ong **T**op **P**art
Here **C**omes **T**he **T**humb

The first letter of each word corresponds to the first letter of a carpal. Notice that this device arranges the carpals in order for the proximal, transverse row of four bones from lateral to medial, followed by the distal, transverse row from medial to lateral, which ends nearest the thumb. This arrangement assumes the anatomical position of the hand.

OPTIONAL ACTIVITY

Use different colored pencils to distinguish the individual bones in figure 15.5.

Web Quest

What are the functions of individual bones and features? Search these and review the anatomy of the skeleton at www.mhhe.com/shieress8

Laboratory Report **15**

Name ______________________

Date ______________________

Section ______________________

PECTORAL GIRDLE AND UPPER LIMB

PART A

Complete the following statements:

1. The pectoral girdle is an incomplete ring because it is open in the back between the ______________ .
2. The medial ends of the clavicles articulate with the ______________ of the sternum.
3. The lateral ends of the clavicles articulate with the ______________ of the scapula.
4. The ______________ divides the posterior side of the scapula into unequal portions.
5. The tip of the shoulder is due to the ______________ of the scapula.
6. At the lateral end of the scapula, the ______________ curves anteriorly and inferiorly from the clavicle.
7. The glenoid cavity of the scapula articulates with the ______________ of the humerus.

PART B

Match the bones in column A with the features in column B. Place the letter of your choice in the space provided.

Column A		Column B
a. Carpals	_______	1. Capitate
b. Humerus	_______	2. Coronoid fossa
c. Metacarpals	_______	3. Deltoid tuberosity
d. Phalanges	_______	4. Greater tubercle
e. Radius	_______	5. Intertubercular groove
f. Ulna	_______	6. Lunate
	_______	7. Olecranon fossa
	_______	8. Five palmar bones
	_______	9. Radial tuberosity
	_______	10. Trapezium
	_______	11. Triquetrum (triangular)
	_______	12. Trochlear notch
	_______	13. Fourteen bones in digits

PART C

Identify the bones and features indicated in the radiographs (X rays) of figures 15.6, 15.7, and 15.8.

Figure 15.6 Identify the bones and features indicated on this radiograph of the elbow, using the terms provided.

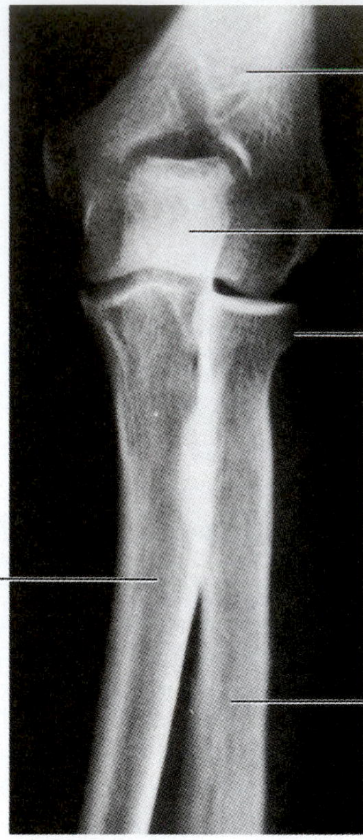

Terms:
Head of radius
Humerus
Olecranon process
Radius
Ulna

Figure 15.7 Identify the bones and features indicated on this radiograph of the shoulder, using the terms provided.

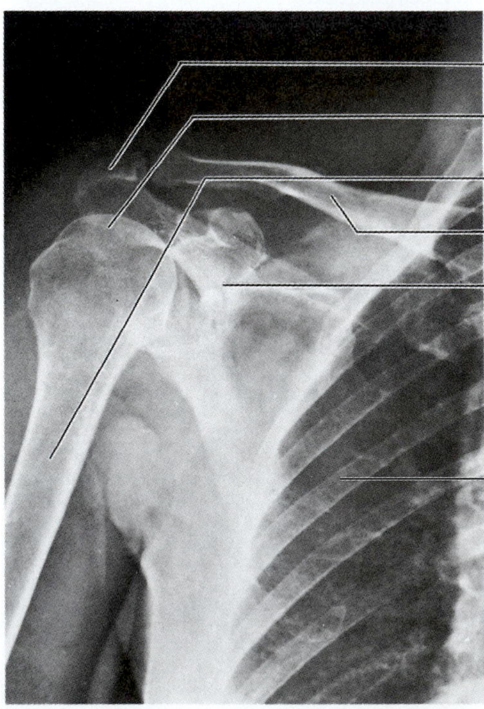

Terms:
Acromion process
Clavicle
Head of humerus
Humerus
Rib
Scapula

Figure 15.8 Identify the bones indicated on this radiograph of the left hand, using the terms provided.

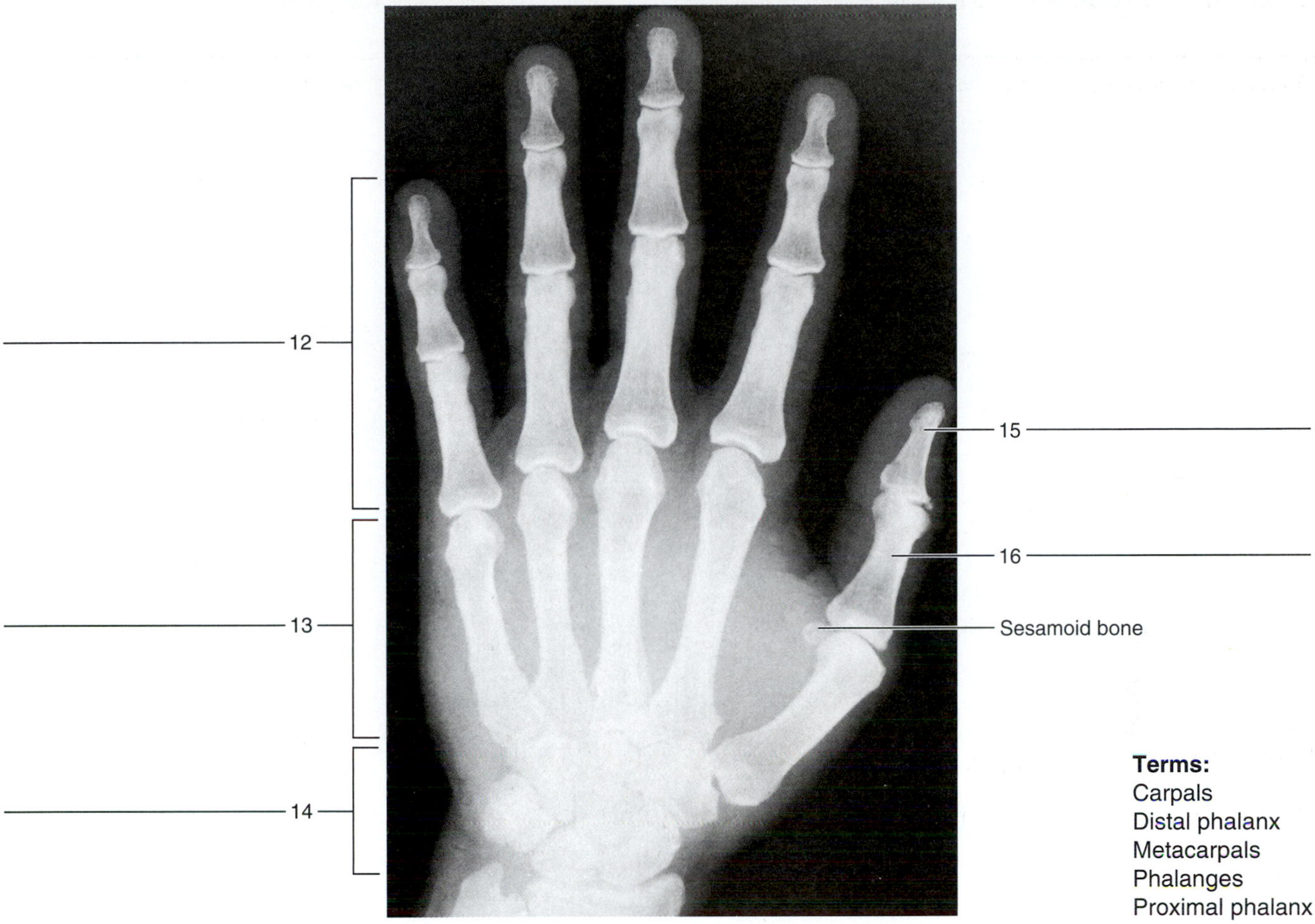

Terms:
Carpals
Distal phalanx
Metacarpals
Phalanges
Proximal phalanx

PELVIC GIRDLE AND LOWER LIMB

MATERIALS NEEDED

Textbook
Human skeleton, articulated
Human skeleton, disarticulated
Male and female pelves

For Optional Activity:

Colored pencils

The pelvic girdle includes two coxae (hipbones) that articulate with each other anteriorly at the symphysis pubis and with the sacrum posteriorly. Together, the pelvic girdle, sacrum, and coccyx comprise the pelvis. The pelvis, in turn, provides support for the trunk of the body and provides attachments for the lower limbs.

The bones of the lower limb form the framework of the thigh, leg, and foot. Each limb includes a femur, a patella, a tibia, a fibula, seven tarsals, five metatarsals, and fourteen phalanges.

PURPOSE OF THE EXERCISE

To examine the bones of the pelvic girdle and lower limb and to identify the major features of these bones.

LEARNING OBJECTIVES

After completing this exercise, you should be able to

1. locate and identify the bones of the pelvic girdle and their major features;
2. locate and identify the bones of the lower limb and their major features.

PROCEDURE A—THE PELVIC GIRDLE

1. Review the section entitled "Pelvic Girdle" in chapter 7 of the textbook.
2. As a review activity, label figures 16.1 and 16.2.
3. Examine the bones of the pelvic girdle and locate the following:

   coxa (hipbone; innominate)

   - acetabulum
   - ilium
   - iliac crest
   - sacroiliac joint
   - anterior superior iliac spine
   - ischium
   - ischial tuberosity
   - ischial spine
   - pubis
   - symphysis pubis
   - pubic arch
   - obturator foramen

4. Complete Part A of Laboratory Report 16.

Critical Thinking Application

Examine the male and female pelves. Look for major differences between them. Note especially the flare of the iliac bones, the angle of the pubic arch, the distance between the ischial spines and ischial tuberosities, and the curve and width of the sacrum. In what ways are the differences you observed related to the function of the female pelvis as a birth canal?

Figure 16.1 Label the bones of the pelvis (posterior view).

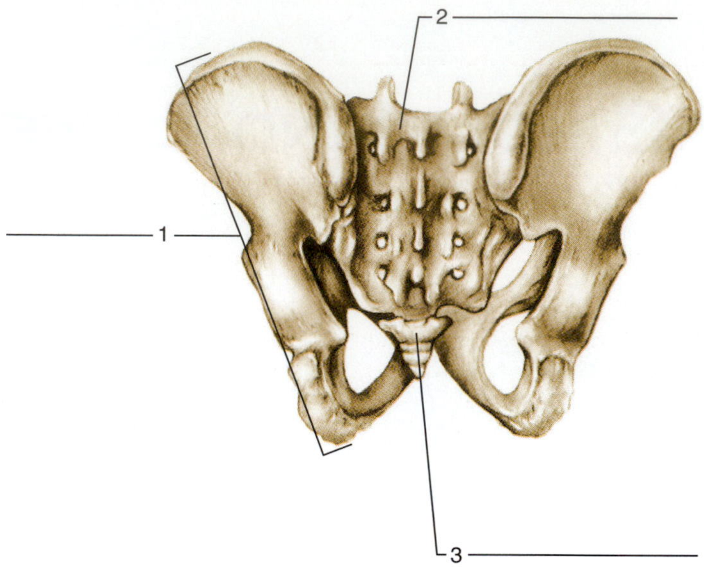

Figure 16.2 Label the lateral features of the left coxa.

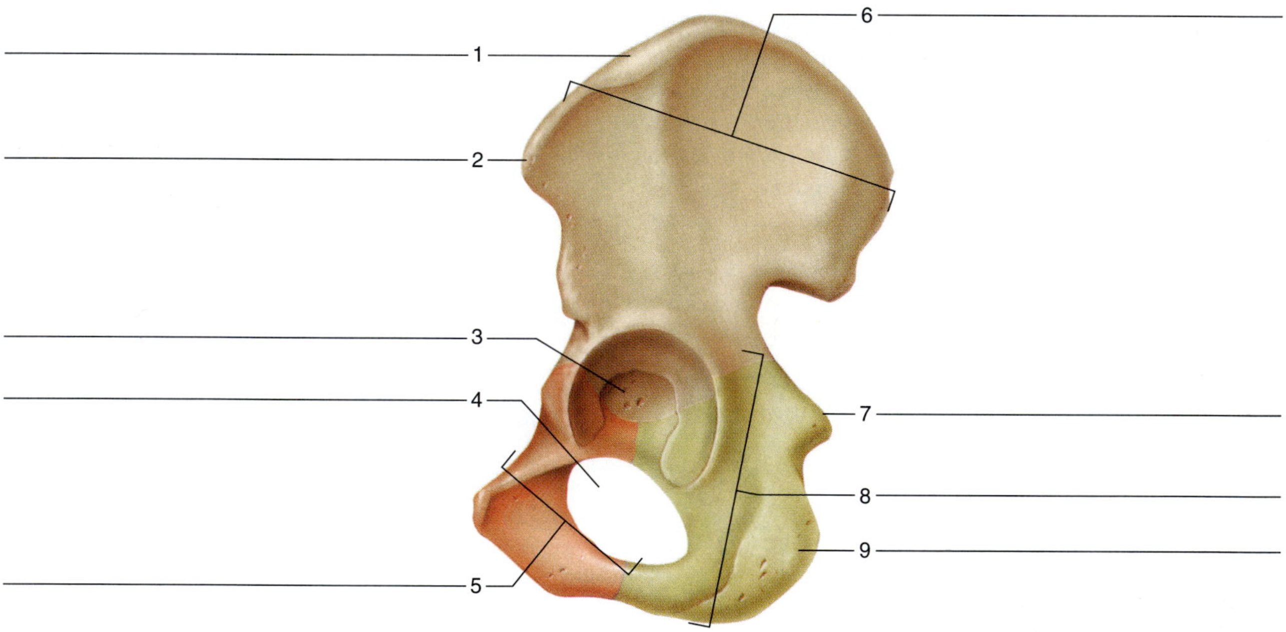

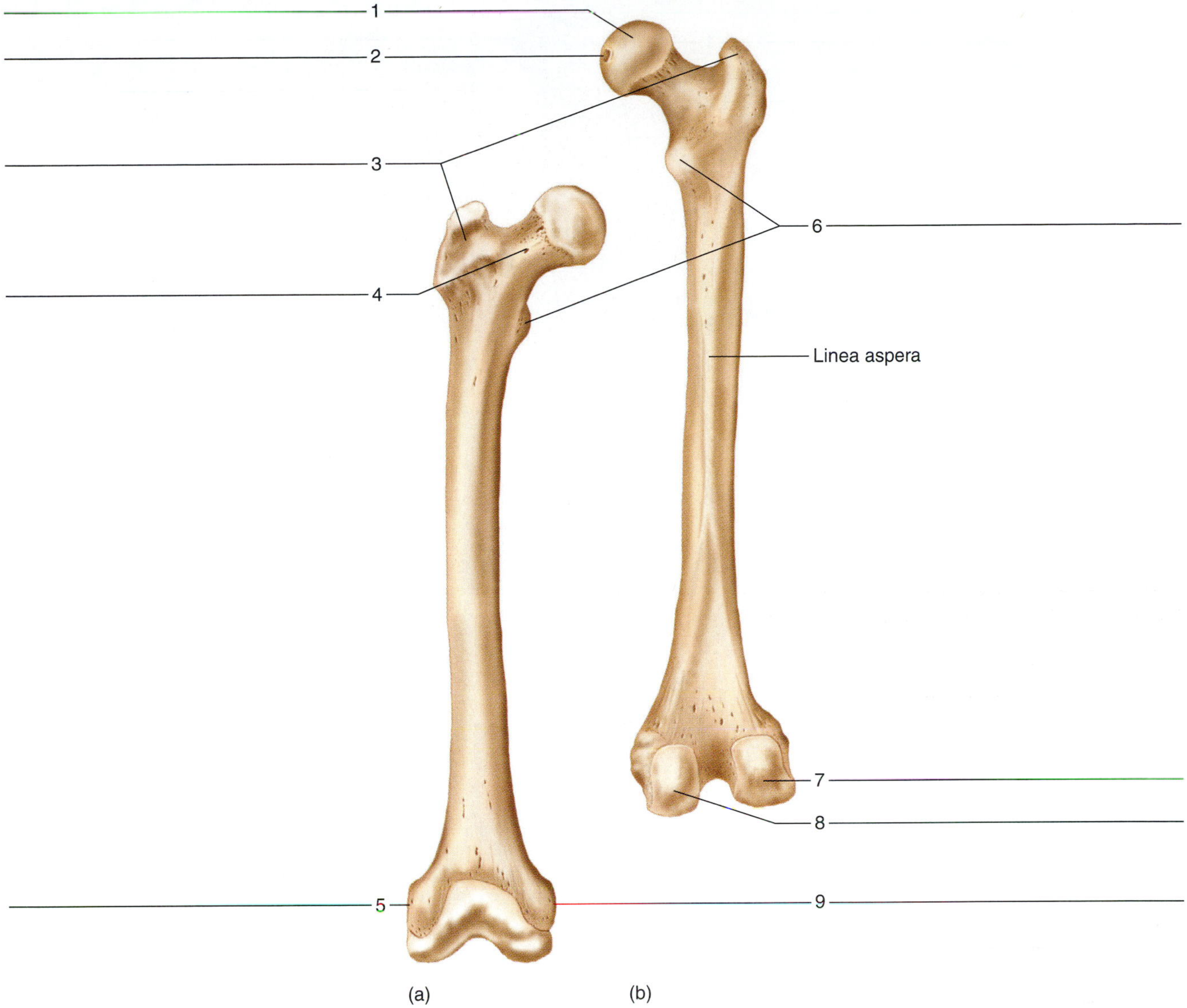

Figure 16.3 Label the features of *(a)* the anterior surface and *(b)* the posterior surface of the right femur.

PROCEDURE B—THE LOWER LIMB

1. Review the section entitled "Lower Limb" in chapter 7 of the textbook.
2. As a review activity, label figures 16.3, 16.4, and 16.5.
3. Examine the bones of the lower limb and locate each of the following:

femur

head

fovea capitis

neck

greater trochanter

lesser trochanter

linea aspera

lateral condyle

medial condyle

lateral epicondyle

medial epicondyle

patella

tibia

medial condyle

lateral condyle

Figure 16.4 Label the bones and features of the right tibia and fibula in this anterior view.

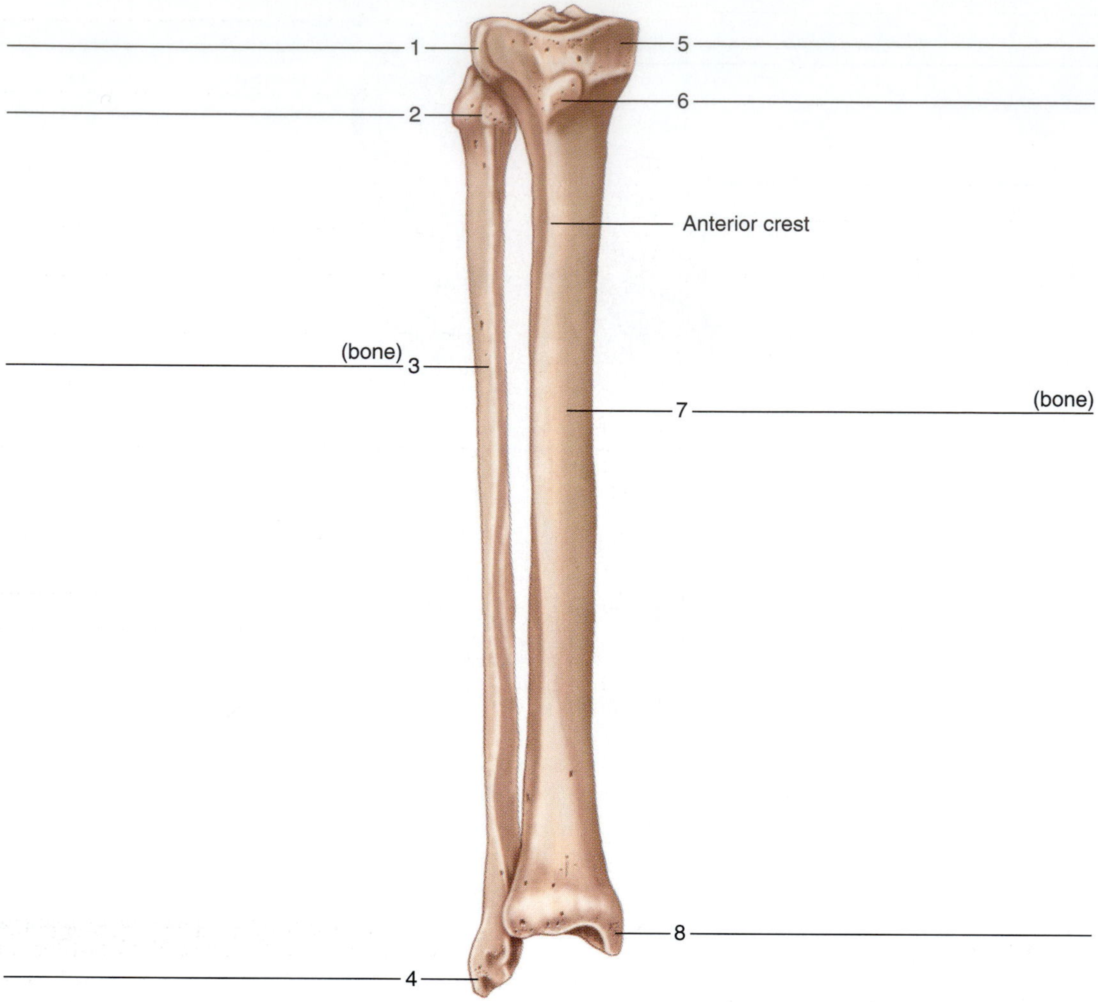

- tibial tuberosity
- anterior crest
- medial malleolus

fibula

- head
- lateral malleolus

tarsal bones

- talus
- calcaneus
- navicular
- cuboid
- lateral cuneiform
- intermediate cuneiform
- medial cuneiform

metatarsal bones

phalanges

- proximal phalanx
- middle phalanx
- distal phalanx

4. Complete Parts B and C of the laboratory report.

Figure 16.5 Label the bones in this superior view of the left foot.

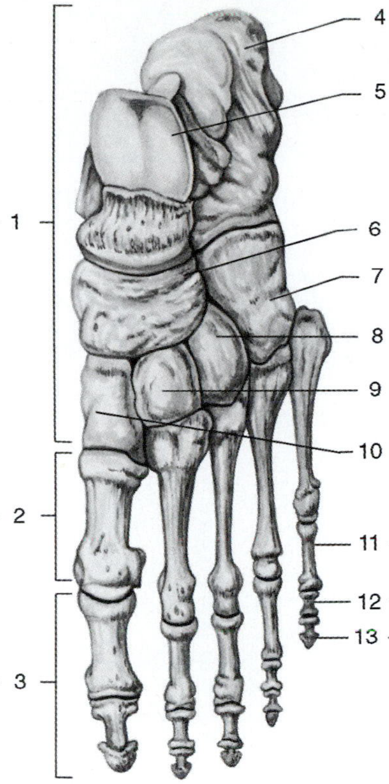

OPTIONAL ACTIVITY

Use different colored pencils to distinguish the individual bones in figure 16.5.

Web Quest

What are the functions of individual bones and features? Search these and review the anatomy of the skeleton at www.mhhe.com/shieress8

Laboratory Report **16**

Name ______________________

Date ______________________

Section ______________________

Pelvic Girdle and Lower Limb

PART A

Complete the following statements:

1. The pelvic girdle consists of two ______________________ .
2. The head of the femur articulates with the ______________________ of the coxa.
3. The ______________________ is the largest portion of the coxa.
4. The pubic bones come together anteriorly to form the joint called the ______________________ .
5. The ______________________ is the portion of the ilium that causes the prominence of the hip.
6. When a person sits, the ______________________ of the ischium supports the weight of the body.
7. The angle formed by the pubic bones below the symphysis pubis is called the ______________________ .
8. ______________________ is the largest foramen in the skeleton.
9. The ilium joins the sacrum at the ______________________ joint.

PART B

Match the bones in column A with the features in column B. Place the letter of your choice in the space provided.

Column A		Column B
a. Femur	_______	1. Middle phalanx
b. Fibula	_______	2. Lesser trochanter
c. Metatarsals	_______	3. Medial malleolus
d. Patella	_______	4. Fovea capitis
e. Phalanges	_______	5. Calcaneus
f. Tarsals	_______	6. Lateral cuneiform
g. Tibia	_______	7. Tibial tuberosity
	_______	8. Talus
	_______	9. Cuboid
	_______	10. Lateral malleolus
	_______	11. Located in a tendon over the knee
	_______	12. Five bones that form the instep

PART C

Identify the bones and features indicated in the radiographs (X rays) of figures 16.6, 16.7, and 16.8.

Figure 16.6 Identify the bones and features indicated on this radiograph of the pelvic region, using the terms provided.

Terms:
Head of femur
Ilium
Obturator foramen
Pubis
Sacrum
Symphysis pubis

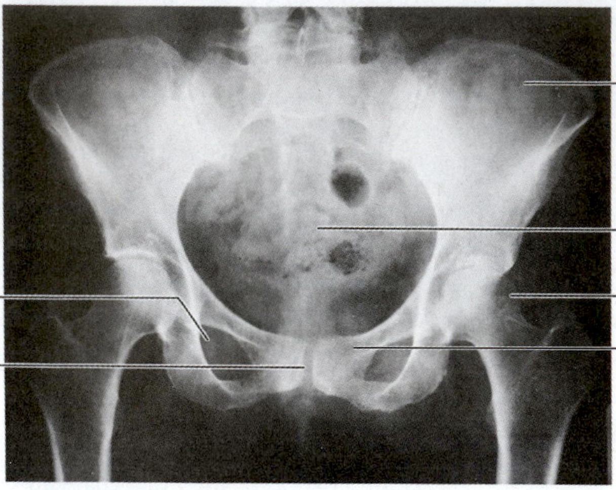

Figure 16.7 Identify the bones and features indicated on this radiograph of the knee, using the terms provided.

Terms:
Femur
Fibula
Head of fibula
Lateral condyle
Lateral epicondyle
Tibia

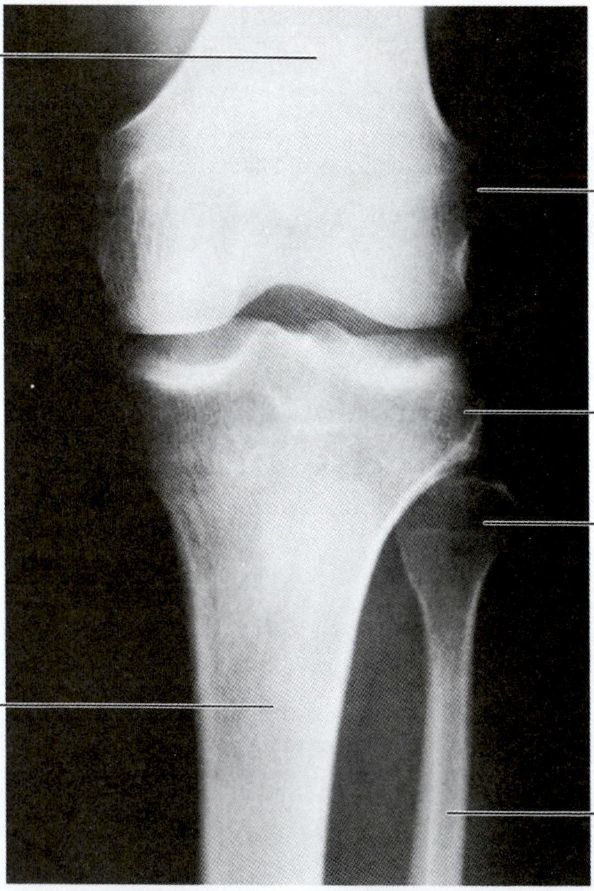

Figure 16.8 Identify the bones indicated on this radiograph of the left foot, using the terms provided.

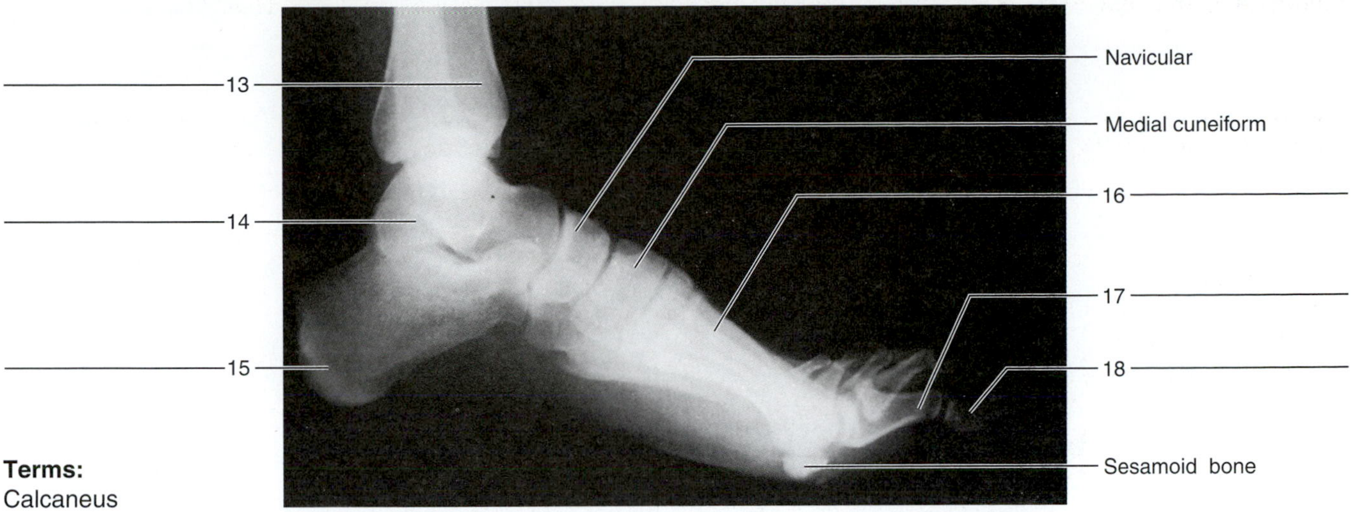

Terms:
Calcaneus
Distal phalanx
Metatarsal
Proximal phalanx
Talus
Tibia

Laboratory Exercise 17

The Joints

MATERIALS NEEDED

Textbook
Human skull
Human skeleton, articulated
Model of knee joint

For Demonstration:

Fresh animal joint (knee joint preferred)
Radiographs of major joints

SAFETY

- Wear disposable gloves when handling the fresh animal joint.
- Wash your hands before leaving the laboratory.

Joints are junctions between bones. Although they vary considerably in structure, they can be classified according to the type of tissue that binds the bones together. Thus, the three groups of joints can be identified as (1) fibrous joints, (2) cartilaginous joints, and (3) synovial joints.

Movements occurring at freely movable synovial joints are due to the contractions of skeletal muscles. In each case, the type of movement depends on the kind of joint involved and the way in which the muscles are attached to the bones on either side of the joint.

PURPOSE OF THE EXERCISE

To examine examples of the three types of joints, to identify the major features of these joints, and to review the types of movements produced at synovial joints.

LEARNING OBJECTIVES

After completing this exercise, you should be able to

1. distinguish among fibrous, cartilaginous, and synovial joints;
2. identify examples of each type of joint;
3. identify the major features of each type of joint;
4. locate and identify examples of each of the six types of synovial joints;
5. identify the types of movements that occur at synovial joints;
6. describe the structure of the knee joint.

PROCEDURE A—TYPES OF JOINTS

1. Review the section entitled "Joints" in chapter 7 of the textbook.
2. Study the knee joint model and locate the structures indicated in figure 17.1.
3. Complete Part A of Laboratory Report 17.
4. Examine the skull and locate examples of *fibrous joints* as described in the textbook.
5. Locate and examine examples of *cartilaginous joints* in the skeleton.
6. Locate examples of the following types of *synovial joints* in the skeleton and in your own body. Experiment with each joint to experience its range of movements.

   ball-and-socket joint

   condyloid joint

   gliding joint

   hinge joint

   pivot joint

   saddle joint

7. Complete Part B of the laboratory report.

Figure 17.1 Anterior view of right knee (patella removed).

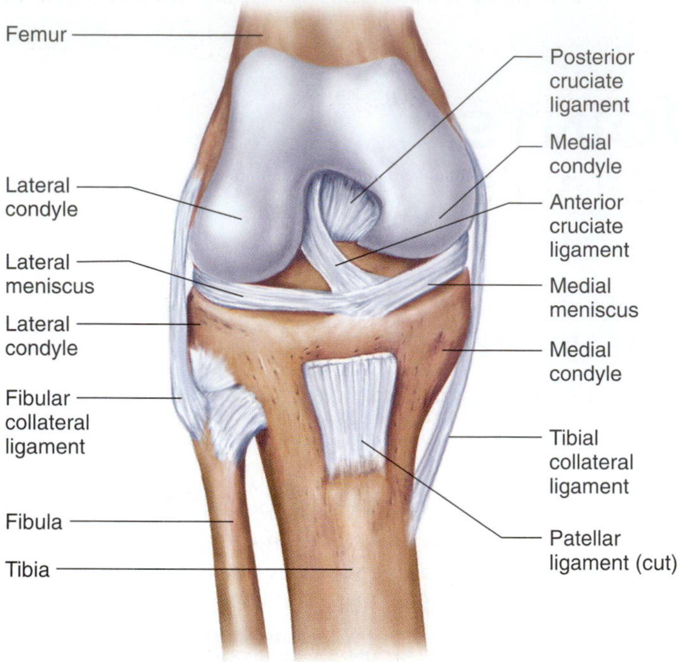

DEMONSTRATION

Examine a longitudinal section of a fresh synovial animal joint. Locate the dense connective tissue that forms the joint capsule and the hyaline cartilage that forms the articular cartilage on the ends of the bones. Locate the synovial membrane on the inside of the joint capsule. Does the joint have any semilunar cartilages (menisci)?

____

____

What is the function of such cartilages? ______________

____

____

PROCEDURE B—JOINT MOVEMENTS

1. Review the section entitled "Types of Joint Movements" in chapter 7 of the textbook.
2. When the body is in anatomical position, most joints are extended and/or adducted. Skeletal muscle action involves the movable end (*insertion*) being pulled toward the stationary end (*origin*). In the limbs, the origin is usually proximal to the insertion; in the trunk, the origin is usually medial to the insertion. Use these concepts as reference points as you move joints. Move various parts of your own body to demonstrate the following joint movements:

   flexion

   extension

   hyperextension

   dorsiflexion

   plantar flexion

   abduction

   adduction

   rotation

   circumduction

   supination

   pronation

   eversion

   inversion

   protraction

   retraction

   elevation

   depression

3. Have your laboratory partner do some of the preceding movements and see if you can identify correctly the movements made.

Critical Thinking Application

Describe a body position that can exist when all major body parts are flexed.

____

____

____

____

4. Complete Part C of the laboratory report.

DEMONSTRATION

Study the available radiographs of joints by holding them in front of a light source. Identify the type of joint and the bones incorporated in the joint. Also identify other major features that are visible.

Web Quest

Why does the shoulder joint allow extensive movement? What is joint fluoroscopy? Search for these answers and review joints at www.mhhe.com/shieress8

Laboratory Report **17**

Name ____________________

Date ____________________

Section ____________________

The Joints

PART A

Complete the following statements:

1. A ____________________ of the skull is an example of an immovable fibrous joint.
2. The joint between the first rib and the sternum is ____________________ .
3. The joint at the distal tibia and fibula is ____________________ .
4. Intervertebral discs are composed of ____________________ .
5. ____________________ are the most common type of joint in the human body.
6. ____________________ acts as a joint lubricant in synovial joints.
7. The discs of fibrocartilage between the articulating surfaces of the knee are called ____________________ .
8. The fluid-filled sacs sometimes associated with synovial joints are called ____________________ .
9. The symphysis pubis is an example of a(n) ____________________ joint.
10. Articular cartilage is composed of ____________________ tissue.

PART B

Match the types of synovial joints in column A with the examples in column B. Place the letter of your choice in the space provided.

Column A		Column B
a. Ball-and-socket	_______	1. Hip joint
b. Condyloid	_______	2. Metacarpal-phalanx
c. Gliding	_______	3. Proximal radius-ulna
d. Hinge	_______	4. Humerus-ulna of the elbow joint
e. Pivot	_______	5. Phalanx-phalanx
f. Saddle	_______	6. Shoulder joint
	_______	7. Tarsal-tarsal
	_______	8. Carpal-metacarpal of the thumb
	_______	9. Carpal-carpal

PART C

Identify the types of joint movements that are numbered in figure 17.2.

1. ______________________ (of head)
2. ______________________ (of shoulder)
3. ______________________ (of shoulder)
4. ______________________ (of hand)
5. ______________________ (of hand)
6. ______________________ (of arm at shoulder)
7. ______________________ (of arm at shoulder)
8. ______________________ (of hand at wrist)
9. ______________________ (of hand at wrist)
10. ______________________ (of thigh at hip)
11. ______________________ (of thigh at hip)
12. ______________________ (of lower limb at hip)
13. ______________________ (of chin/mandible)
14. ______________________ (of chin/mandible)
15. ______________________ (of vertebral column/trunk)
16. ______________________ (of vertebral column/trunk)
17. ______________________ (of head and neck)
18. ______________________ (of head and neck)
19. ______________________ (of arm at shoulder)
20. ______________________ (of arm at shoulder)
21. ______________________ (of forearm at elbow)
22. ______________________ (of forearm at elbow)
23. ______________________ (of thigh at hip)
24. ______________________ (of thigh at hip)
25. ______________________ (of leg at knee)
26. ______________________ (of leg at knee)
27. ______________________ (of foot at ankle)
28. ______________________ (of foot at ankle)

Figure 17.2 Identify each of the types of movements that are numbered and illustrated: (*a*) anterior view; (*b*) lateral view of head; (*c*) lateral view. (Original illustration drawn by Ross Martin)

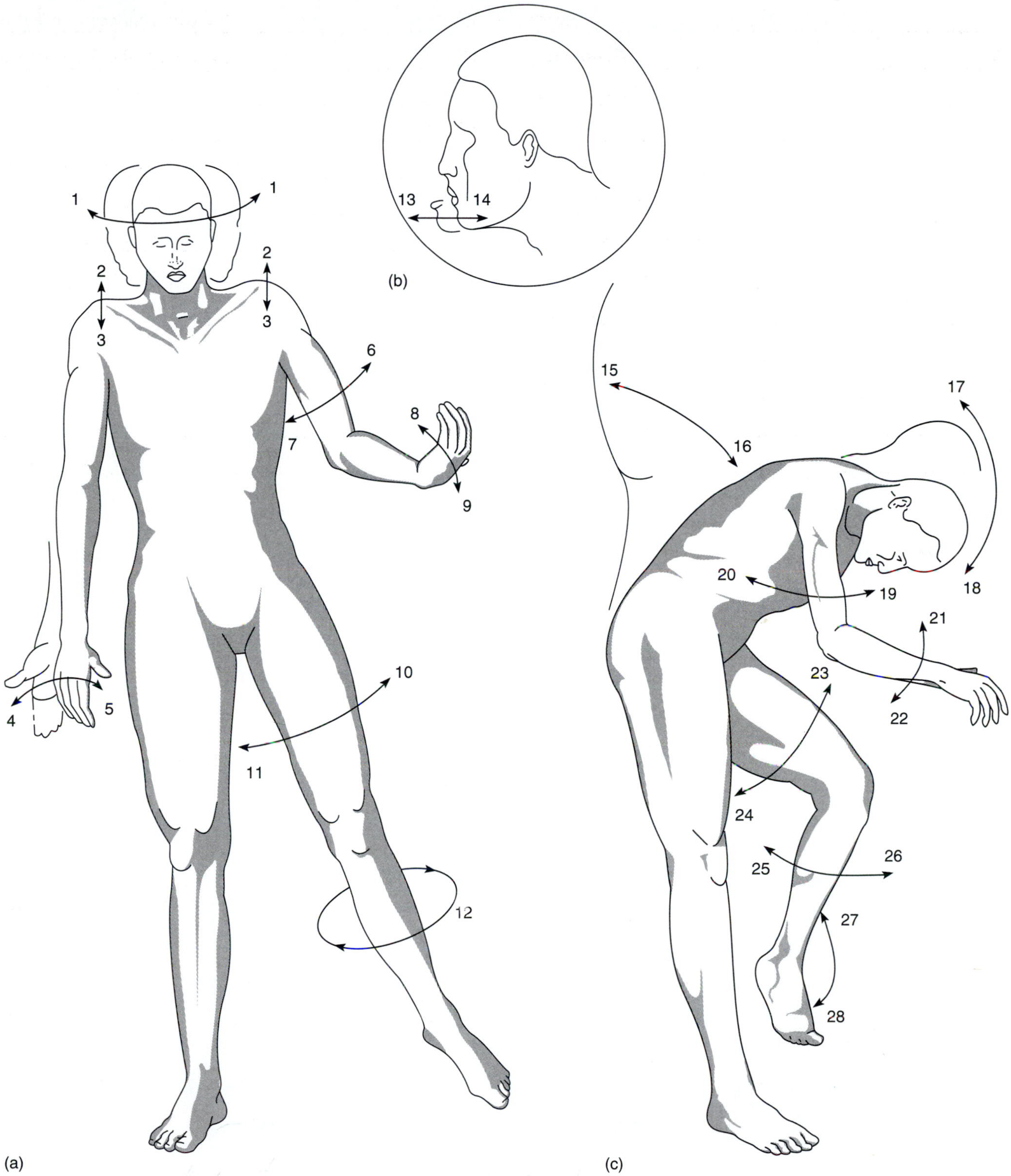

Laboratory Exercise 18

Skeletal Muscle Structure

MATERIALS NEEDED

Textbook
Compound microscope
Prepared microscope slide of skeletal muscle tissue
Torso with musculature
Model of skeletal muscle fiber

For Demonstration:

Fresh round beefsteak
Virtual Physiology Lab: 3, Frog Muscle

SAFETY

- Wear disposable gloves when handling the fresh beefsteak.
- Wash your hands before leaving the laboratory.

A skeletal muscle represents an organ of the muscular system and is composed of several kinds of tissues. These tissues include skeletal muscle tissue, nervous tissue, blood, and various connective tissues.

Each skeletal muscle is surrounded by connective tissue. The connective tissue often extends beyond the end of a muscle, providing an attachment to other muscles or to bones. Connective tissue also extends into the structure of a muscle and separates it into compartments.

PURPOSE OF THE EXERCISE

To review the structure of a skeletal muscle.

LEARNING OBJECTIVES

After completing this exercise, you should be able to

1. describe how connective tissue is associated with muscle tissue within a skeletal muscle;
2. name and locate the major structures of a skeletal muscle fiber on a model;
3. distinguish between the origin and insertion of a muscle;
4. describe the general actions of prime movers, synergists, and antagonists.

PROCEDURE

1. Review the section entitled "Skeletal Muscle Tissue" in chapter 5 of the textbook.
2. Reexamine the microscopic structure of skeletal muscle (fig. 18.1) by observing a prepared microscope slide of this tissue. Locate the following features:

   skeletal muscle fiber

   nuclei

   striations

3. Review the section entitled "Structure of a Skeletal Muscle" in chapter 8 of the textbook.
4. As a review activity, label figures 18.2 and 18.3.
5. Examine the torso and locate examples of fascia, tendons, and aponeuroses. Locate examples of tendons in your own body.
6. Complete Part A of Laboratory Report 18.

DEMONSTRATION

Examine the fresh round beefsteak. It represents a cross section through the beef thigh muscles. Note the white lines of connective tissue that separate the individual skeletal muscles. Also note how the connective tissue extends into the structure of a muscle and separates it into small compartments of muscle tissue. Locate the epimysium and the perimysium of the fascia.

Figure 18.1 Structures found in skeletal muscle fibers (cells) (250× micrograph enlarged to 700×).

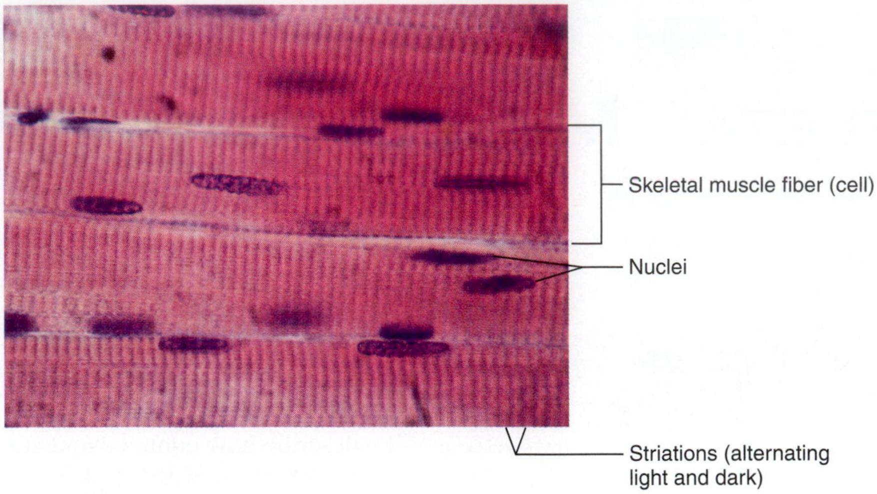

Figure 18.2 Label the structures of a skeletal muscle.

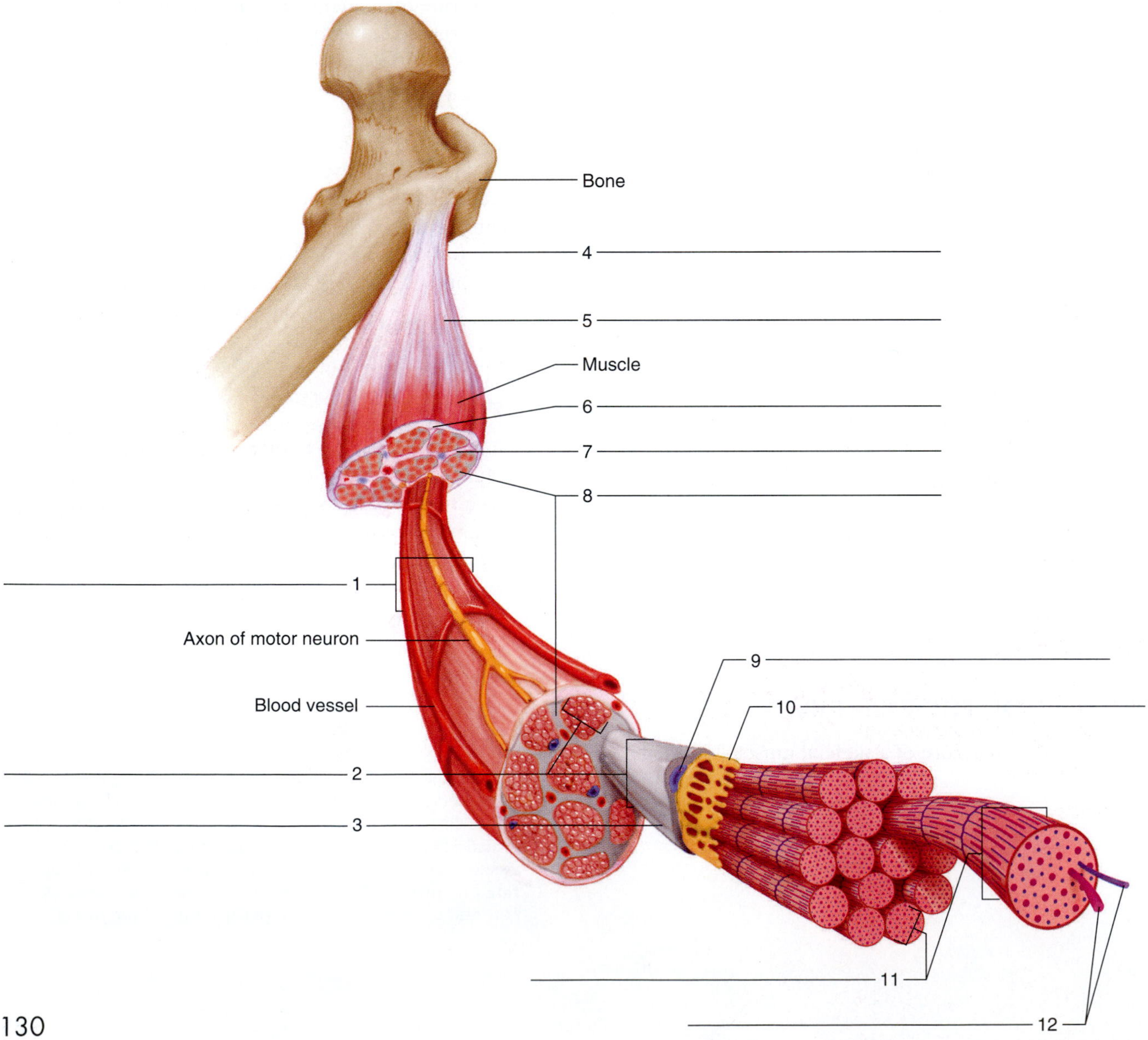

Figure 18.3 Label the structures of this skeletal muscle fiber.

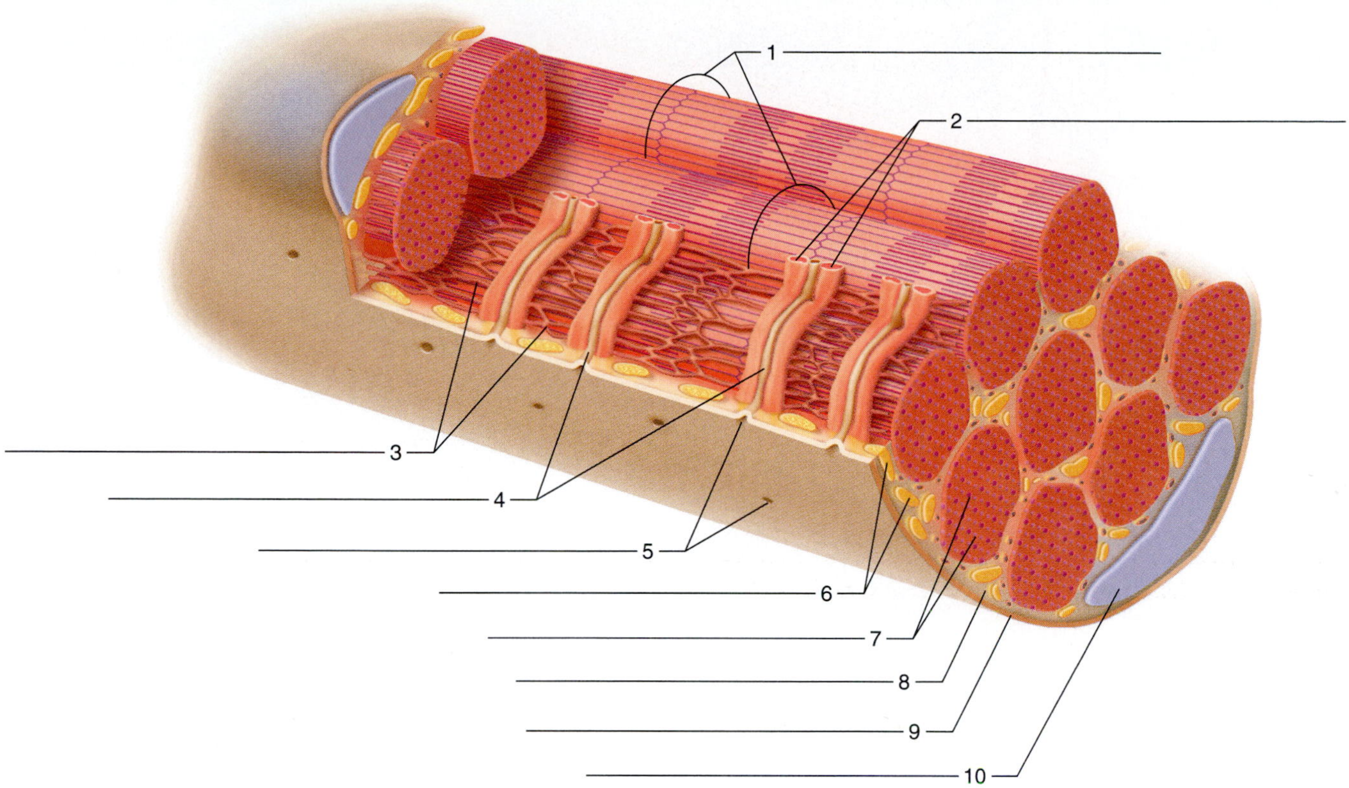

7. Examine the model of the skeletal muscle fiber and locate the following:

   sarcolemma

   sarcoplasm

   myofibril

   myosin filament

   actin filament

   sarcomere

   A band

   I band

   Z line

   sarcoplasmic reticulum

   cisternae

   transverse tubules

8. Complete Part B of the laboratory report.
9. Review the section entitled "Skeletal Muscle Actions" in chapter 8 of the textbook.
10. Locate the biceps brachii muscle and its origin and insertion in the torso and in your own body.
11. Make various movements with your upper limb at the shoulder and elbow. For each movement, determine the location of the muscles functioning as prime movers and as antagonists. **(Remember, when a prime mover contracts, its antagonist must relax.)**
12. Complete Part C of the laboratory report.

Laboratory Report **18**

Name ______________________

Date ______________________

Section ______________________

Skeletal Muscle Structure

PART A

Match the terms in column A with the definitions in column B. Place the letter of your choice in the space provided.

Column A	Column B
a. Aponeurosis	_______ 1. Membranous channel extending inward from muscle fiber membrane
b. Cisterna	_______ 2. Cytoplasm of a muscle fiber
c. Endomysium	_______ 3. Network of connective tissue that extends throughout the muscular system
d. Epimysium	_______ 4. Layer of connective tissue that separates a muscle into small bundles called fascicles
e. Fascia	_______ 5. Enlarged portion of sarcoplasmic reticulum on either side of a transverse tubule
f. Fascicle	_______ 6. Broad sheet of connective tissue that attaches coverings of adjacent muscles
g. Myosin	_______ 7. Cell membrane of a muscle fiber
h. Perimysium	_______ 8. Layer of connective tissue that surrounds a skeletal muscle
i. Sarcolemma	_______ 9. Unit of alternating light and dark striations between Z lines
j. Sarcomere	_______ 10. Layer of connective tissue that surrounds an individual muscle fiber
k. Sarcoplasm	_______ 11. Cellular organelle in muscle fiber corresponding to the endoplasmic reticulum
l. Sarcoplasmic reticulum	_______ 12. Cordlike part that attaches a muscle to a bone
m. Tendon	_______ 13. Protein found within thick myofibril
n. Transverse tubule	_______ 14. A small bundle of muscle fibers

PART B

Provide the labels for the electron micrograph in figure 18.4.

1. ______________________
2. ______________________
3. ______________________
4. ______________________

Figure 18.4 Identify the bands and lines of the striations in this transmission electron micrograph of relaxed sarcomeres (8,400×), using the terms provided.

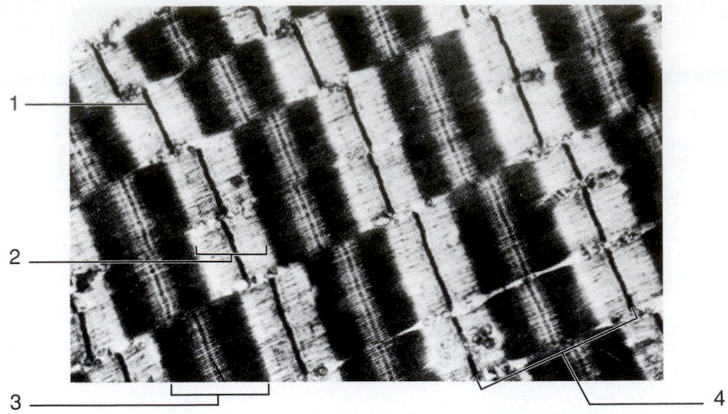

Terms:
A band
I band
Sarcomere
Z line

PART C

Complete the following statements:

1. The ______________________________ of a muscle is usually attached to a fixed part.
2. The ______________________________ of a muscle is usually attached to a movable part.
3. The name *biceps* means ______________________________ .
4. The forearm is flexed at the elbow when the ______________________________ muscle contracts.
5. A muscle responsible for most of a movement is called a(n) ______________________________ .
6. Assisting muscles are called ______________________________ .
7. Antagonists are muscles that resist the actions of ______________________________ and cause movement in the opposite direction.

Laboratory Exercise 19

Muscles of the Face, Head, and Neck

MATERIALS NEEDED

Textbook
Torso with musculature
Human skull
Human skeleton, articulated

The skeletal muscles of the face and head include the muscles of facial expression, which lie just beneath the skin, the muscles of mastication, which are attached to the mandible, and the muscles that move the head, which are located in the neck.

PURPOSE OF THE EXERCISE

To review the locations, actions, origins, and insertions of the muscles of the face, head, and neck.

LEARNING OBJECTIVES

After completing this exercise, you should be able to

1. locate and identify the muscles of facial expression, the muscles of mastication, and the muscles that move the head;
2. describe and demonstrate the action of each of these muscles;
3. locate the origin and insertion of each of these muscles in a human skeleton and the musculature of the torso.

PROCEDURE

1. Review the sections entitled "Muscles of Facial Expression," "Muscles of Mastication," and "Muscles That Move the Head" in chapter 8 of the textbook.
2. As a review activity, label figures 19.1 and 19.2.
3. Locate the following muscles in the torso and in your own body whenever possible:

   epicranius (frontalis and occipitalis)

   orbicularis oculi

   orbicularis oris

   buccinator

   zygomaticus

   platysma

   masseter

   temporalis

   sternocleidomastoid

   splenius capitis

   semispinalis capitis

4. Demonstrate the actions of these muscles in your own body.
5. Locate the origins and insertions of these muscles in the human skull and skeleton.
6. Complete Parts A, B, and C of Laboratory Report 19.

Web Quest

Determine the origin, insertion, action, nerve innervation, and blood supply of all the major muscles.

Identify muscles and detailed explanations from an interactive site.

Find information about these topics at www.mhhe.com/shieress8

Figure 19.1 Label the muscles of expression and mastication.

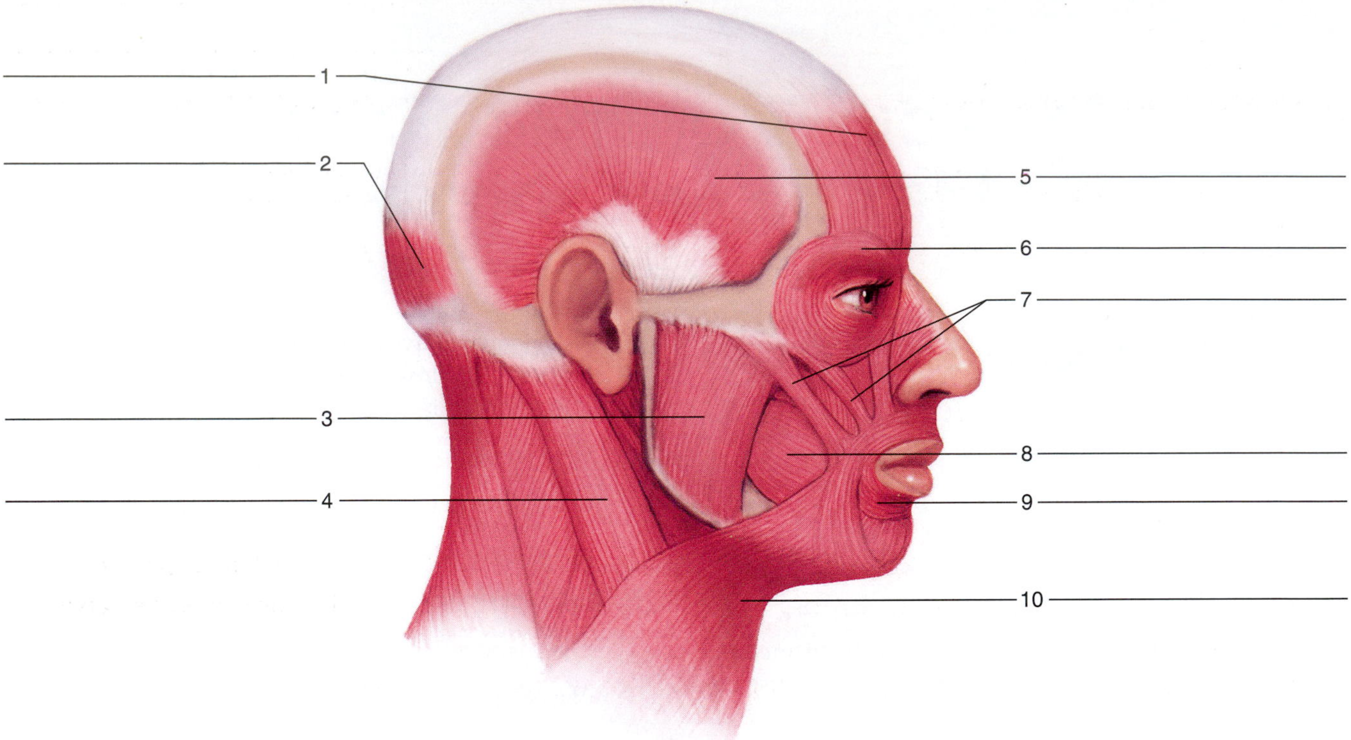

Figure 19.2 Label these deep muscles of the posterior neck (trapezius removed).

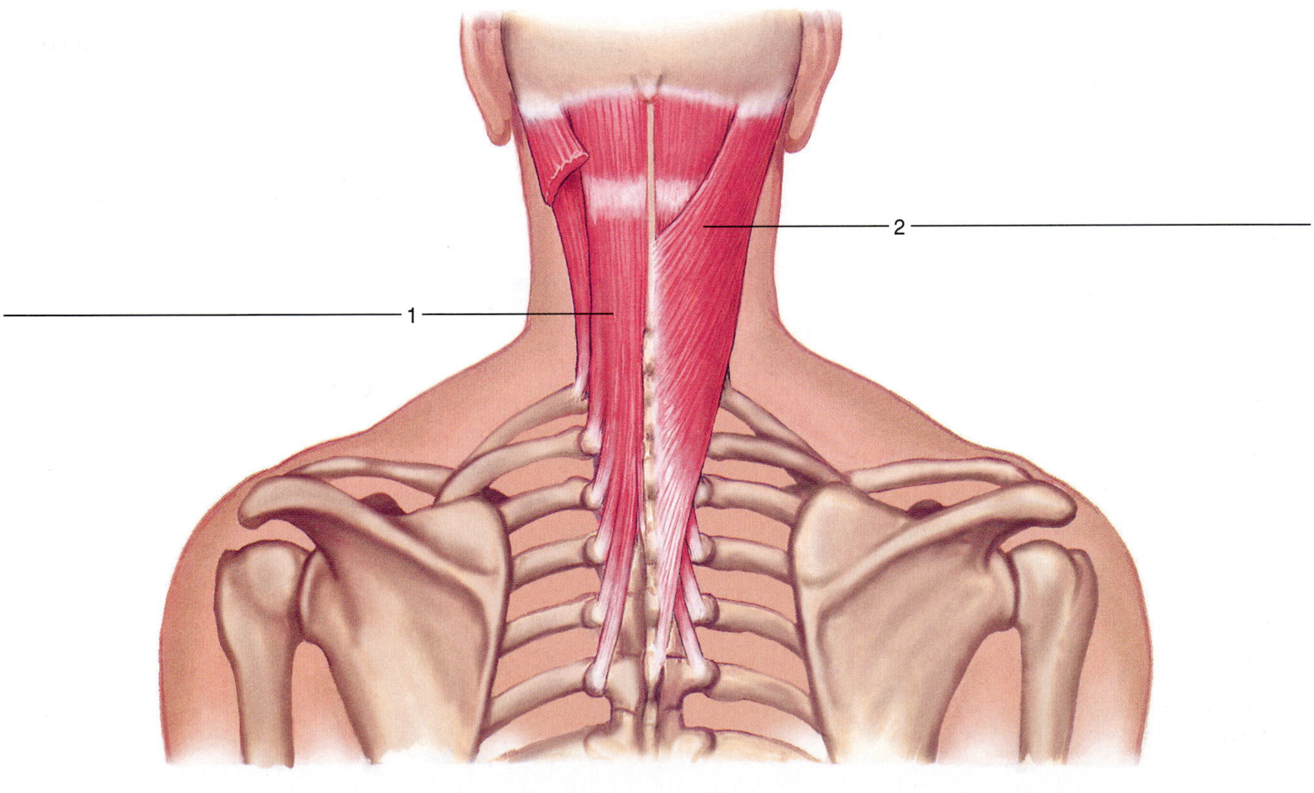

Laboratory Report **19**

Name ______________________________

Date ______________________________

Section ______________________________

Muscles of the Face, Head, and Neck

PART A

Complete the following statements:

1. When the ______________________________ contracts, the corner of the mouth is drawn upward.
2. The ______________________________ acts to compress the wall of the cheeks when air is blown out of the mouth.
3. The ______________________________ causes the lips to protrude or close.
4. The temporalis acts to ______________________________ .
5. The ______________________________ can close the eye, as in blinking.
6. The ______________________________ can pull the head toward the chest.
7. The ______________________________ can pull the corners of the mouth downward, as when pouting.

PART B

Name the muscle indicated by the following combinations of origin and insertion.

Origin	Insertion	Muscle
1. Occipital bone	Skin and muscle around eye	______________
2. Zygomatic bone	Orbicularis oris	______________
3. Zygomatic arch	Lateral surface of mandible	______________
4. Anterior surface of sternum and upper clavicle	Mastoid process of temporal bone	______________
5. Outer surfaces of mandible and maxilla	Orbicularis oris	______________
6. Fascia in upper chest	Lower border of mandible and skin around corner of mouth	______________
7. Temporal bone	Coronoid process and lateral surface of mandible	______________
8. Spinous processes of cervical and thoracic vertebrae	Mastoid process of temporal bone	______________
9. Processes of cervical and thoracic vertebrae	Occipital bone	______________

PART C

Critical Thinking Application

Identify the muscles of various facial expressions in the photographs of figure 19.3.

1. ____________________
2. ____________________
3. ____________________
4. ____________________
5. ____________________

Figure 19.3 Identify the muscles of facial expression being contracted in each of these photographs (*a–c*), using the terms provided.

(a)

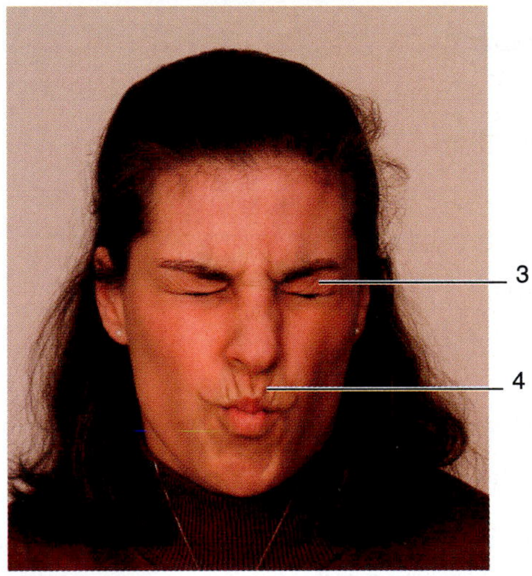

(b)

Terms:
Epicranius/frontalis
Orbicularis oculi
Orbicularis oris
Platysma
Zygomaticus

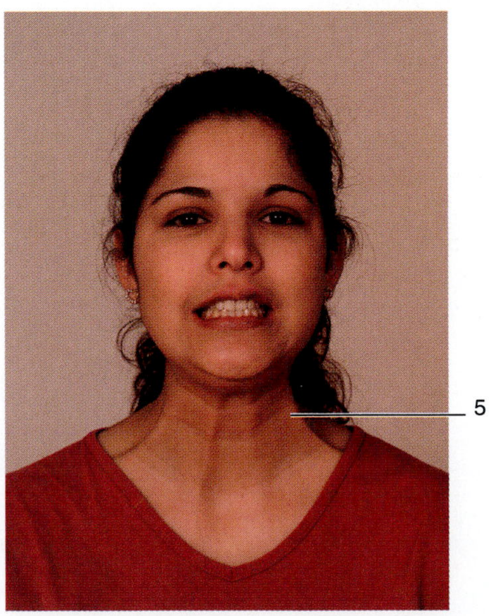

(c)

MUSCLES OF THE CHEST, SHOULDER, AND UPPER LIMB

MATERIALS NEEDED

Textbook
Torso
Human skeleton, articulated
Muscular models of the upper limb

The muscles of the chest and shoulder are responsible for moving the scapula and arm, whereas those within the arm and forearm act to move joints in the elbow and hand.

PURPOSE OF THE EXERCISE

To review the locations, actions, origins, and insertions of the muscles in the chest, shoulder, and upper limb.

LEARNING OBJECTIVES

After completing this exercise, you should be able to

1. locate and identify the muscles of the chest, shoulder, and upper limb;
2. describe and demonstrate the action of each of these muscles;
3. locate the origin and insertion of each of these muscles.

PROCEDURE

1. Review the sections entitled "Muscles That Move the Pectoral Girdle," "Muscles That Move the Arm," "Muscles That Move the Forearm," and "Muscles That Move the Hand" in chapter 8 of the textbook.
2. As a review activity, label figures 20.1, 20.2, 20.3, and 20.4.
3. Locate the following muscles in the torso and models of the upper limb. Also locate in your own body as many of the muscles as you can.

muscles that move the pectoral girdle

trapezius
rhomboideus major
levator scapulae
serratus anterior
pectoralis minor

muscles that move the arm

coracobrachialis
pectoralis major
teres major
latissimus dorsi
supraspinatus
deltoid
subscapularis
infraspinatus
teres minor

muscles that move the forearm

biceps brachii
brachialis
brachioradialis
triceps brachii
supinator
pronator teres
pronator quadratus

Figure 20.1 Label the muscles of the posterior shoulder. Superficial muscles are illustrated on the left side and deep muscles on the right side.

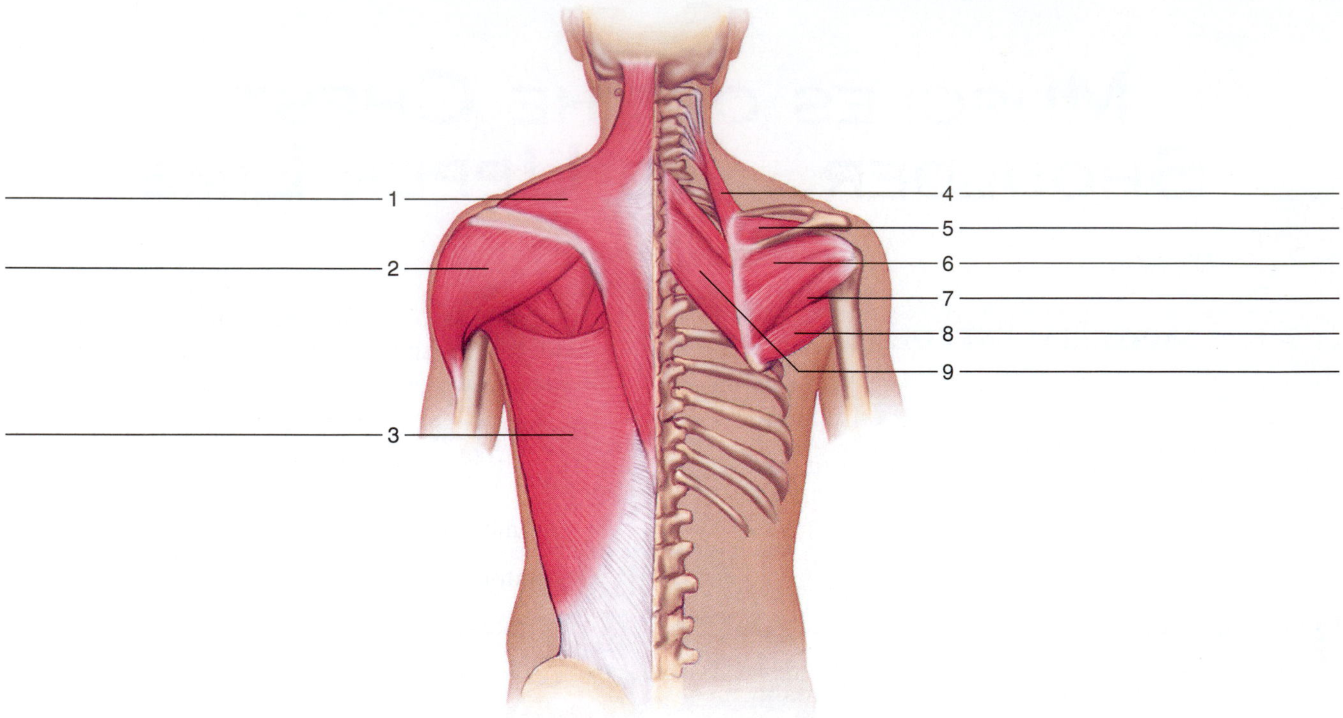

Figure 20.2 Label the muscles of the anterior chest. Superficial muscles are illustrated on the left side and deep muscles on the right side.

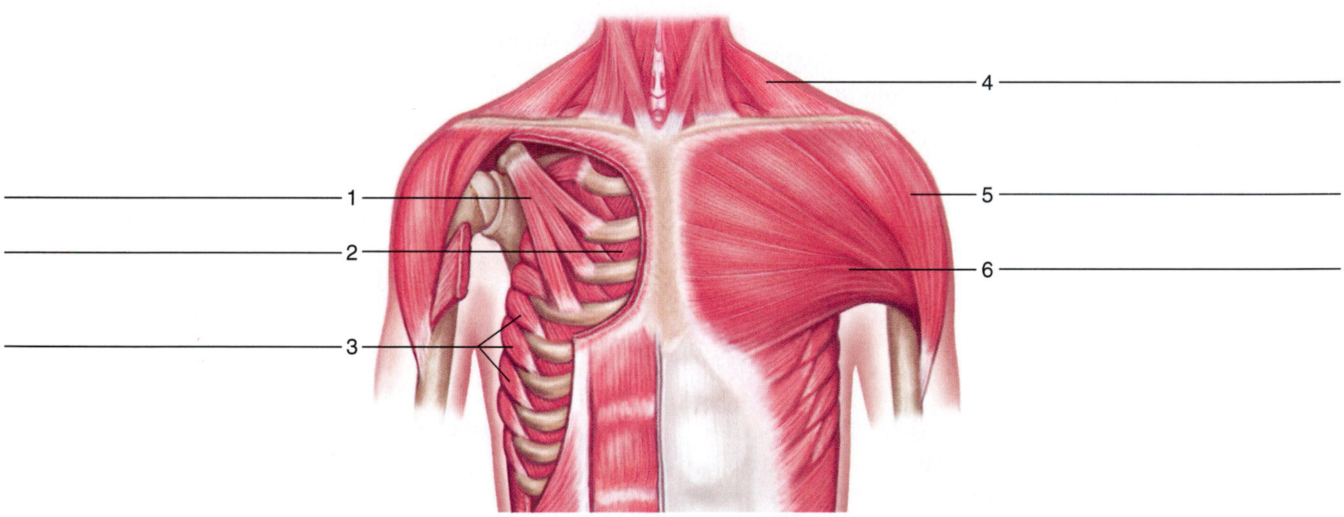

Figure 20.3 Label (*a*) the muscles of the posterior shoulder and arm and (*b*) the muscles of the anterior shoulder and arm.

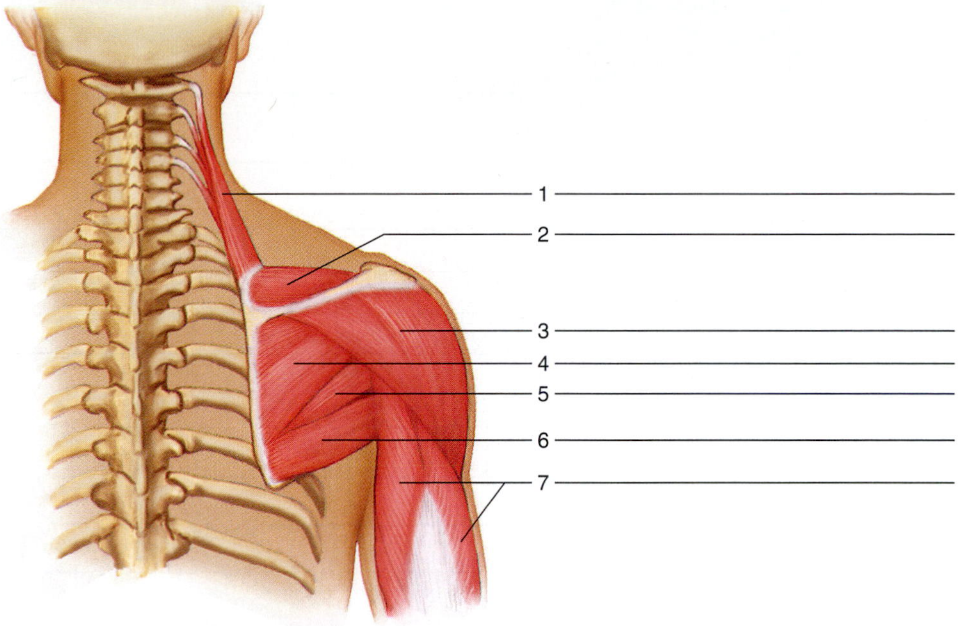

(a)

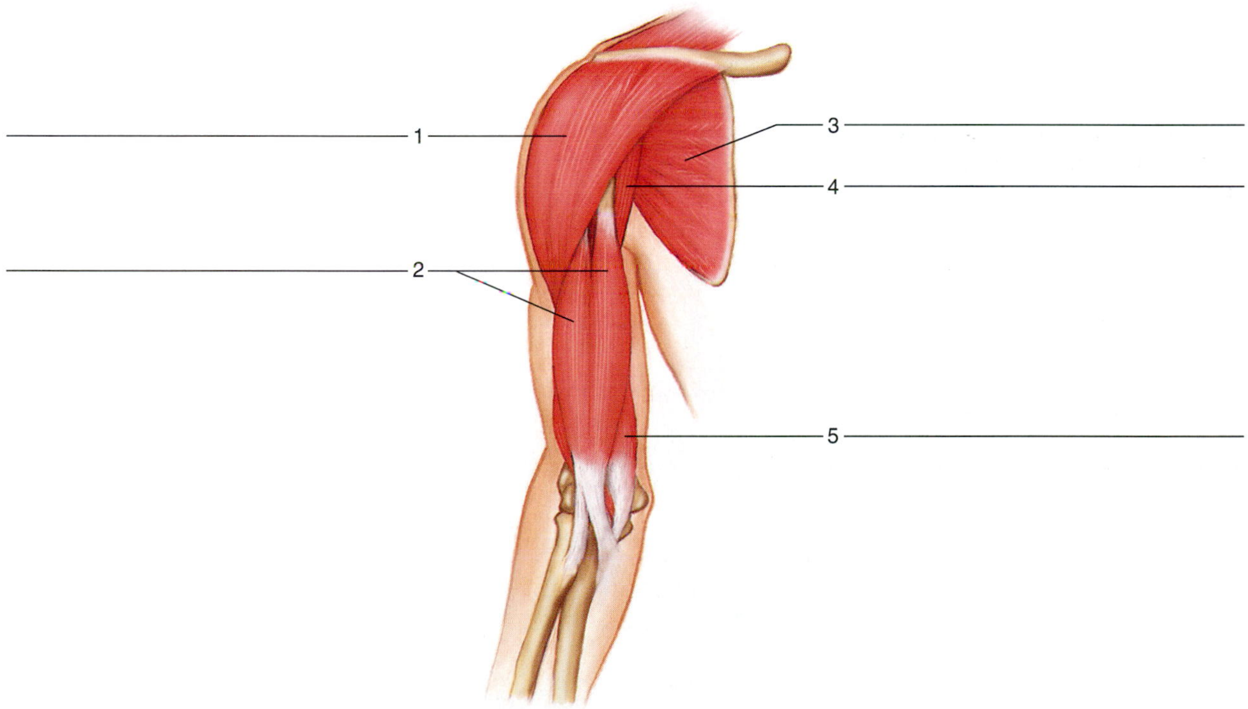

(b)

Figure 20.4 Label (*a*) the muscles of the anterior forearm and (*b*) the muscles of the posterior forearm.

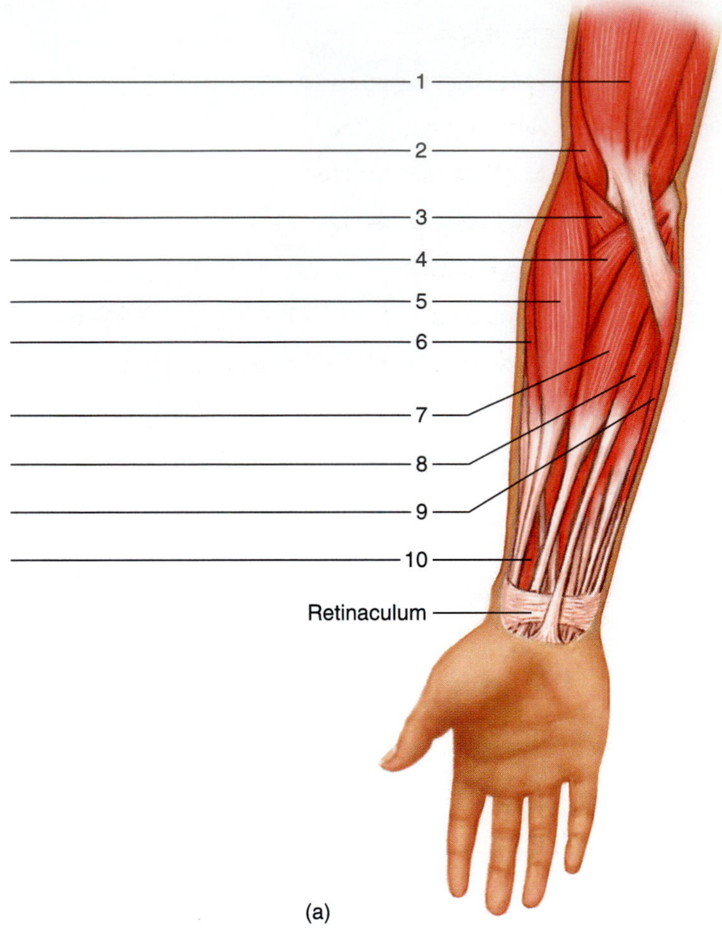

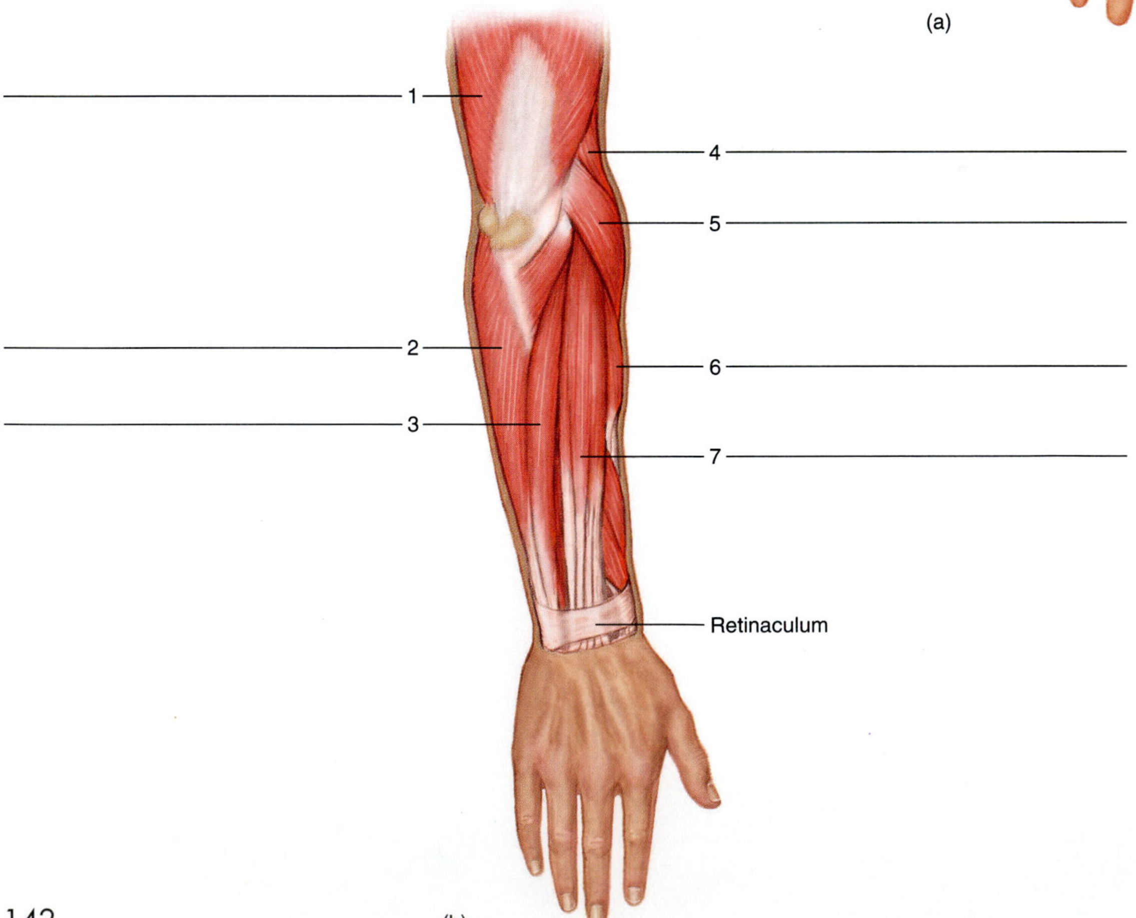

muscles that move the hand

- flexor carpi radialis
- flexor carpi ulnaris
- palmaris longus
- flexor digitorum profundus
- extensor carpi radialis longus
- extensor carpi radialis brevis
- extensor carpi ulnaris
- extensor digitorum

4. Demonstrate the actions of these muscles in your own body.
5. Locate the origins and insertions of these muscles in the human skeleton.
6. Complete Parts A, B, and C of Laboratory Report 20.

Web Quest

Determine the origin, insertion, action, nerve innervation, and blood supply of all the major muscles.

Identify muscles and detailed explanations from an interactive site.

Find information about these topics at www.mhhe.com/shieress8

Laboratory Report **20**

Name ______________________________

Date ______________________________

Section ______________________________

Muscles of the Chest, Shoulder, and Upper Limb

PART A

Match the muscles in column A with the actions in column B. Place the letter of your choice in the space provided.

Column A

a. Brachialis
b. Coracobrachialis
c. Deltoid
d. Extensor carpi ulnaris
e. Flexor carpi ulnaris
f. Flexor digitorum profundus
g. Infraspinatus
h. Pectoralis major
i. Pectoralis minor
j. Pronator teres
k. Rhomboideus major
l. Serratus anterior
m. Supinator
n. Teres major
o. Triceps brachii

Column B

_______ 1. Abducts arm

_______ 2. Pulls arm forward and across chest

_______ 3. Flexes and adducts hand at the wrist

_______ 4. Raises and adducts scapula

_______ 5. Rotates forearm medially

_______ 6. Raises ribs in forceful inhalation or pulls scapula forward and downward

_______ 7. Rotates forearm laterally

_______ 8. Used to thrust shoulder anteriorly, as when pushing something

_______ 9. Flexes the forearm at the elbow

_______ 10. Flexes and adducts arm at the shoulder along with pectoralis major

_______ 11. Extends the forearm at the elbow

_______ 12. Extends, adducts, and rotates arm medially

_______ 13. Extends and adducts hand at the wrist

_______ 14. Rotates arm laterally

_______ 15. Flexes distal joints of fingers 2–5

PART B

Name the muscle indicated by the following combinations of origin and insertion.

	Origin	Insertion	Muscle
1.	Spines of upper thoracic vertebrae	Medial border of scapula	____________
2.	Outer surfaces of upper ribs	Ventral surfaces of scapula	____________
3.	Sternal ends of upper ribs	Coracoid process of scapula	____________
4.	Coracoid process of scapula	Shaft of humerus	____________
5.	Lateral border of scapula	Intertubercular groove of humerus	____________
6.	Anterior surface of scapula	Lesser tubercle of humerus	____________
7.	Lateral border of scapula	Greater tubercle of humerus	____________
8.	Anterior shaft of humerus	Coronoid process of ulna	____________
9.	Medial epicondyle of humerus and coronoid process of ulna	Lateral surface of radius	____________
10.	Anterior distal end of ulna	Anterior distal end of radius	____________
11.	Distal lateral end of humerus	Lateral surface of radius above styloid process	____________
12.	Medial epicondyle of humerus	Base of second and third metacarpals	____________
13.	Medial epicondyle of humerus	Fascia of palm	____________
14.	Distal end of humerus	Base of second metacarpal	____________
15.	Lateral epicondyle of humerus	Base of fifth metacarpal	____________

PART C

Critical Thinking Application

Identify the muscles indicated in figure 20.5.

1. ______________________
2. ______________________
3. ______________________
4. ______________________
5. ______________________
6. ______________________
7. ______________________
8. ______________________
9. ______________________
10. ______________________
11. ______________________
12. ______________________
13. ______________________
14. ______________________
15. ______________________
16. ______________________
17. ______________________
18. ______________________
19. ______________________
20. ______________________
21. ______________________

Figure 20.5 Identify the muscles that appear as body surface features in these photographs (*a, b,* and *c*), using the terms provided.

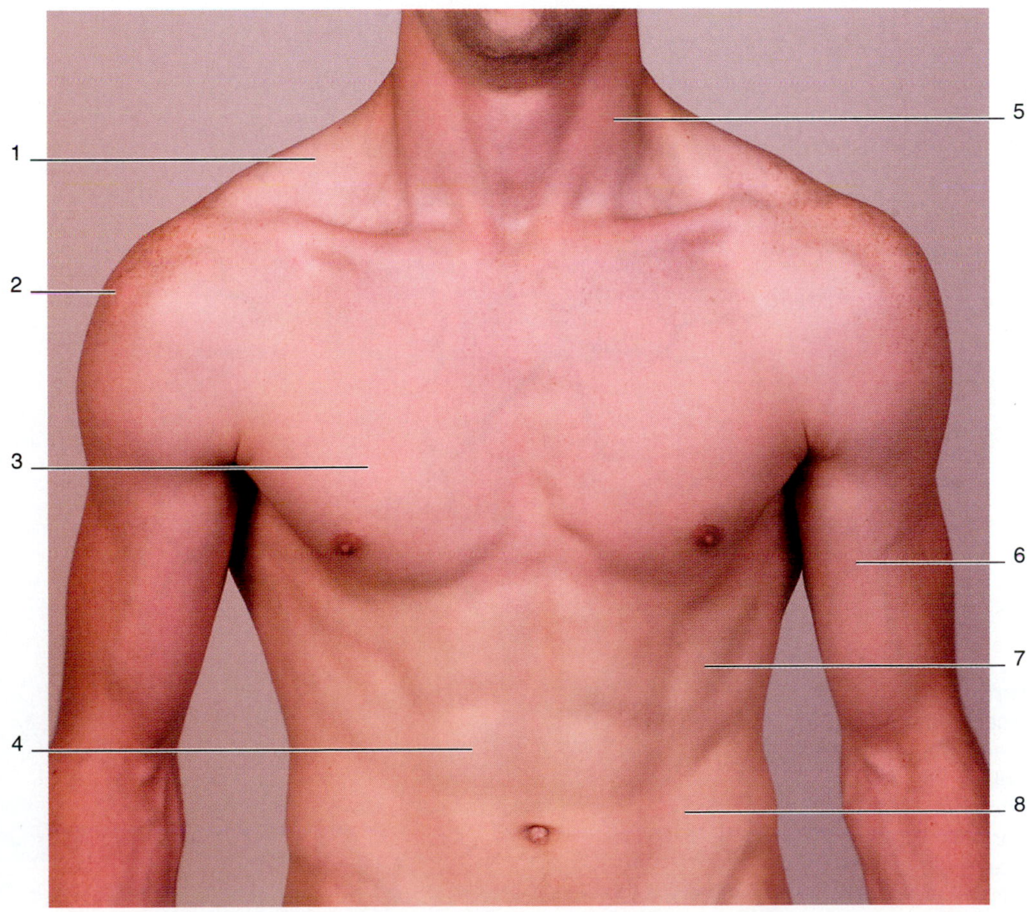

(a)

Terms:
Biceps brachii
Deltoid
External oblique
Pectoralis major
Rectus abdominis
Serratus anterior
Sternocleidomastoid
Trapezius

Figure 20.5 *Continued.*

Terms:
Biceps brachii
Deltoid
Infraspinatus
Latissimus dorsi
Trapezius
Triceps brachii

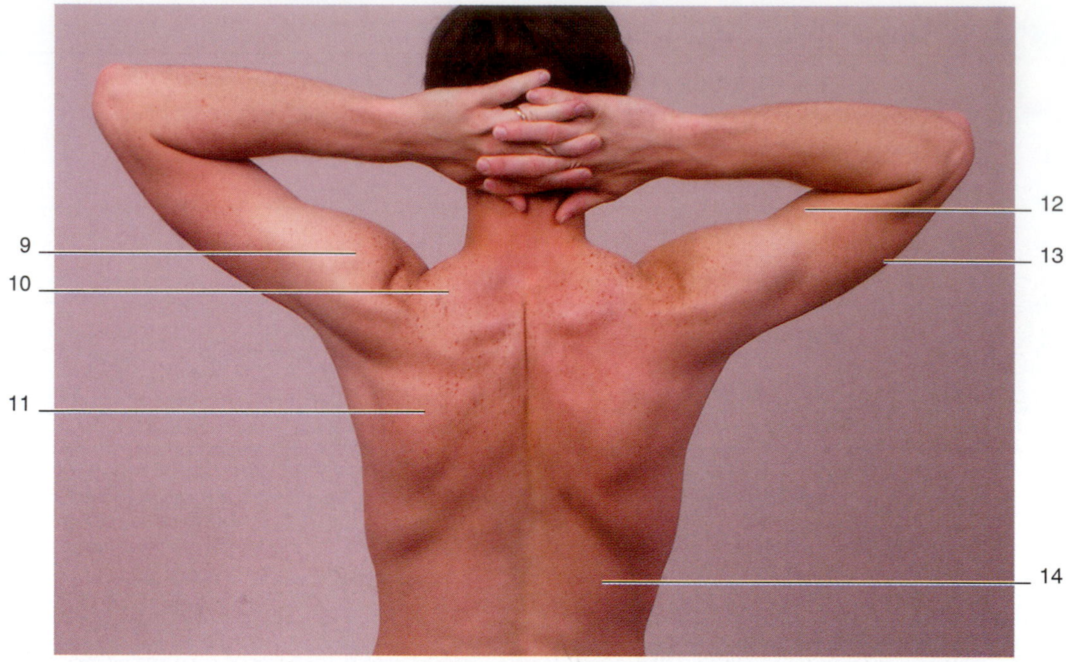

(b)

Terms:
Biceps brachii
Brachioradialis
Deltoid
Pectoralis major
Serratus anterior
Trapezius
Triceps brachii

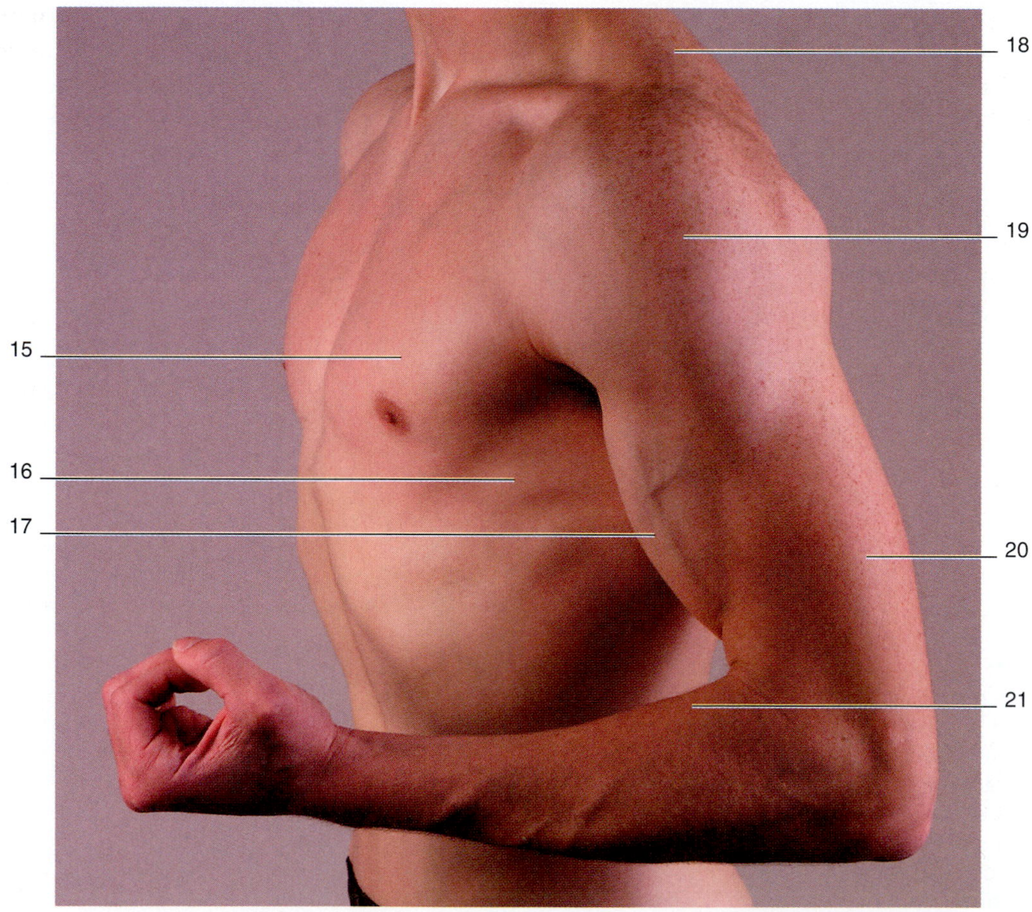

(c)

Laboratory Exercise 21

Muscles of the Abdominal Wall and Pelvic Outlet

MATERIALS NEEDED

Textbook
Torso with musculature
Human skeleton, articulated
Muscular models of male and female pelves

The anterior and lateral walls of the abdomen contain broad, flattened muscles arranged in layers. These muscles connect the rib cage and vertebral column to the pelvic girdle.

The muscles of the pelvic outlet are arranged in two muscular sheets: (1) a deeper pelvic diaphragm that forms the floor of the pelvic cavity and (2) a urogenital diaphragm that fills the space within the pubic arch.

PURPOSE OF THE EXERCISE

To review the actions, origins, and insertions of the muscles of the abdominal wall and pelvic outlet.

LEARNING OBJECTIVES

After completing this exercise, you should be able to

1. locate and identify the muscles of the abdominal wall and pelvic outlet;
2. describe the action of each of these muscles;
3. locate the origin and insertion of each of these muscles.

PROCEDURE

1. Review the sections entitled "Muscles of the Abdominal Wall" and "Muscles of the Pelvic Outlet" in chapter 8 of the textbook.
2. As a review activity, label figures 21.1, 21.2, and 21.3.
3. Locate the following muscles in the torso:

   external oblique

   internal oblique

   transversus abdominis

   rectus abdominis

4. Demonstrate the actions of these muscles in your own body.
5. Locate the origin and insertion of each of these muscles in the human skeleton.
6. Complete Part A of Laboratory Report 21.
7. Locate the following muscles in the models of the male and female pelves:

   levator ani

   superficial transversus perinei

   bulbospongiosus

   ischiocavernosus

8. Locate the origin and insertion of each of these muscles in the human skeleton.
9. Complete Part B of the laboratory report.

Critical Thinking Application

List the muscles from superficial to deep for an appendectomy incision.

Figure 21.1 Label the muscles of the abdominal wall.

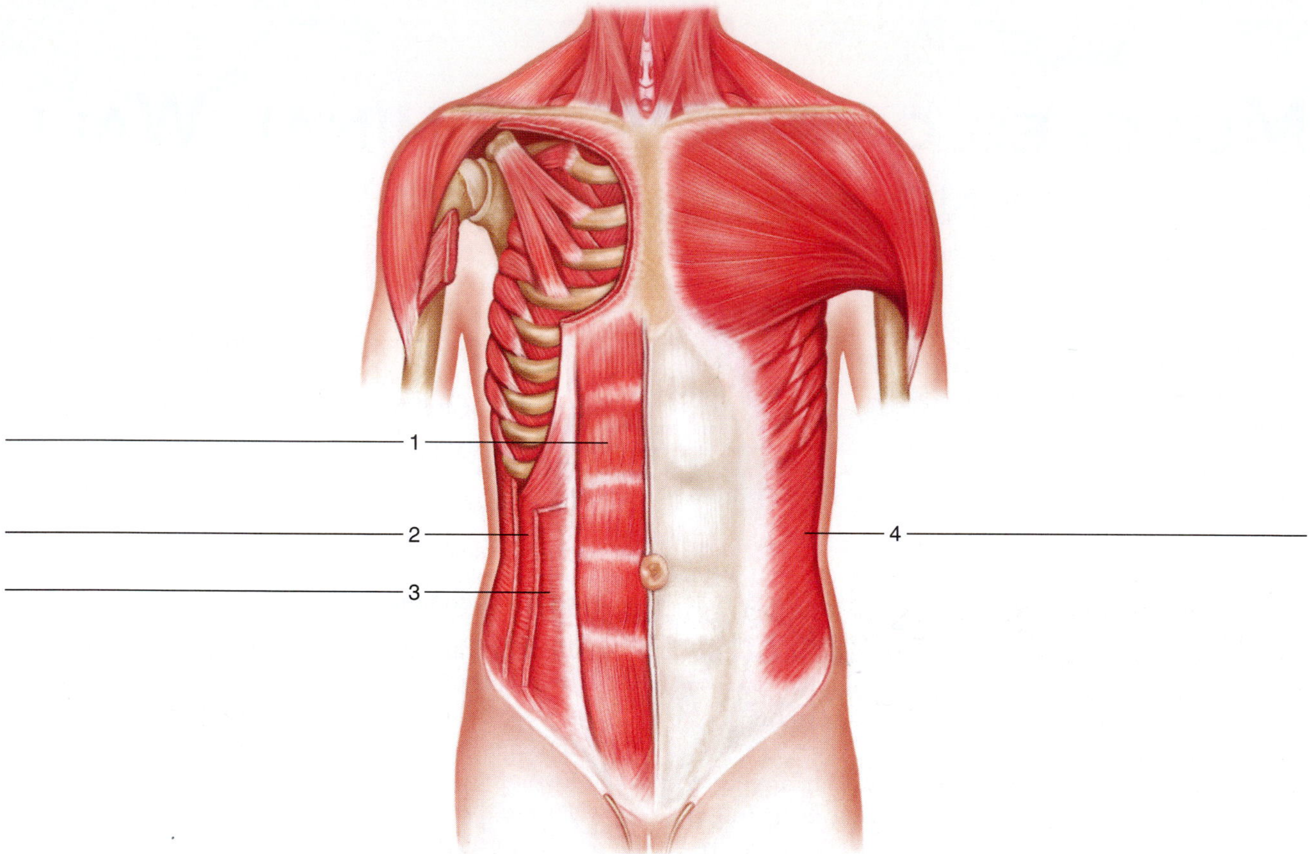

Figure 21.2 Label the muscles of the male pelvic outlet.

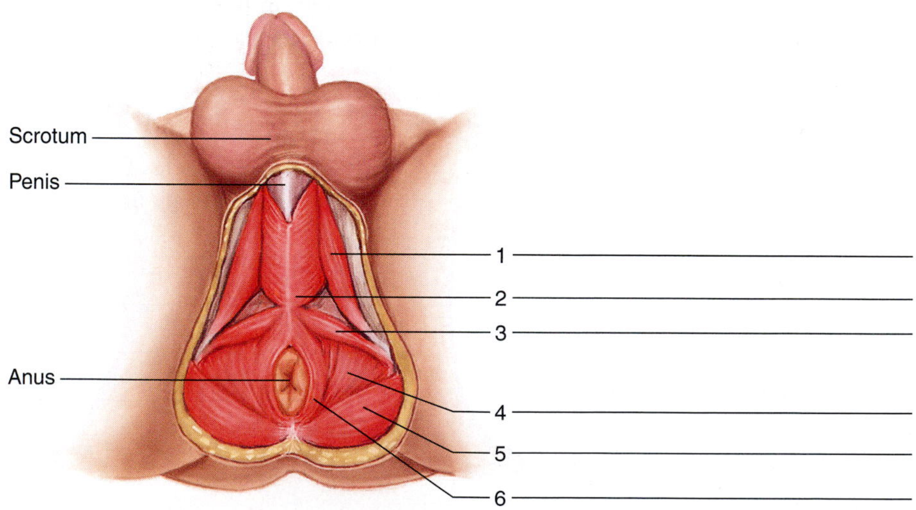

Figure 21.3 Label the muscles of the female pelvic outlet.

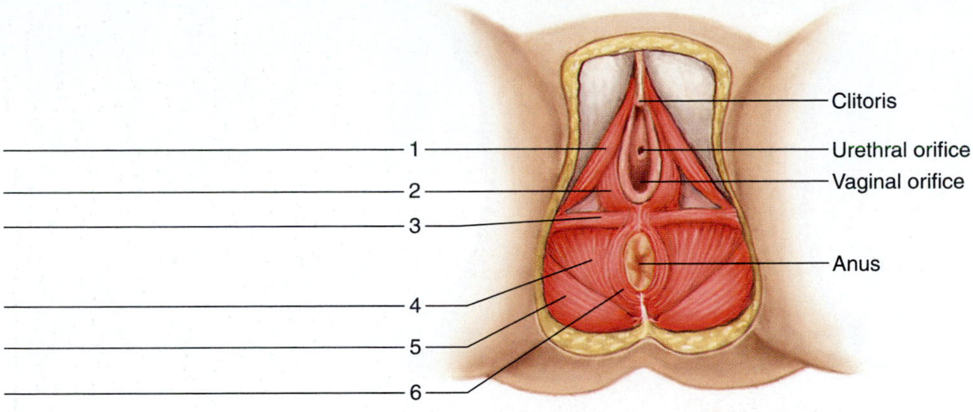

Web Quest

Determine the origin, insertion, action, nerve innervation, and blood supply of all the major muscles.

Identify muscles and detailed explanations from an interactive site.

Find information about these topics at www.mhhe.com/shieress8

Laboratory Report 21

Name ____________________

Date ____________________

Section ____________________

Muscles of the Abdominal Wall and Pelvic Outlet

PART A

Complete the following statements:

1. A band of tough connective tissue in the midline of the anterior abdominal wall called the ____________ serves as a muscle attachment.
2. The ____________ muscle spans from the ribs and sternum to the pubic bones.
3. The ____________ forms the third layer (deepest layer) of the abdominal wall muscles.
4. The action of the external oblique muscle is to ____________.
5. The action of the rectus abdominis is to ____________.

PART B

Complete the following statements:

1. The levator ani forms the ____________ diaphragm.
2. The levator ani provides a sphincterlike action in the ____________.
3. The action of the superficial transversus perinei is to ____________.
4. The ____________ assists in emptying the male urethra.
5. In females, the bulbospongiosus acts to ____________.
6. The ischiocavernosus extends from the margin of the pubic arch to the ____________.

Laboratory Exercise 22

Muscles of the Hip and Lower Limb

MATERIALS NEEDED

Textbook
Torso with musculature
Human skeleton, articulated
Muscular models of the lower limb

The muscles that move the thigh are attached to the femur and to some part of the pelvic girdle. Those attached anteriorly primarily act to flex the thigh at the hip, whereas those attached posteriorly act to extend, abduct, or rotate the thigh.

The muscles that move the leg connect the tibia or fibula to the femur or to the pelvic girdle. They function to flex or extend the leg at the knee. Other muscles, located in the leg, act to move the foot.

PURPOSE OF THE EXERCISE

To review the actions, origins, and insertions of the muscles that move the thigh, leg, and foot.

LEARNING OBJECTIVES

After completing this exercise, you should be able to

1. locate and identify the muscles that move the thigh, leg, and foot;
2. describe and demonstrate the actions of each of these muscles;
3. locate the origin and insertion of each of these muscles.

PROCEDURE

1. Review the sections entitled "Muscles That Move the Thigh," "Muscles That Move the Leg," and "Muscles That Move the Foot" in chapter 8 of the textbook.
2. As a review activity, label figures 22.1, 22.2, 22.3, 22.4, 22.5, and 22.6.
3. Locate the following muscles in the torso and in the lower limb models. Also locate as many of them as possible in your own body.

muscles that move the thigh

- iliopsoas group
  - psoas major
  - iliacus
- gluteus maximus
- gluteus medius
- gluteus minimus
- tensor fasciae latae
- adductor longus
- adductor magnus
- gracilis

muscles that move the leg

- hamstring group
  - biceps femoris
  - semitendinosus
  - semimembranosus
- sartorius
- quadriceps femoris group
  - rectus femoris
  - vastus lateralis
  - vastus medialis
  - vastus intermedius

Figure 22.1 Label the muscles of the anterior right hip and thigh.

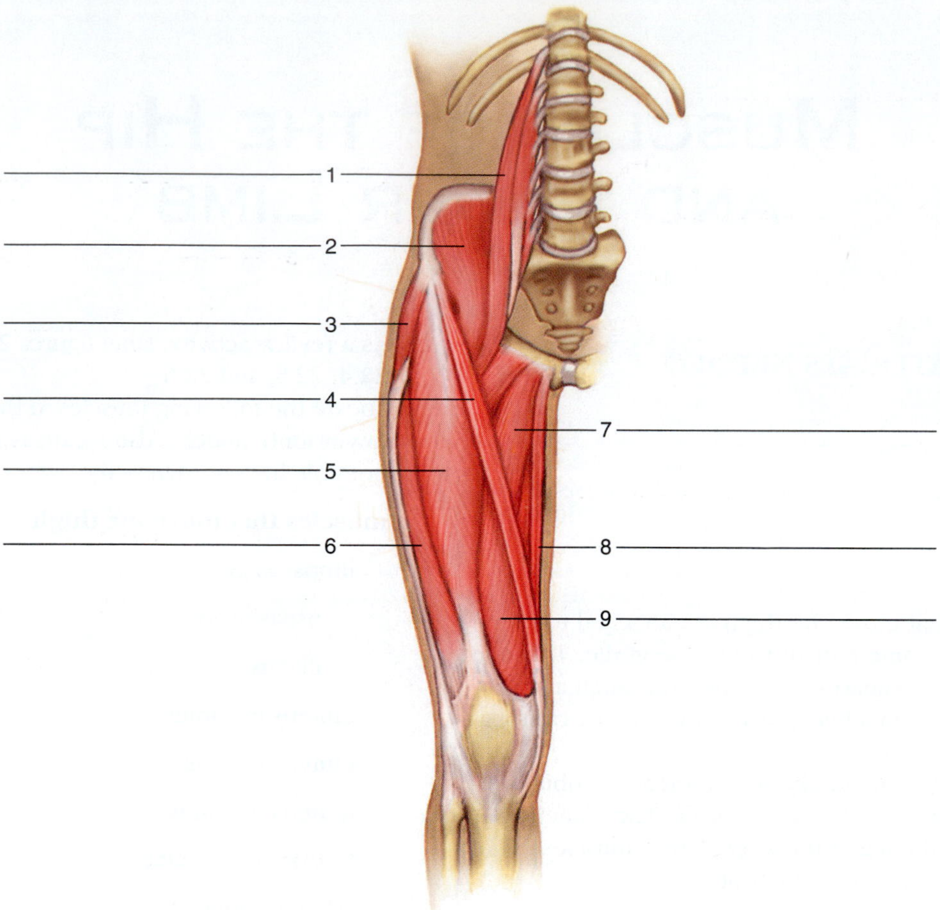

Figure 22.2 Label the muscles of the lateral right hip and thigh.

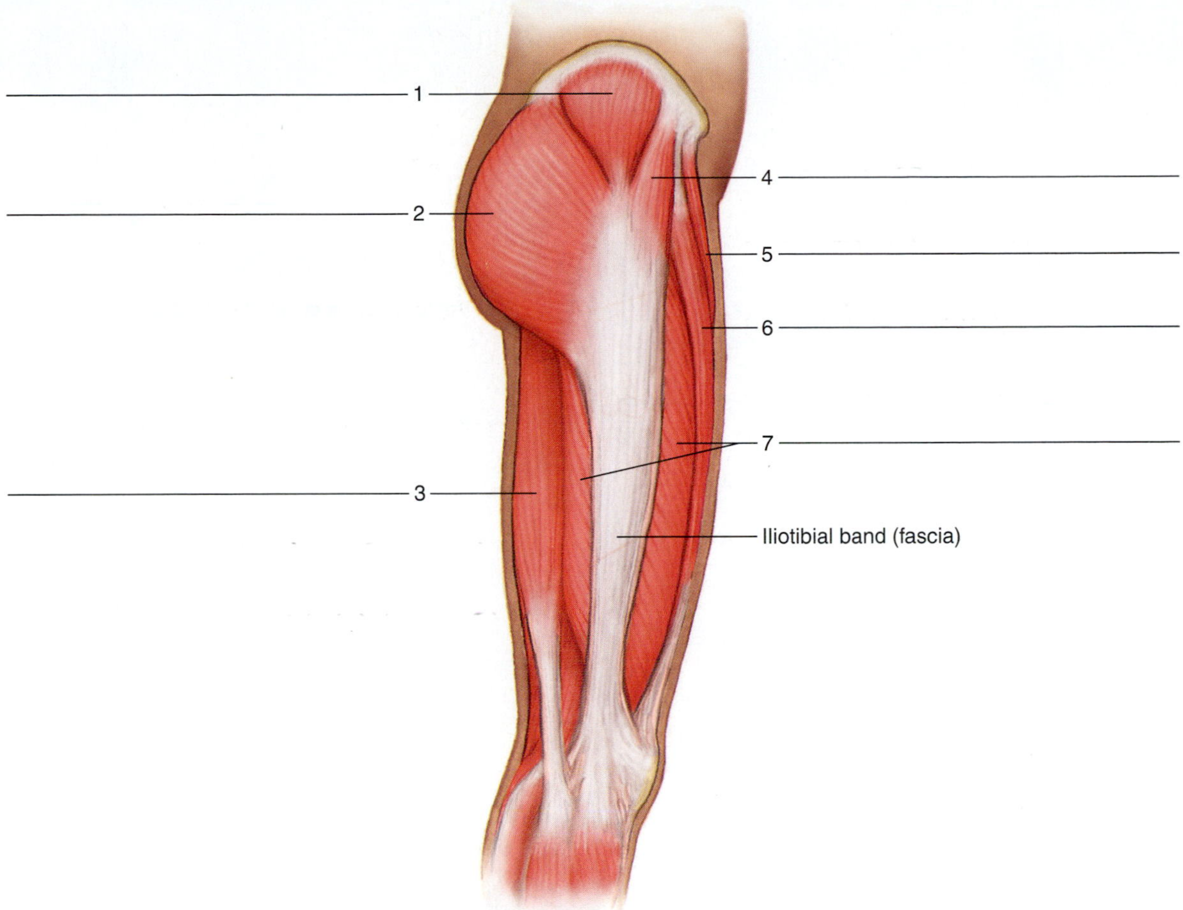

Figure 22.3 Label the muscles of the posterior right hip and thigh.

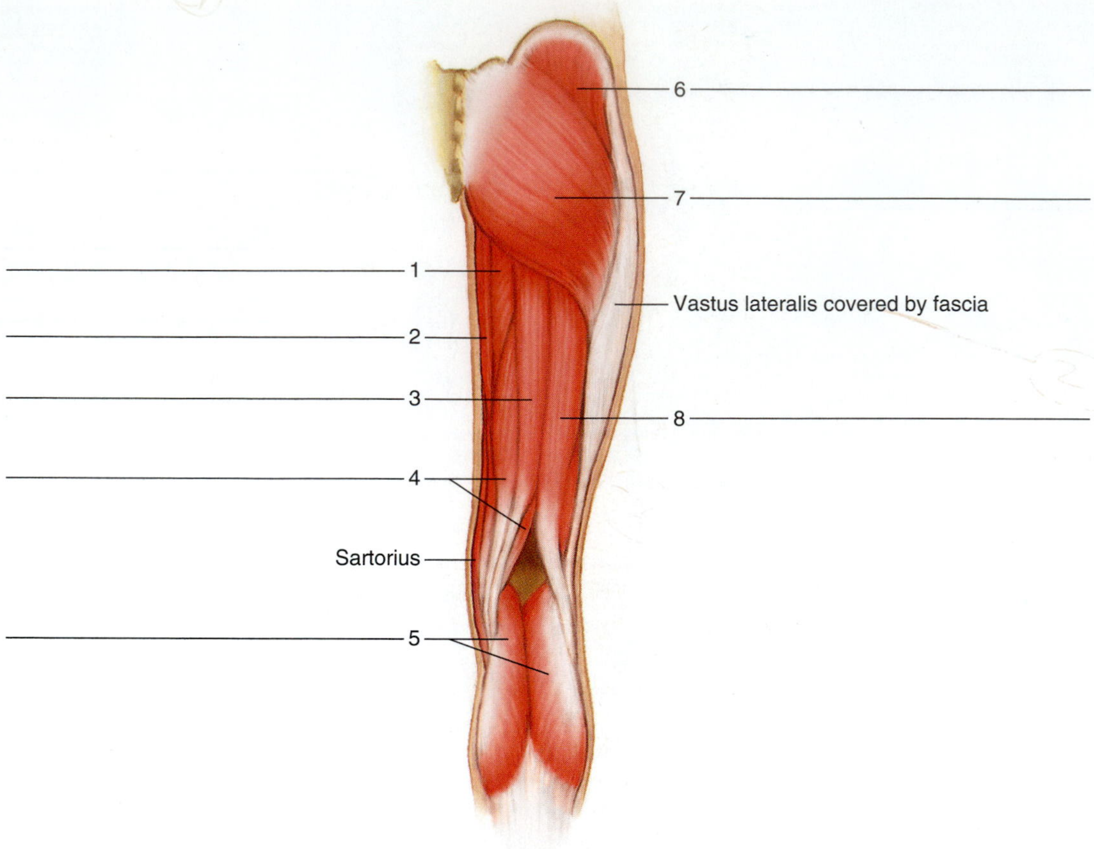

Figure 22.4 Label the muscles of the anterior right leg.

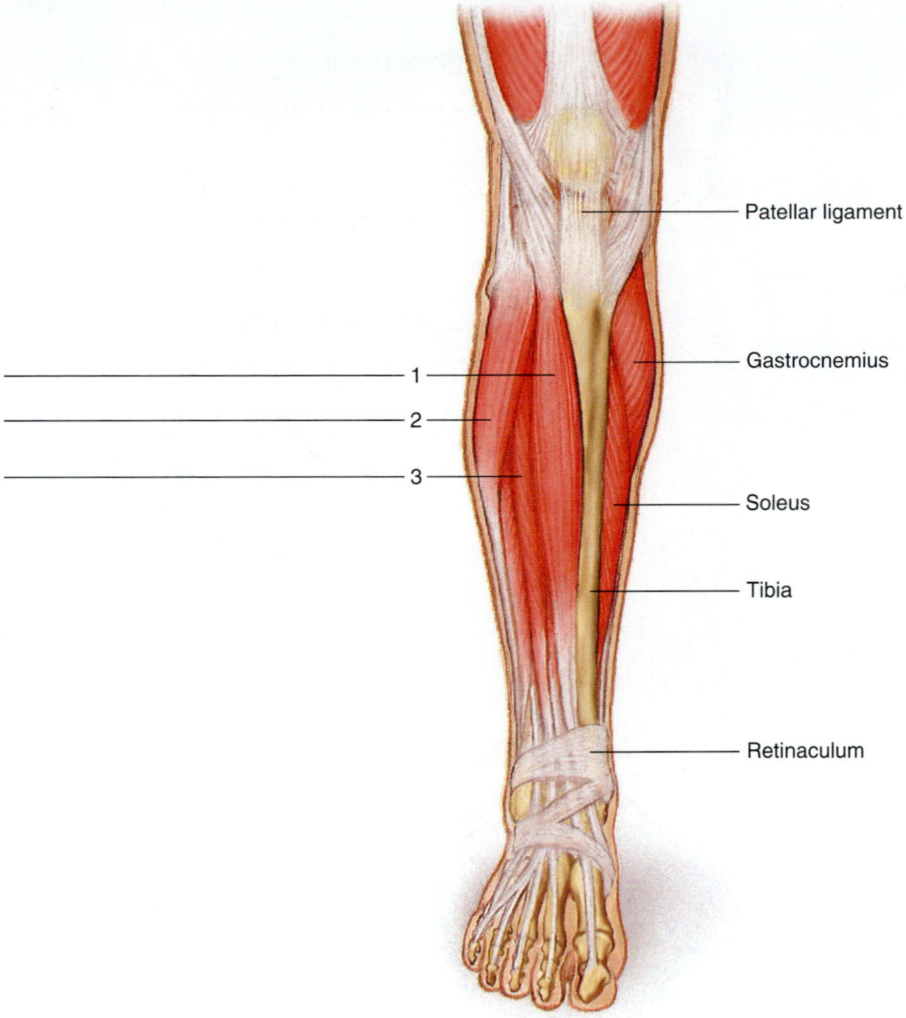

Figure 22.5 Label the muscles of the lateral right leg.

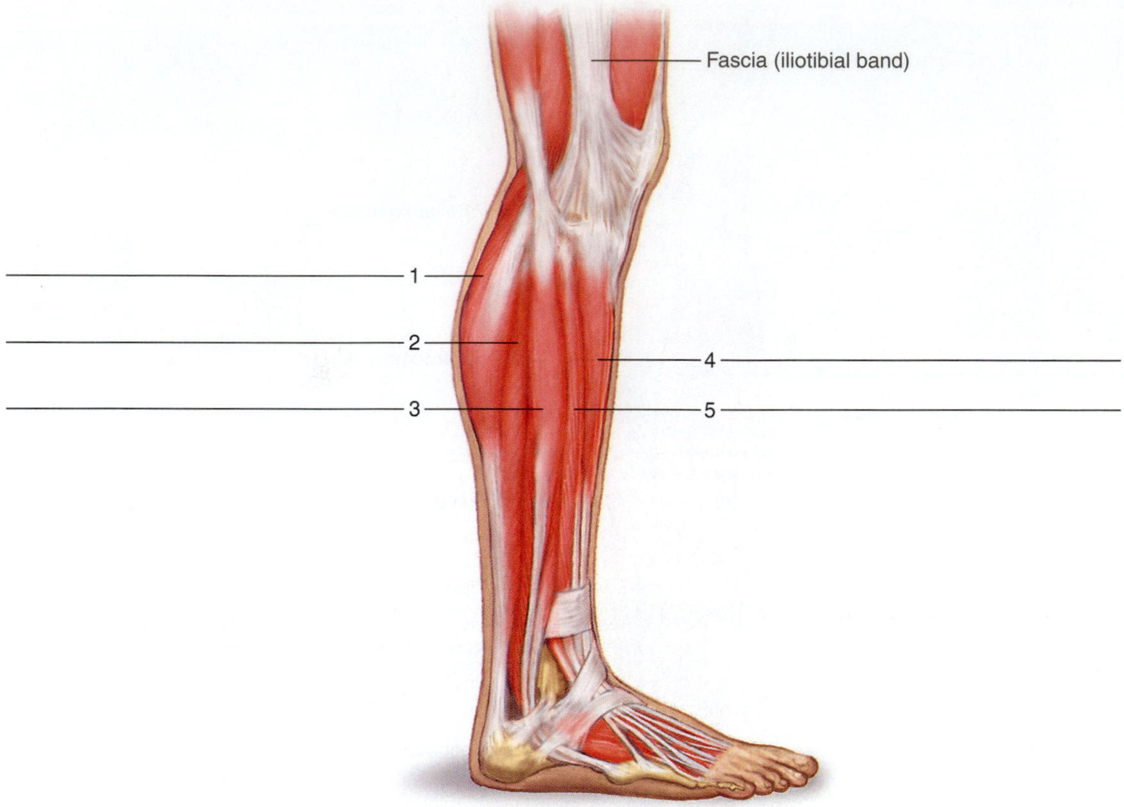

Figure 22.6 Label the muscles of the posterior right leg.

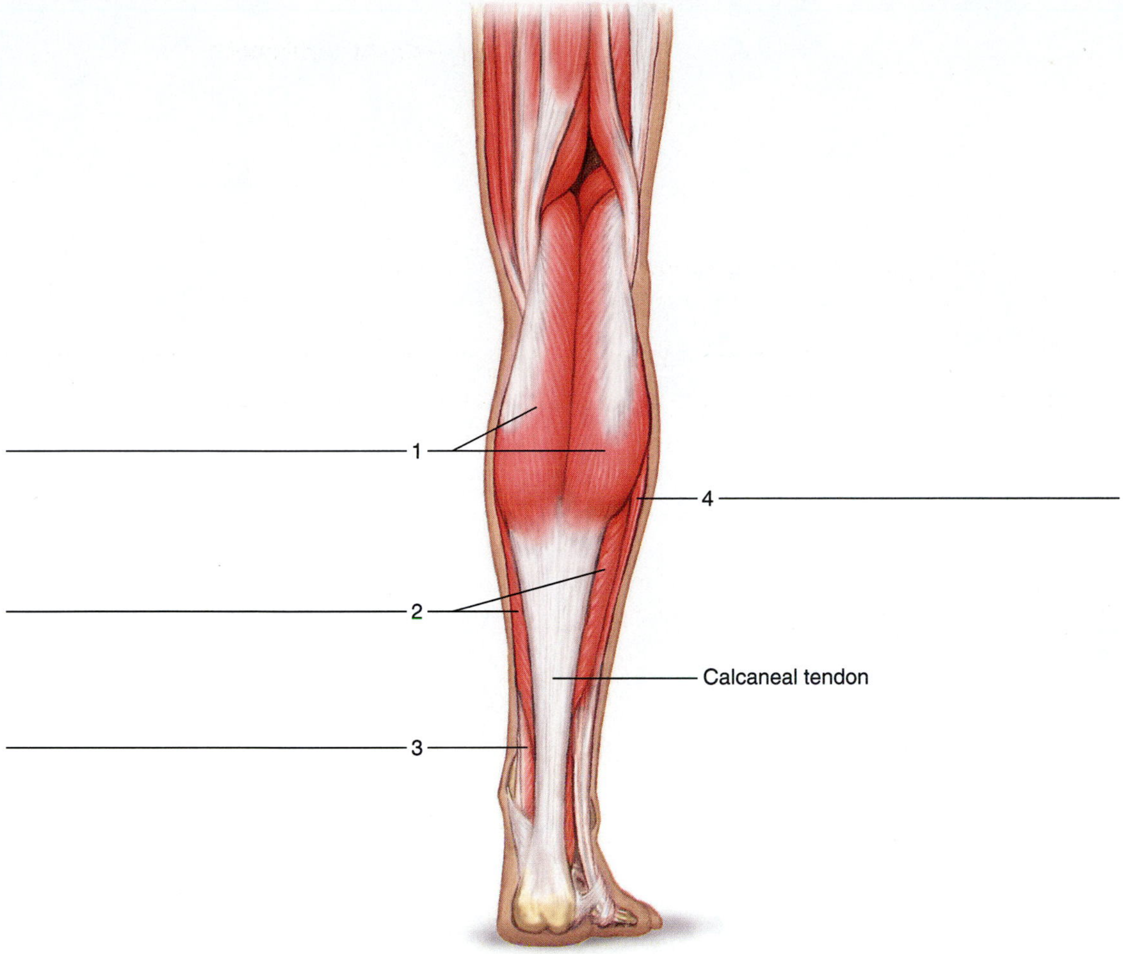

muscles that move the foot

tibialis anterior

peroneus (fibularis) tertius

extensor digitorum longus

gastrocnemius

soleus

flexor digitorum longus

peroneus (fibularis) longus

4. Demonstrate the action of each of these muscles in your own body.
5. Locate the origin and insertion of each of these muscles in the human skeleton.
6. Complete Parts A, B, and C of Laboratory Report 22.

Web Quest

Determine the origin, insertion, action, nerve innervation, and blood supply of all the major muscles.

Identify muscles and detailed explanations from an interactive site.

Find information about these topics at www.mhhe.com/shieress8

Laboratory Report **22**

Name ______________________

Date ______________________

Section ______________________

Muscles of the Hip and Lower Limb

PART A

Match the muscles in column A with the actions in column B. Place the letter of your choice in the space provided.

Column A		Column B
a. Biceps femoris	______	1. Adducts thigh
b. Gluteus medius	______	2. Plantar flexion and eversion of foot
c. Gracilis	______	3. Flexes thigh at the hip
d. Peroneus (fibularis) longus	______	4. Abducts thigh and rotates it laterally
e. Psoas major and iliacus	______	5. Abducts thigh and rotates it medially
f. Quadriceps femoris group	______	6. Flexes the leg at the knee
g. Sartorius	______	7. Extends the leg at the knee
h. Tibialis anterior	______	8. Dorsiflexion and inversion of foot

PART B

Name the muscle indicated by the following combinations of origin and insertion.

Origin	Insertion	Muscle
1. Lateral surface of ilium	Greater trochanter of femur	______
2. Ischial tuberosity	Posterior surface of femur	______
3. Anterior superior iliac spine	Medial surface of tibia	______
4. Lateral and medial condyles of femur	Posterior surface of calcaneus	______
5. Anterior iliac crest	Fascia (iliotibial band) of the thigh	______
6. Greater trochanter and posterior surface of femur	Patella to tibial tuberosity	______
7. Ischial tuberosity	Medial surface of tibia	______
8. Medial surface of femur	Patella to tibial tuberosity	______
9. Posterior surface of tibia	Distal phalanges of four lateral toes	______
10. Lateral condyle and lateral surface of tibia	Cuneiform and first metatarsal	______

PART C

Critical Thinking Application

Identify the muscles indicated in figure 22.7.

1. ______________________________
2. ______________________________
3. ______________________________
4. ______________________________
5. ______________________________
6. ______________________________
7. ______________________________
8. ______________________________

Figure 22.7 Identify the muscles that appear as lower limb surface features in these photographs (*a* and *b*), using the terms provided.

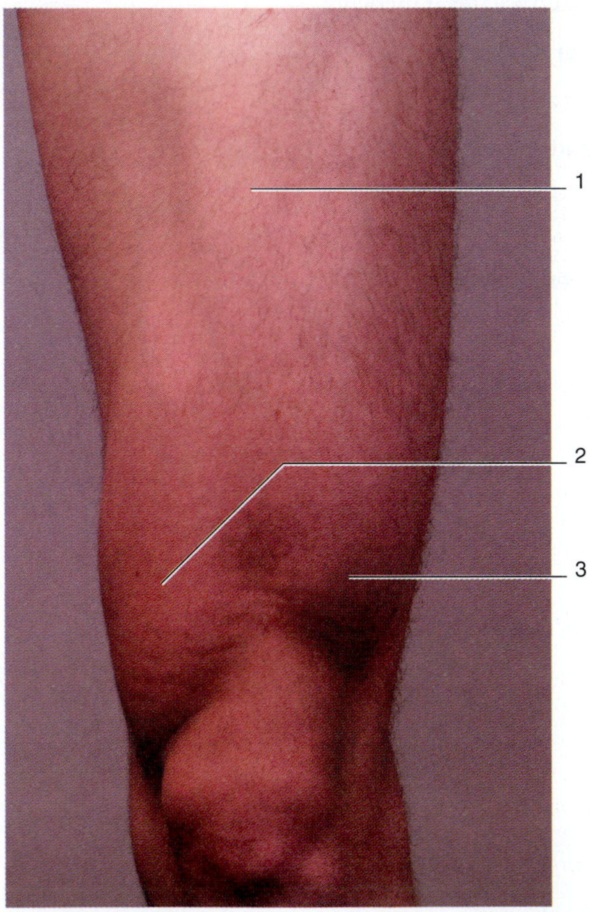

(a) Left thigh, anterior view

Terms:
Gastrocnemius
Rectus femoris
Sartorius
Soleus
Tibialis anterior
Vastus lateralis
Vastus medialis
Vastus medialis

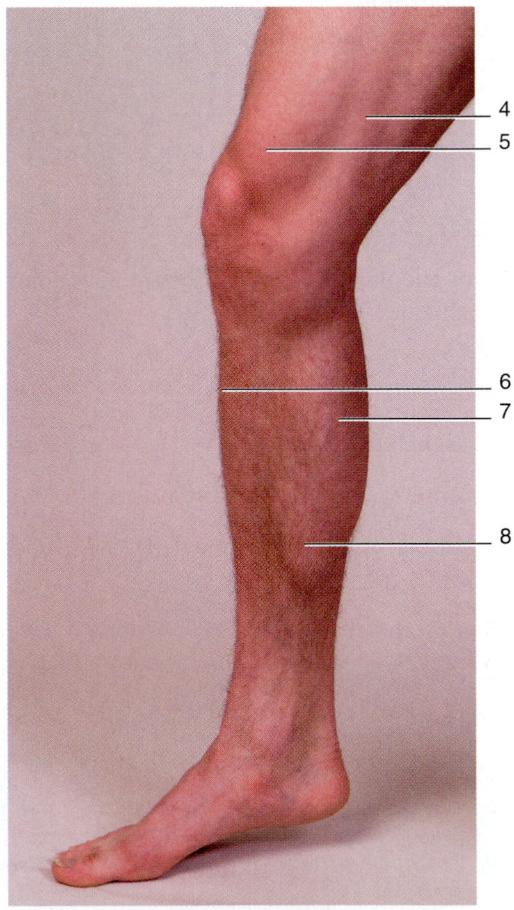

(b) Right lower limb, medial view

NERVOUS TISSUE AND NERVES

MATERIALS NEEDED

Textbook
Compound microscope
Prepared microscope slides of the following:
- Spinal cord (smear)
- Dorsal root ganglion (section)
- Neuroglial cells (astrocytes)
- Peripheral nerve (cross section and longitudinal section)

Neuron model

For Optional Activity:

Prepared microscope slide of Purkinje cells from cerebellum

Nervous tissue, which occurs in the brain, spinal cord, and nerves, contains neurons and neuroglial cells. The neurons are the basic structural and functional units of the nervous system involved in decision-making processes, detecting stimuli, and conducting messages. The neuroglial cells perform various supportive and protective functions for neurons.

PURPOSE OF THE EXERCISE

To review the characteristics of nervous tissue and to observe neurons, neuroglial cells, and various features of the nerves.

LEARNING OBJECTIVES

After completing this exercise, you should be able to

1. describe the general characteristics of nervous tissue;
2. distinguish between neurons and neuroglial cells;
3. identify the major structures of a neuron and a nerve.

PROCEDURE

1. Review the sections entitled "Neurons," "Classification of Neurons," and "Neuroglial Cells" in chapter 9 of the textbook.
2. As a review activity, label figures 23.1 and 23.2.
3. Complete Part A of Laboratory Report 23.
4. Obtain a prepared microscope slide of a spinal cord smear. Using low-power magnification, search the slide and locate the relatively large, deeply stained cell bodies of motor neurons (multipolar neurons).
5. Observe a single motor neuron using high-power magnification, and note the following features:

   cell body

   nucleus

   nucleolus

   neurofibrils (threadlike structures extending into the nerve fibers)

   dendrites

   axon (nerve fiber)

   Compare the slide to the neuron model and to figure 23.3. You also may note small, darkly stained nuclei of neuroglial cells around the motor neuron.
6. Sketch and label a single motor (efferent) neuron in the space provided in Part B of the laboratory report.
7. Obtain a prepared microscope slide of a dorsal root ganglion. Search the slide and locate a cluster of sensory neuron cell bodies. You also may note bundles of nerve fibers passing among groups of neuron cell bodies (fig. 23.4).
8. Sketch and label a single sensory (afferent) neuron cell body in the space provided in Part B of the laboratory report.
9. Obtain a prepared microscope slide of neuroglial cells. Search the slide and locate some darkly stained astrocytes with numerous long, slender processes (fig. 23.5).

Figure 23.1 Label this diagram of a motor neuron.

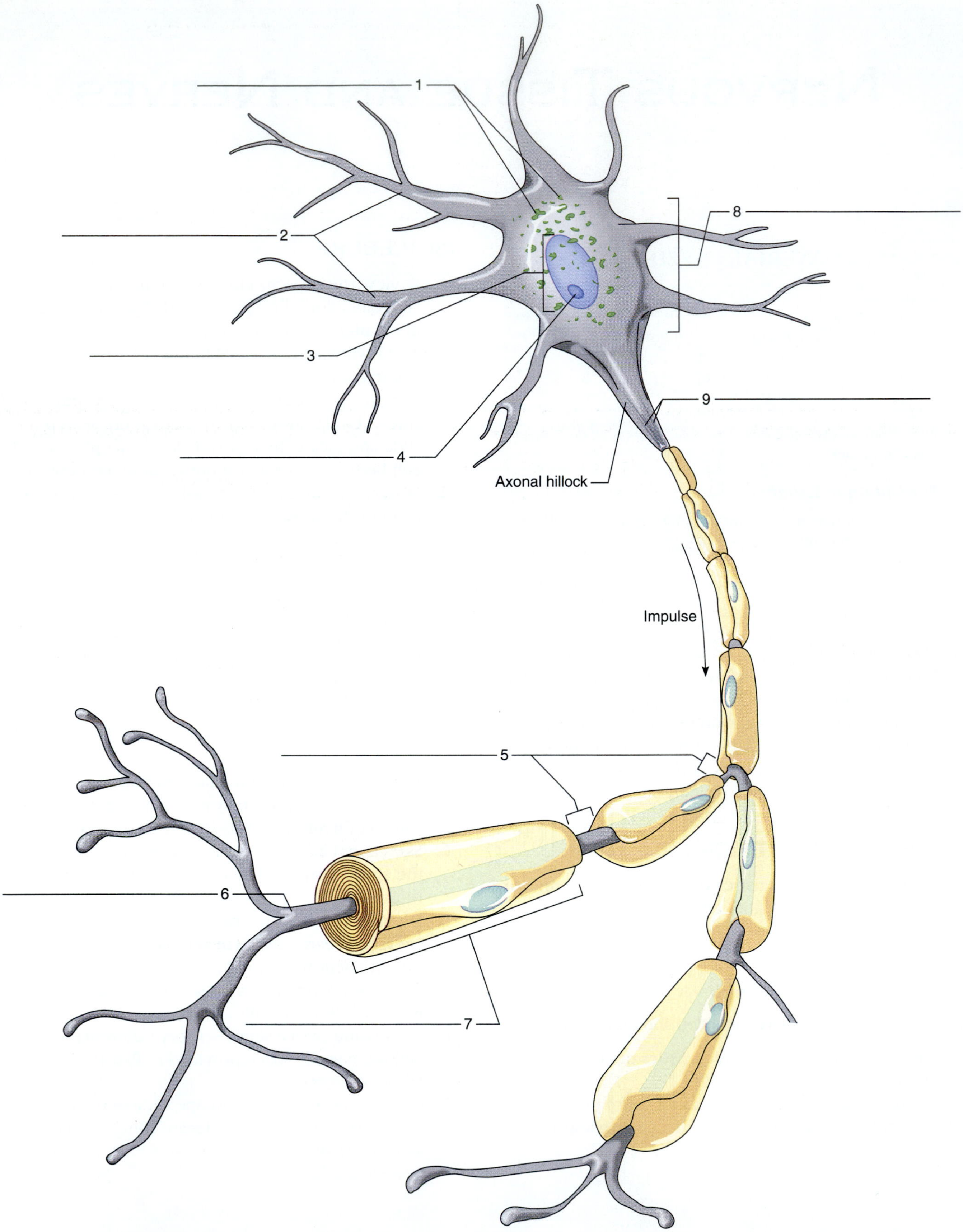

Figure 23.2 Label the features of a myelinated nerve fiber.

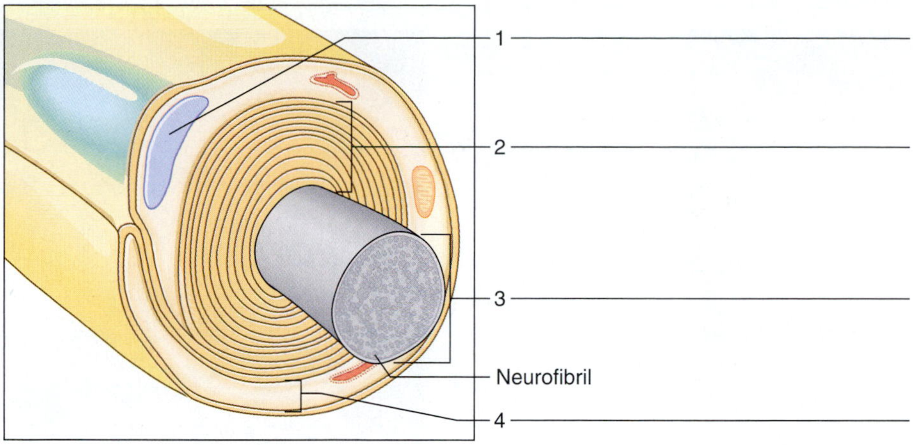

Figure 23.3 Micrograph of a multipolar neuron and neuroglia from a spinal cord smear (100× micrograph enlarged to 600×).

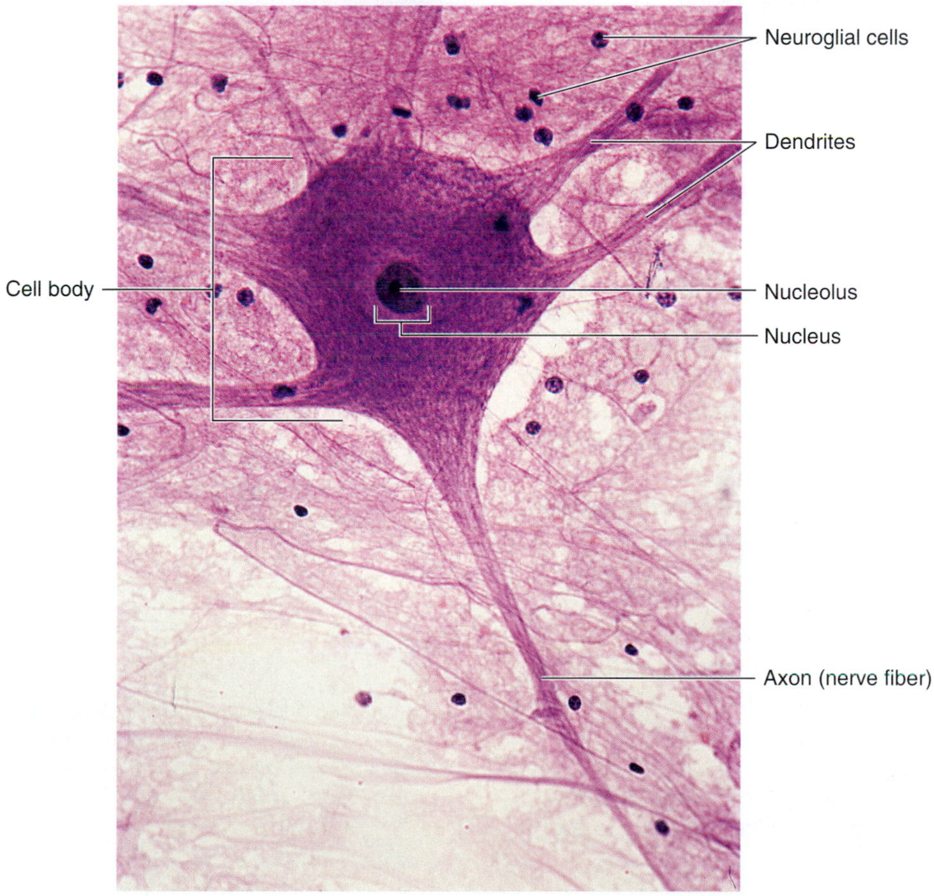

Figure 23.4 Micrograph of a dorsal root ganglion (50× micrograph enlarged to 100×).

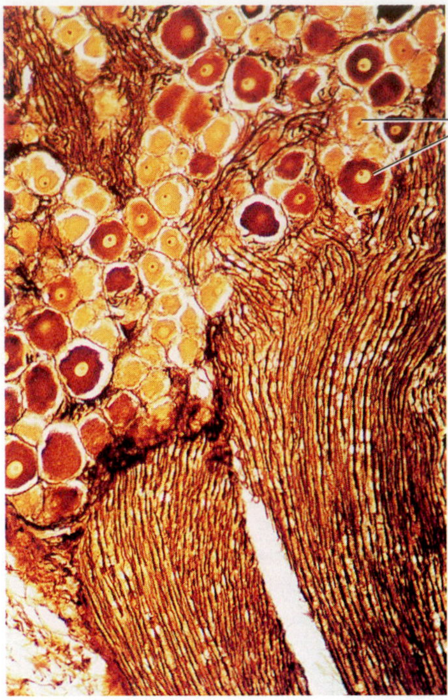

Figure 23.5 Micrograph of astrocytes (250× micrograph enlarged to 1,000×).

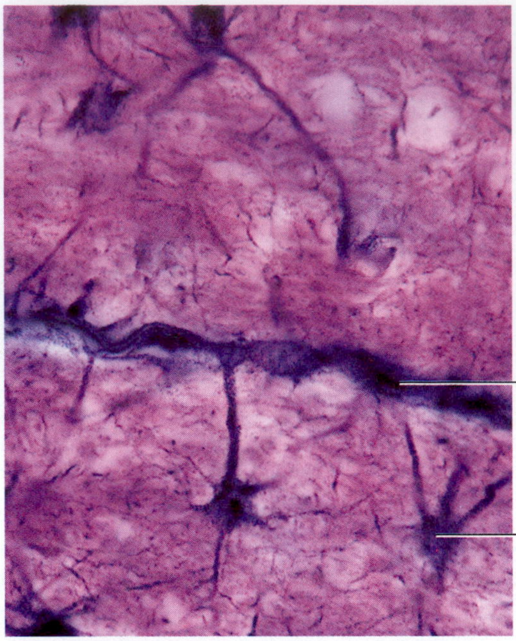

10. Sketch a single neuroglial cell in the space provided in Part B of the laboratory report.
11. Obtain a prepared microscope slide of a nerve. Locate the cross section of the nerve and note the many round nerve fibers inside. Also note the dense layer of connective tissue (perineurium) that encircles the nerve fibers and holds them together in a bundle. The individual nerve fibers are surrounded by a layer of more delicate connective tissue (endoneurium) (fig. 23.6).
12. Using high-power magnification, observe a single nerve fiber. Note the following features:

    central axon

    myelin around the axon (actually, most of the myelin may have been dissolved and lost during the slide preparation)

    neurilemma
13. Sketch and label a single nerve fiber with Schwann cell (cross section) in the space provided in Part B of the laboratory report.
14. Locate the longitudinal section of the nerve on the slide (fig. 23.7). Note the following:

    central axons

    myelin sheaths of Schwann cells

    neurilemmal sheaths of Schwann cells

    nodes of Ranvier
15. Sketch and label a single nerve fiber with Schwann cell (longitudinal section) in the space provided in Part B of the laboratory report.

OPTIONAL ACTIVITY

Obtain a prepared microscope slide of Purkinje cells. To locate these neurons, search the slide for large, flask-shaped cell bodies. Note that each cell body has one or two large, thick dendrites that give rise to branching networks of fibers. These cells are found in a particular region of the brain (cerebellar cortex).

Figure 23.6 Cross section of a bundle of neurons within a nerve (400×).

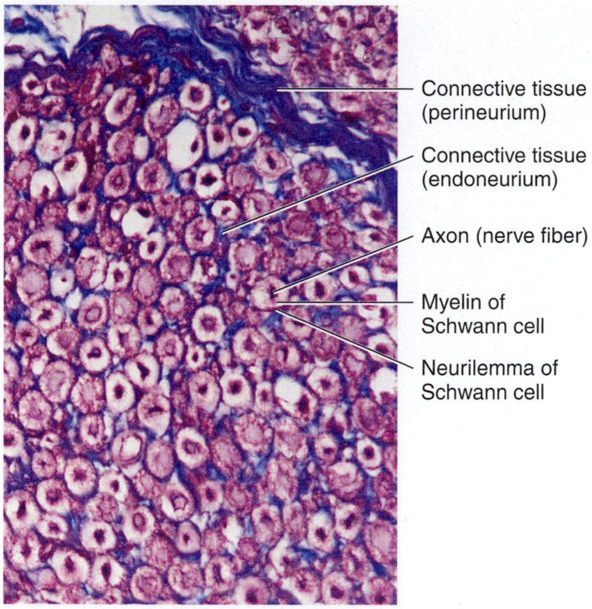

Figure 23.7 Longitudinal section of a nerve (250× micrograph enlarged to 2,000×).

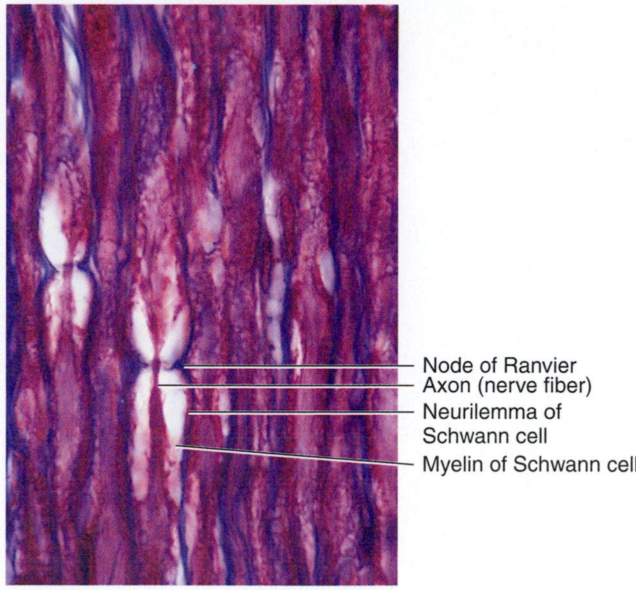

Laboratory Report **23**

Name ______________________

Date ______________________

Section ______________________

Nervous Tissue and Nerves

PART A

Match the terms in column A with the descriptions in column B. Place the letter of your choice in the space provided.

Column A

a. Astrocyte
b. Axon
c. Dendrite
d. Interneurons
e. Microglial cell
f. Motor (efferent) neuron
g. Myelin
h. Neurilemma
i. Neurofibrils
j. Oligodendrocyte
k. Sensory (afferent) neuron

Column B

_______ 1. Sheath of Schwann cell containing cytoplasm and nucleus that encloses myelin

_______ 2. Network of fine threads within a nerve fiber

_______ 3. Substance of Schwann cell composed of lipoprotein

_______ 4. Neuron process with many branches that conducts impulse toward the cell body

_______ 5. Star-shaped neuroglial cell between neurons and blood vessels

_______ 6. Nerve fiber arising from slight elevation of the cell body that conducts impulse away from the cell body

_______ 7. Transmits impulse from sensory to motor neuron

_______ 8. Transmits impulse out of brain or spinal cord to effectors

_______ 9. Transmits impulse into brain or spinal cord from receptors

_______ 10. Myelin-forming neuroglial cell in brain and spinal cord

_______ 11. Phagocytic neuroglial cell

PART B

1. Sketch and label a single motor neuron.

2. Sketch and label a single sensory neuron cell body.

3. Sketch a single neuroglial cell.

4. Sketch and label a single nerve fiber with Schwann cell (cross section).

5. Sketch and label a single nerve fiber with Schwann cell (longitudinal section).

Laboratory Exercise 24

The Reflex Arc and Reflexes

MATERIALS NEEDED

Textbook
Rubber percussion hammer
Virtual Physiology Lab: 1, Action Potential

A reflex arc represents the simplest type of nerve pathway found in the nervous system. This pathway begins with a receptor at the dendrite end of a sensory (afferent) neuron. The sensory neuron leads into the central nervous system and may communicate with one or more interneurons. Some of these interneurons, in turn, communicate with motor (efferent) neurons, whose axons (nerve fibers) lead outward to effectors.

Thus, when a sensory receptor is stimulated by some kind of change occurring inside or outside the body, nerve impulses may pass through a reflex arc, and, as a result, effectors may respond. Such an automatic, subconscious response is called a *reflex.*

Most reflexes demonstrated in this lab are stretch reflexes. When a tendon is stretched by a tap over its tendon, stretch receptors called *muscle spindles* are stretched within the muscle, which initiates an impulse over a reflex arc. A sensory (afferent) neuron conducts an impulse from the muscle spindle into the gray matter of the spinal cord, where it synapses with a motor (efferent) neuron, which conducts the impulse to the effector muscle. The stretched muscle responds by contracting to resist or reverse further stretching. These stretch reflexes are important to maintain proper posture, balance, and movements. Observation of many of these reflexes in clinical tests on patients may indicate damage to a level of the spinal cord or peripheral nerves of the particular reflex arc.

PURPOSE OF THE EXERCISE

To review the characteristics of reflex arcs and reflex behavior and to demonstrate some of the reflexes that occur in the human body.

LEARNING OBJECTIVES

After completing this exercise, you should be able to

1. describe a reflex arc;
2. distinguish between a reflex arc and a reflex;
3. identify and demonstrate several reflex actions that occur in humans.

PROCEDURE

1. Review the section entitled "Nerve Pathways" in chapter 9 of the textbook.
2. As a review activity, label figure 24.1.
3. Complete Part A of Laboratory Report 24.
4. Work with a laboratory partner to demonstrate each of the reflexes listed. (See fig. 24.2 also.) *It is important that muscles involved in the reflexes be totally relaxed in order to observe proper responses.* After each demonstration, record your observations in the table provided in Part B of the laboratory report.
   a. *Knee-jerk reflex (patellar reflex).* Have your laboratory partner sit on a table (or sturdy chair) with legs relaxed and hanging freely over the edge without touching the floor. Gently strike your partner's patellar ligament (just below the patella) with the blunt side of a rubber percussion hammer. The normal response is a moderate extension of the leg.
   b. *Ankle-jerk reflex.* Have your partner kneel on a chair with back toward you and with feet over the edge and relaxed. Gently strike the calcaneal tendon (just above its insertion on the calcaneus) with the blunt side of the rubber hammer. The normal response is plantar flexion of the foot.

Figure 24.1 Label this diagram of a withdrawal (polysynaptic) reflex arc by placing the correct numbers in the spaces provided. Most reflexes demonstrated in this lab are stretch (monosynaptic) reflex arcs and lack the interneuron.

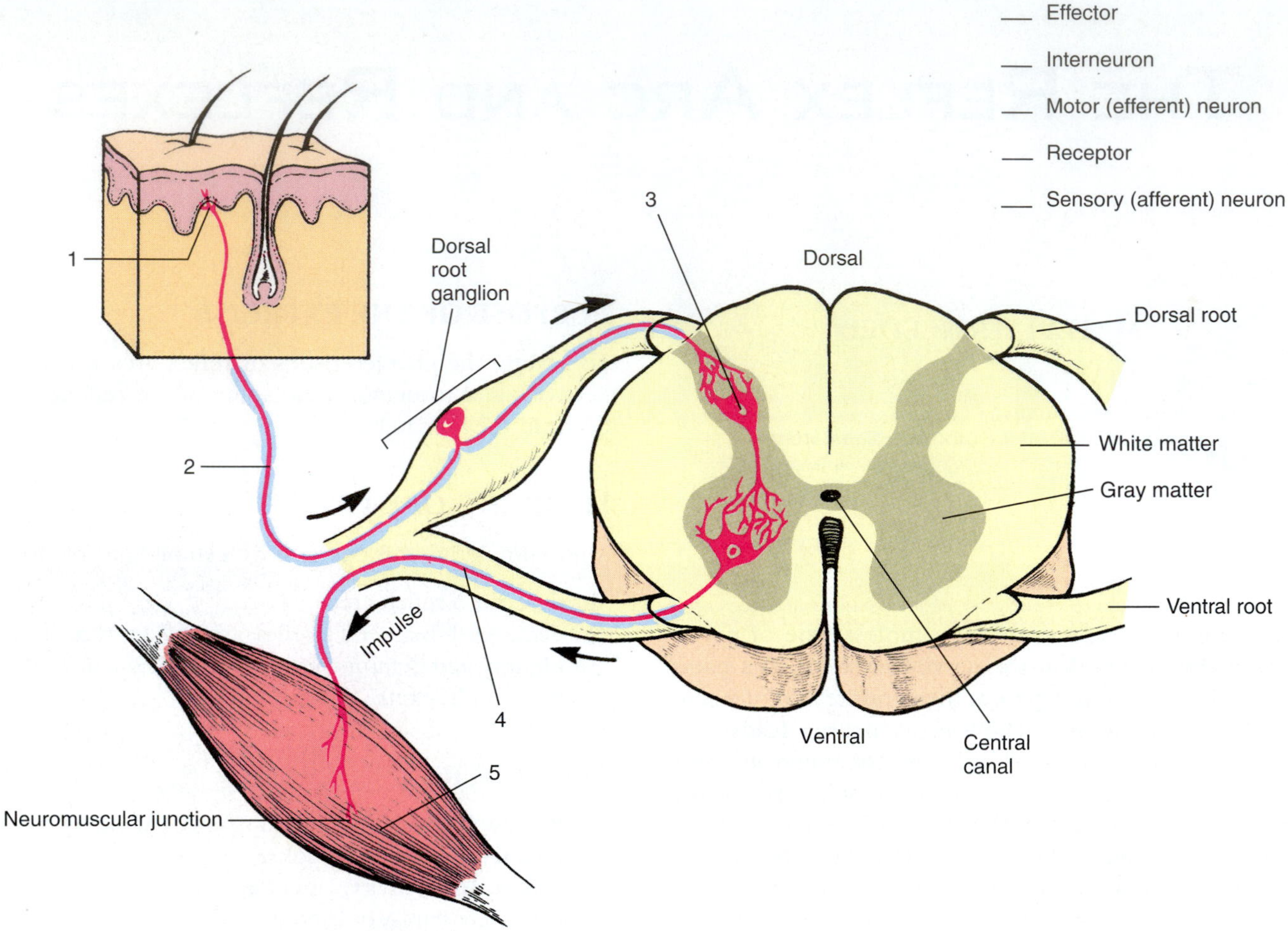

c. *Biceps-jerk reflex.* Have your partner place a bare arm bent about 90° at the elbow on the table. Press your thumb on the inside of the elbow over the tendon of the biceps brachii, and gently strike your finger with the rubber hammer. Watch the biceps brachii for a response. The response might be a slight twitch of the muscle or flexion of the forearm at the elbow joint.

d. *Triceps-jerk reflex.* Have your partner lie supine with an upper limb bent about 90° across the abdomen. Gently strike the tendon of the triceps brachii near its insertion just proximal to the olecranon process at the tip of the elbow. Watch the triceps brachii for a response. The response might be a slight twitch of the muscle or extension of the forearm at the elbow joint.

e. *Plantar reflex.* Have your partner remove a shoe and sock and lie supine with the lateral surface of the foot resting on the table. Draw the metal tip of the rubber hammer, applying firm pressure, over the sole from the heel to the base of the large toe. The normal response is flexion of the toes and plantar flexion of the foot. If the toes spread apart and dorsiflexion occurs, the reflex is the abnormal *Babinski reflex* response (normal in infants until the nerve fibers have completed myelinization).

5. Complete Part B of the laboratory report.

Figure 24.2 Demonstrate each of the following reflexes: (*a*) knee-jerk reflex; (*b*) ankle-jerk reflex; (*c*) biceps-jerk reflex; (*d*) triceps-jerk reflex; and (*e*) plantar reflex.

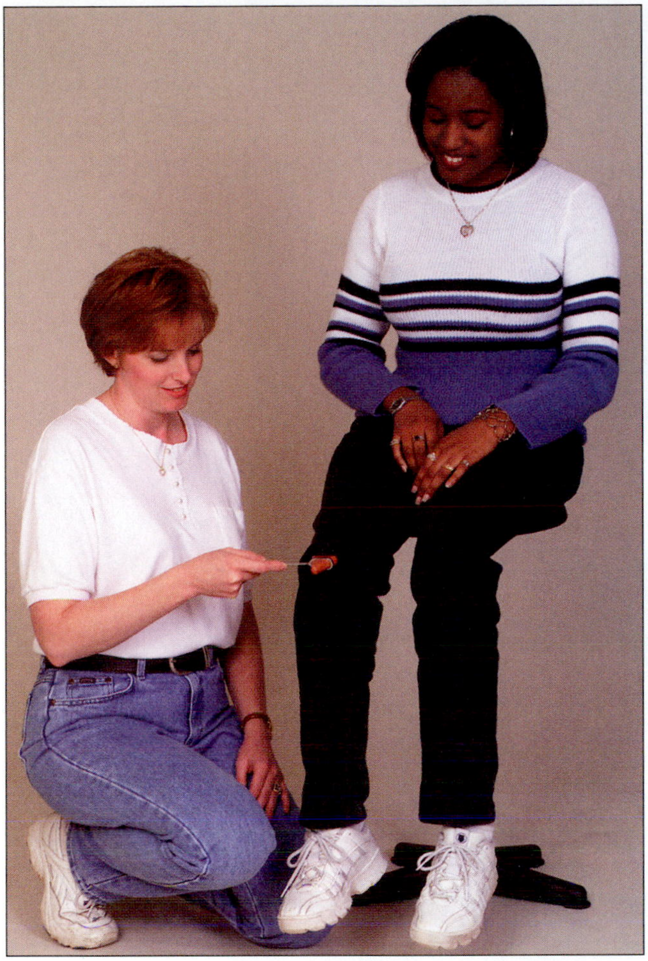

(a) Knee-jerk reflex

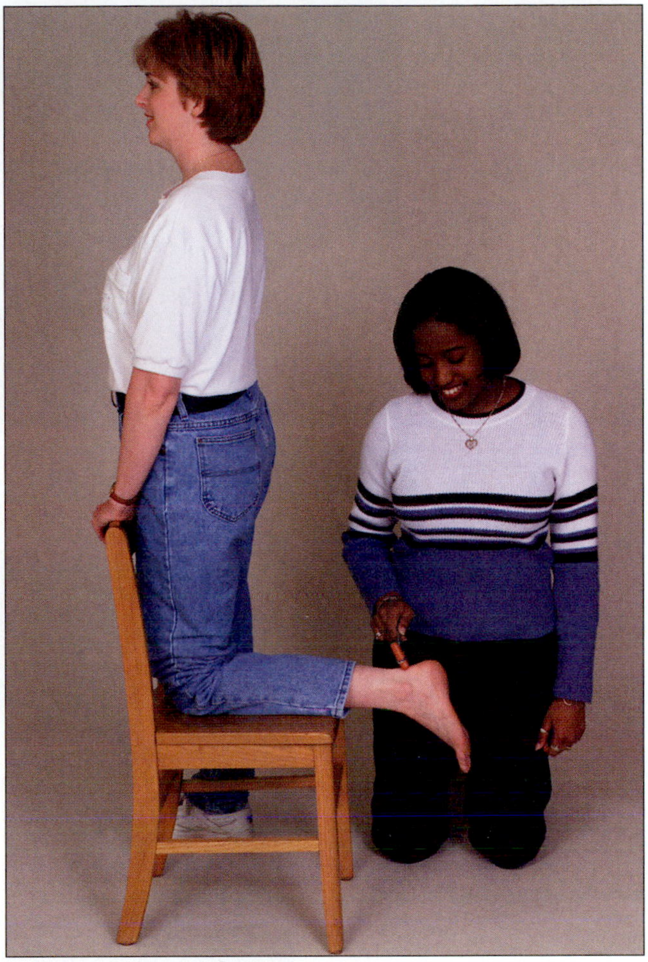

(b) Ankle-jerk reflex

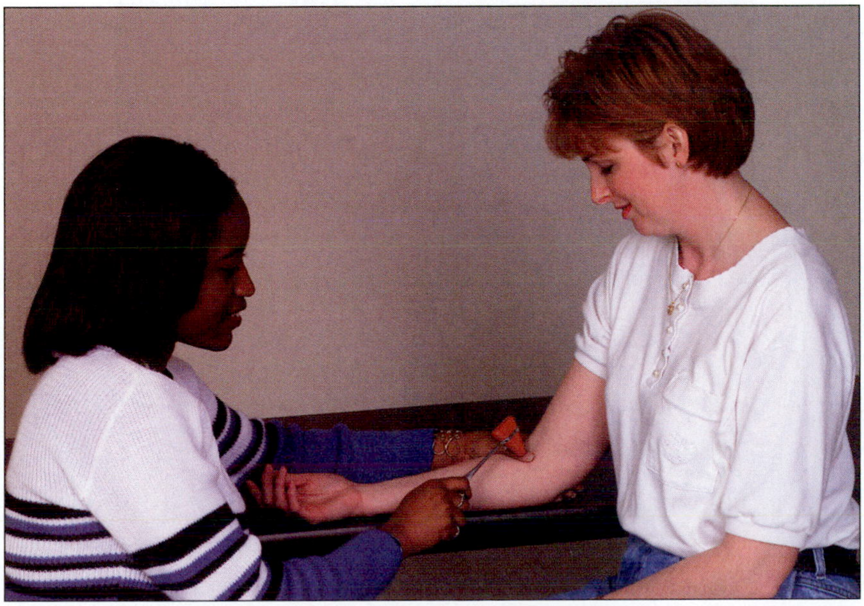

(c) Biceps-jerk reflex

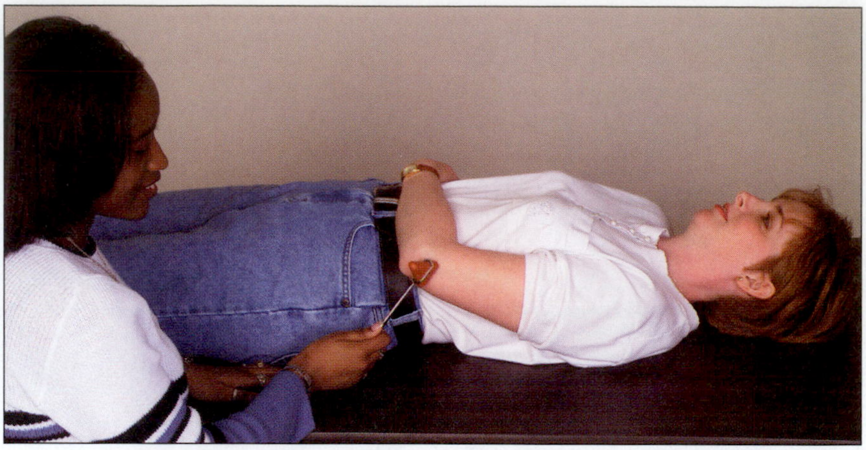

(d) Triceps-jerk reflex

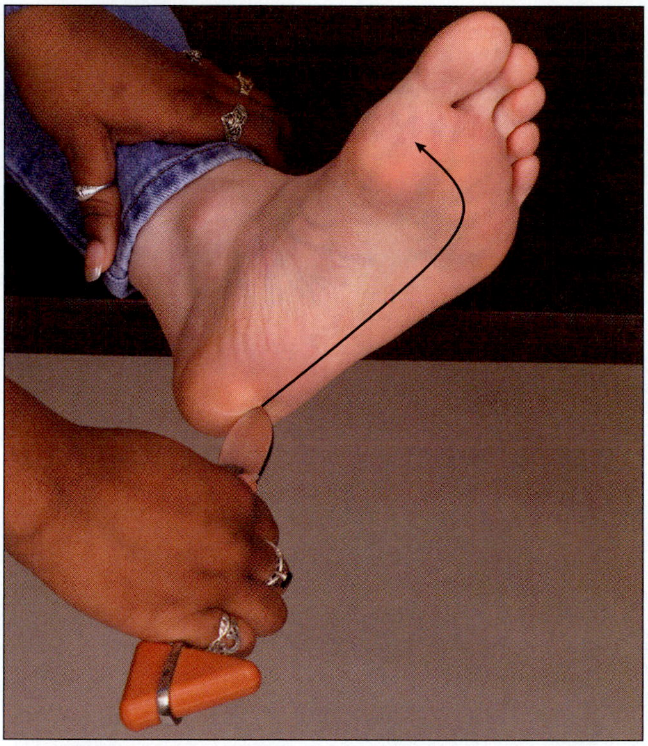

(e) Plantar reflex

Laboratory Report **24**

Name ______________________________

Date ______________________________

Section ______________________________

The Reflex Arc and Reflexes

PART A

Complete the following statements:

1. ______________________ are routes followed by nerve impulses as they pass through the nervous system.
2. Interneurons in a withdrawal reflex are located in the ______________________ .
3. ______________________ are automatic subconscious responses to stimuli within or outside the body.
4. Effectors of a reflex arc are glands and ______________________ .
5. A knee-jerk reflex employs only ______________________ and motor neurons.
6. The effector of the knee-jerk reflex is the ______________________ muscle.
7. The sensory stretch receptors of the knee-jerk reflex are located in the ______________________ .
8. The knee-jerk reflex helps the body to maintain ______________________ .
9. The sensory receptors of a withdrawal reflex are located in the ______________________ .
10. ______________________ muscles in the limbs are the effectors of a withdrawal reflex.

PART B

1. Complete the following table.

Reflex Tested	Response Observed	Effector Involved
Knee-jerk		
Ankle-jerk		
Biceps-jerk		
Triceps-jerk		
Plantar		

2. List the major events that occur in the knee-jerk reflex, from the striking of the patellar ligament to the resulting response. ______________________________

____

____

____

____

____

Critical Thinking Application

What characteristics do the reflexes you demonstrated have in common?

____

____

____

____

THE MENINGES AND SPINAL CORD

MATERIALS NEEDED

Textbook
Compound microscope
Prepared microscope slide of a spinal cord cross section with spinal nerve roots
Spinal cord model

For Demonstration:

Preserved spinal cord with meninges intact

The meninges consist of layers of membranes located between the bones of the skull and vertebral column and the soft tissues of the central nervous system. They include the dura mater, the arachnoid mater, and the pia mater.

The spinal cord is a column of nerve fibers that extends down through the vertebral canal. Together with the brain, it makes up the central nervous system.

Neurons within the spinal cord provide a two-way communication system between the brain and the body parts outside the central nervous system. The cord also contains the processing centers for the spinal reflexes.

PURPOSE OF THE EXERCISE

To review the characteristics of the meninges and spinal cord and to observe the major features of these structures.

LEARNING OBJECTIVES

After completing this exercise, you should be able to

1. name the layers of the meninges and describe the structure of each;
2. identify the major features of the spinal cord.

PROCEDURE A—MENINGES

1. Review the section entitled "Meninges" in chapter 9 of the textbook.
2. Complete Part A of Laboratory Report 25.

DEMONSTRATION

Observe the preserved section of a spinal cord. Note the heavy covering of dura mater, which is firmly attached to the cord on each side by ligaments (denticulate ligaments) originating in the pia mater. The intermediate layer of meninges, the arachnoid mater, is devoid of blood vessels, but in a live human being, the space beneath this layer contains cerebrospinal fluid. The pia mater, which is closely attached to the surface of the spinal cord, contains many blood vessels. What are the functions of these layers?

____

____

PROCEDURE B—STRUCTURE OF THE SPINAL CORD

1. Review the section entitled "Spinal Cord" in chapter 9 of the textbook.
2. As a review activity, label figures 25.1 and 25.2.
3. Complete Part B of the laboratory report.
4. Obtain a prepared microscope slide of a spinal cord cross section. Use the low power of the microscope to locate the following features:

posterior median sulcus

anterior median fissure

central canal

gray matter

- gray commissure
- posterior horn
- lateral horn
- anterior horn

white matter

- posterior funiculus
- lateral funiculus
- anterior funiculus

Figure 25.1 Label the features of the spinal cord and the surrounding structures.

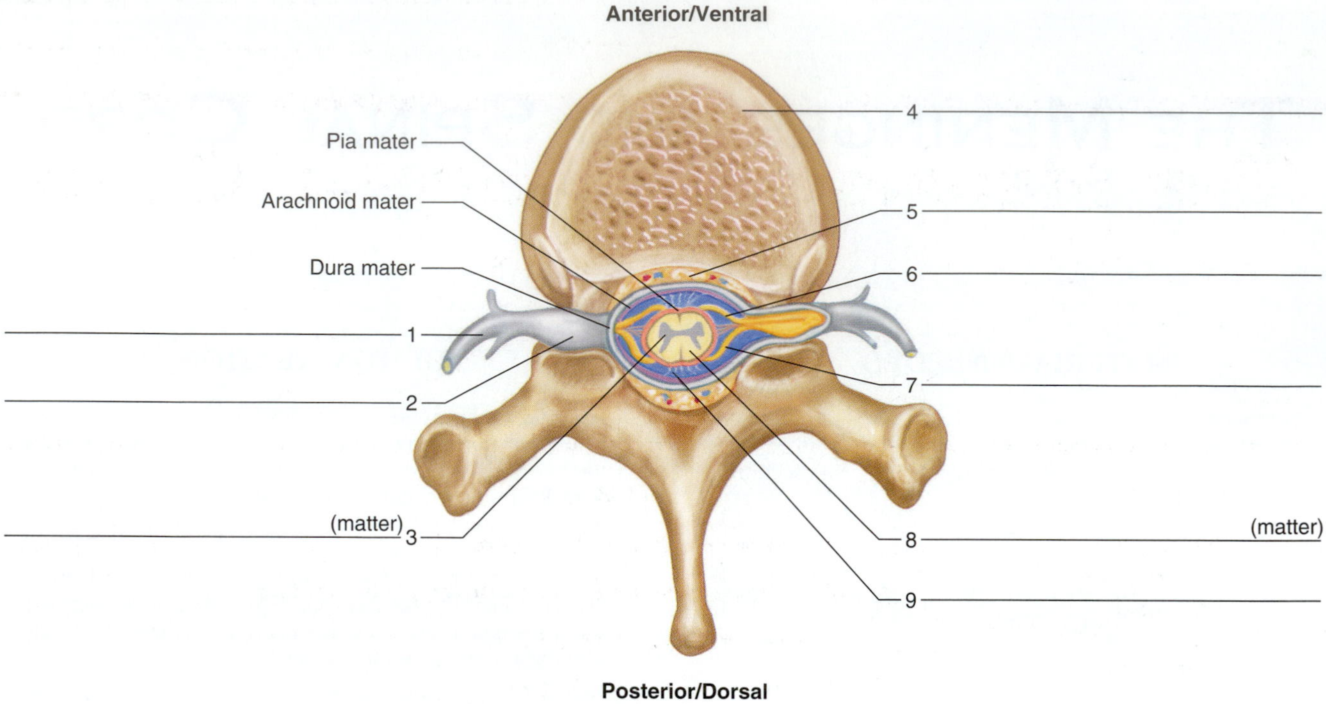

Figure 25.2 Label this cross section of the spinal cord, including the features of the white and gray matter.

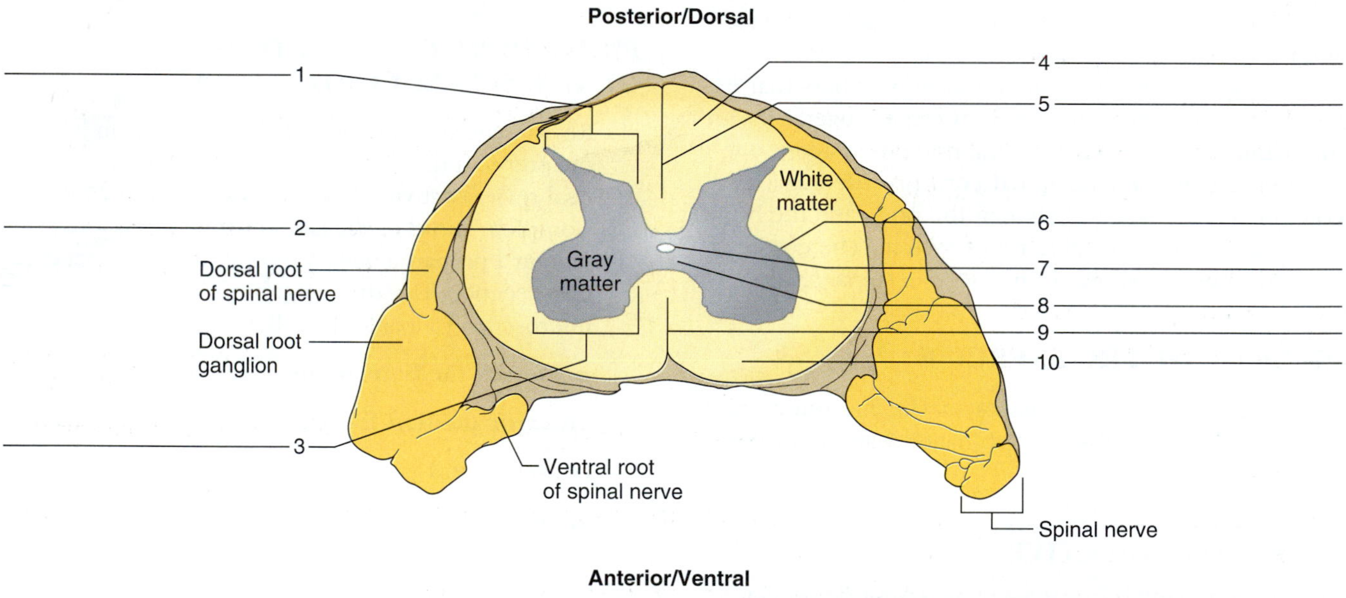

roots of spinal nerve

- dorsal roots
- dorsal root ganglia
- ventral roots

5. Observe the spinal cord model and locate the features listed in step 4.
6. Complete Part C of the laboratory report.

Web Quest

Describe the development of the nervous system, and review the brain, cranial nerves, spinal cord, and CSF. Search these at www.mhhe.com/shieress8

Laboratory Report 25

Name ______________________________

Date ______________________________

Section ______________________________

THE MENINGES AND SPINAL CORD

PART A

Match the terms in column A with the descriptions in column B. Place the letter of your choice in the space provided.

Column A		Column B
a. Arachnoid mater	_______	1. Outermost layer of meninges
b. Dura mater	_______	2. Follows irregular contours of spinal cord surface
c. Epidural space	_______	3. Contains cerebrospinal fluid
d. Pia mater	_______	4. Thin, weblike membrane
e. Subarachnoid space	_______	5. Separates dura mater from bone of vertebra

PART B

Complete the following statements:

1. Each of the thirty-one segments of the spinal cord gives rise to a pair of ______________________________ .
2. The bulge in the spinal cord that gives off nerves to the upper limbs is called the ______________________________ .
3. The bulge in the spinal cord that gives off nerves to the lower limbs is called the ______________________________ .
4. The ______________________________ is a groove that extends the length of the spinal cord posteriorly.
5. In a spinal cord cross section, the posterior ______________________ of the gray matter appear like the upper wings of a butterfly.
6. The cell bodies of motor neurons are found in the ______________________________ horns of the spinal cord.
7. The ______________________________ connects the gray matter on the left and right sides of the spinal cord.
8. The ______________________________ in the gray commissure of the spinal cord contains cerebrospinal fluid.
9. The white matter of the spinal cord is divided into anterior, lateral, and posterior ______________________ .
10. The longitudinal bundles of nerve fibers within the spinal cord comprise major nerve pathways called ______________________________ .
11. Collectively, the dura mater, arachnoid mater, and pia mater are called the ______________________________ .

PART C

Identify the features indicated in the spinal cord section of figure 25.3.

1. ______________________
2. ______________________
3. ______________________
4. ______________________
5. ______________________
6. ______________________

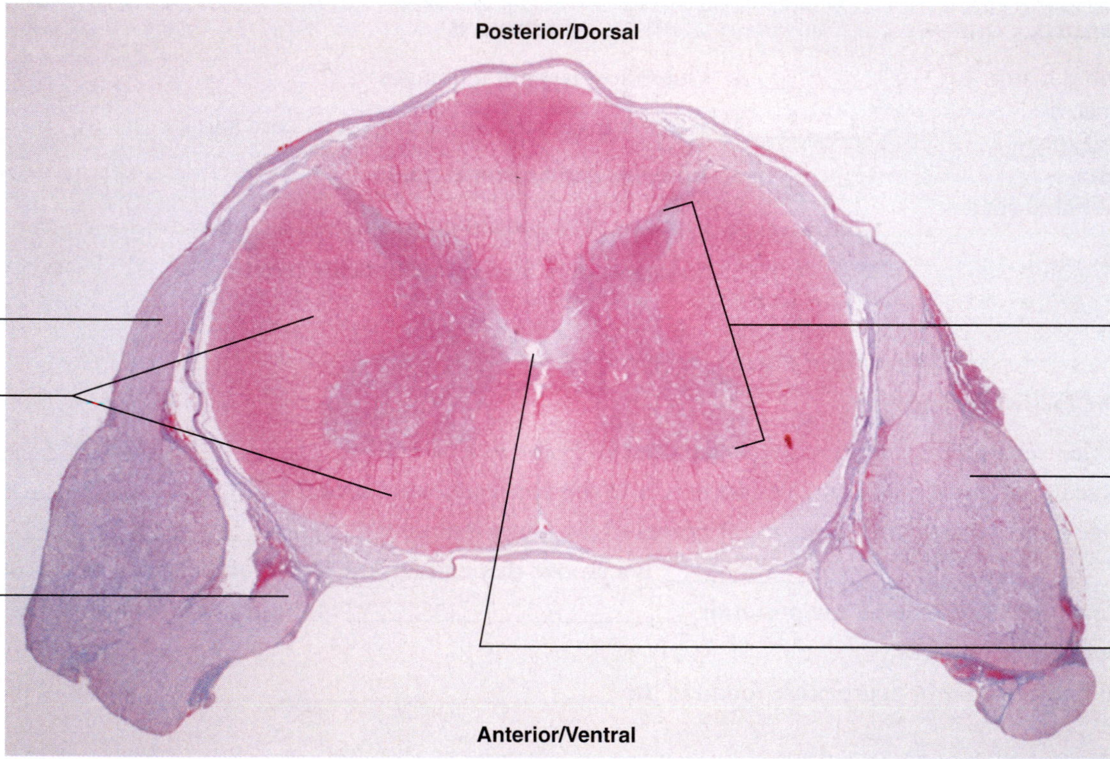

Figure 25.3 Micrograph of a spinal cord cross section with spinal nerve roots (35×). Identify the features, using the terms provided.

Terms:
Central canal
Dorsal root ganglion
Dorsal root of spinal nerve
Gray matter
Ventral root of spinal nerve
White matter

LABORATORY EXERCISE 26

THE BRAIN AND CRANIAL NERVES

MATERIALS NEEDED

Textbook
Dissectible model of the human brain
Preserved human brain
Anatomic charts of the human brain
Virtual Physiology Lab: 2, Synaptic Transmission

The brain, the largest and most complex part of the nervous system, contains nerve centers associated with sensory functions and is responsible for sensations and perceptions. It issues motor commands to skeletal muscles and carries on higher mental activities. It also functions to coordinate muscular movements, and it contains centers and nerve pathways necessary for the regulation of internal organs.

Twelve pairs of cranial nerves arise from the ventral surface of the brain and are designated by number and name. Although most of these nerves conduct both sensory and motor impulses, some contain only sensory fibers associated with special sense organs. Others are primarily composed of motor fibers and are involved with the activities of muscles and glands.

PURPOSE OF THE EXERCISE

To review the structural and functional characteristics of the human brain and cranial nerves.

LEARNING OBJECTIVES

After completing this exercise, you should be able to

1. identify the major structures in the human brain;
2. locate the major functional regions of the brain;
3. identify each of the cranial nerves;
4. list the functions of each cranial nerve.

PROCEDURE A—HUMAN BRAIN

1. Review the section entitled "Brain" in chapter 9 of the textbook.
2. As a review activity, label figures 26.1 and 26.2.
3. Complete Part A of Laboratory Report 26.
4. Observe the dissectible model and the preserved specimen of the human brain. Locate each of the following features:

cerebrum

cerebral hemispheres

corpus callosum

convolutions (gyri)

sulci

central sulcus

lateral sulcus

fissures

longitudinal fissure

transverse fissure

lobes

frontal lobe

parietal lobe

temporal lobe

occipital lobe

cerebral cortex

ventricles

lateral ventricles

third ventricle

fourth ventricle

choroid plexuses

cerebral aqueduct

Figure 26.1 Label this diagram by placing the correct numbers in the spaces provided.

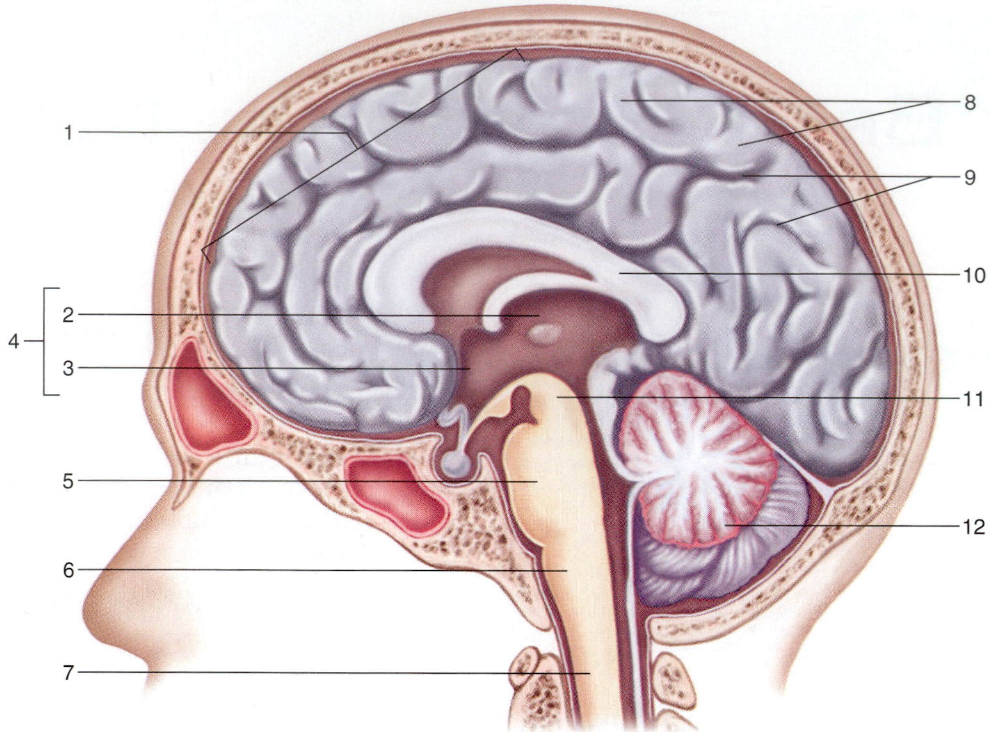

___ Cerebellum	___ Medulla oblongata
___ Cerebrum	___ Midbrain
___ Convolutions/gyri	___ Pons
___ Corpus callosum	___ Spinal cord
___ Diencephalon	___ Sulci
___ Hypothalamus	___ Thalamus

diencephalon

thalamus

hypothalamus

optic chiasma

mammillary bodies

pineal gland

cerebellum

lateral hemispheres

vermis

cerebellar cortex

brain stem

midbrain

pons

medulla oblongata

5. Complete Part B of the laboratory report.
6. Label the areas in figure 26.3 that represent the following functional regions of the cerebrum:

motor area for voluntary muscle control

motor speech area (Broca's area)

cutaneous sensory area

auditory area

visual area

general interpretative area

Figure 26.2 Label the lobes of the cerebrum, using the terms provided.

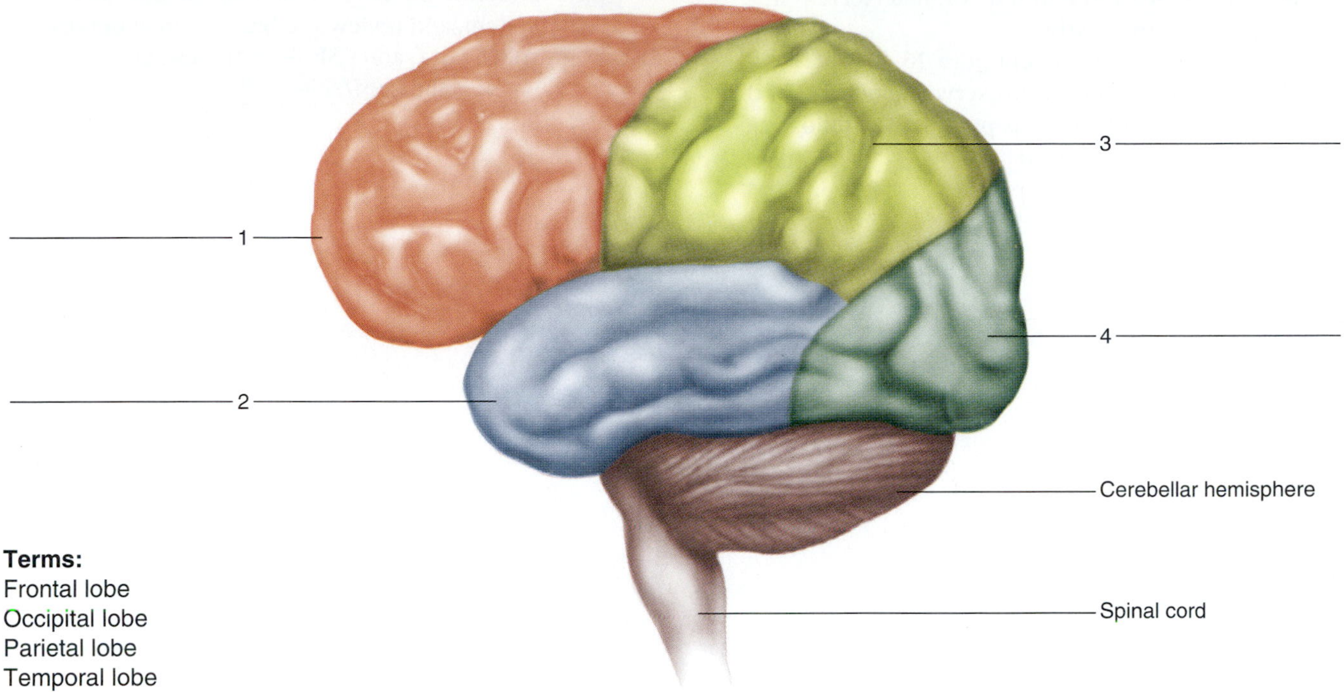

Terms:
Frontal lobe
Occipital lobe
Parietal lobe
Temporal lobe

Figure 26.3 Label the functional areas of the cerebrum, using the terms provided. (*Note:* These areas are not visible as distinct parts of the brain.)

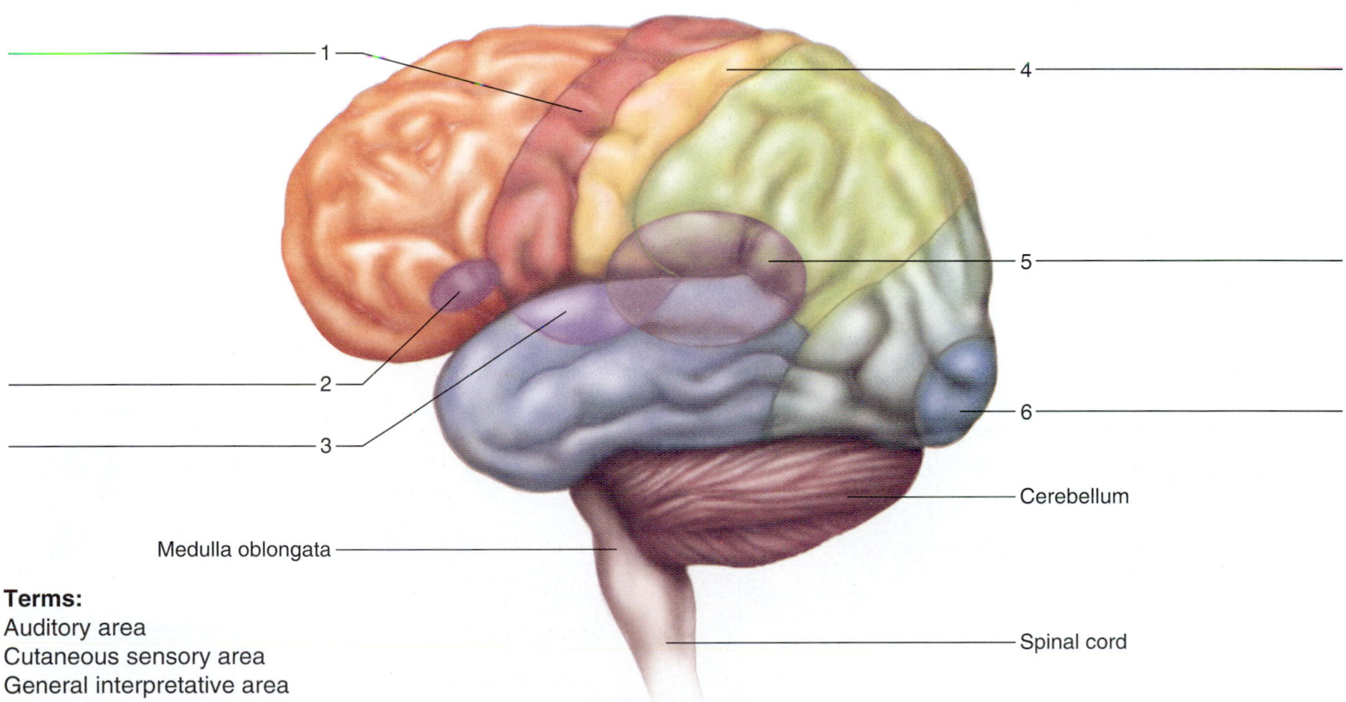

Terms:
Auditory area
Cutaneous sensory area
General interpretative area
Motor area for voluntary muscle control
Motor speech area (Broca's area)
Visual area

PROCEDURE B—CRANIAL NERVES

1. Review the section entitled "Cranial Nerves" in chapter 9 of the textbook.
2. As a review activity, label figure 26.4.
3. Observe the model and preserved specimen of the human brain and locate as many of the following cranial nerves as possible:

   olfactory nerves

   optic nerves

   oculomotor nerves

   trochlear nerves

   trigeminal nerves

   abducens nerves

   facial nerves

   vestibulocochlear nerves

   glossopharyngeal nerves

   vagus nerves

   accessory nerves

   hypoglossal nerves

4. Complete Part C of the laboratory report.

Web Quest

Describe the development of the nervous system, and review the brain, cranial nerves, spinal cord, and CSF. Search these at www.mhhe.com/shieress8

Figure 26.4 Provide the names of the cranial nerves in this ventral view.

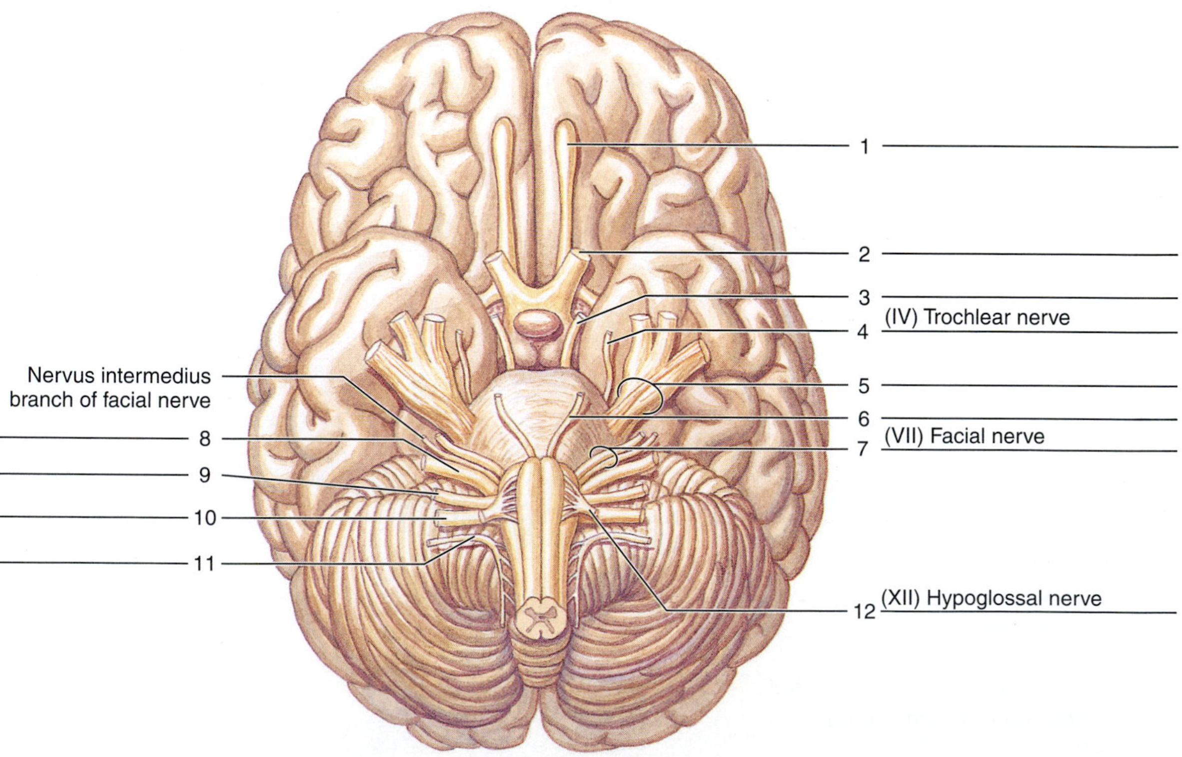

Laboratory Report **26**

Name ______________________________

Date ______________________________

Section ______________________________

The Brain and Cranial Nerves

PART A

Match the terms in column A with the descriptions in column B. Place the letter of your choice in the space provided.

Column A		Column B
a. Central sulcus	_______	1. Cone-shaped structure attached to upper posterior portion of diencephalon
b. Cerebral cortex	_______	2. Connects cerebral hemispheres
c. Convolution (gyrus)	_______	3. Ridge on surface of cerebrum
d. Corpus callosum	_______	4. Separates frontal and parietal lobes
e. Diencephalon	_______	5. Part of brain stem between diencephalon and pons
f. Medulla oblongata	_______	6. Rounded bulge on underside of brain stem
g. Midbrain	_______	7. Part of brain stem continuous with the spinal cord
h. Pineal gland	_______	8. Connects cerebellar hemispheres at midline
i. Pons	_______	9. Contains thalamus and hypothalamus
j. Vermis	_______	10. Thin layer of gray matter on surface of cerebrum

PART B

Identify the indicated features on the midsagittal section of the right half of the human brain in figure 26.5.

1. ______________________________
2. ______________________________
3. ______________________________
4. ______________________________
5. ______________________________
6. ______________________________
7. ______________________________
8. ______________________________
9. ______________________________
10. ______________________________

Figure 26.5 Identify the features on this midsagittal section of the right half of the human brain, using the terms provided.

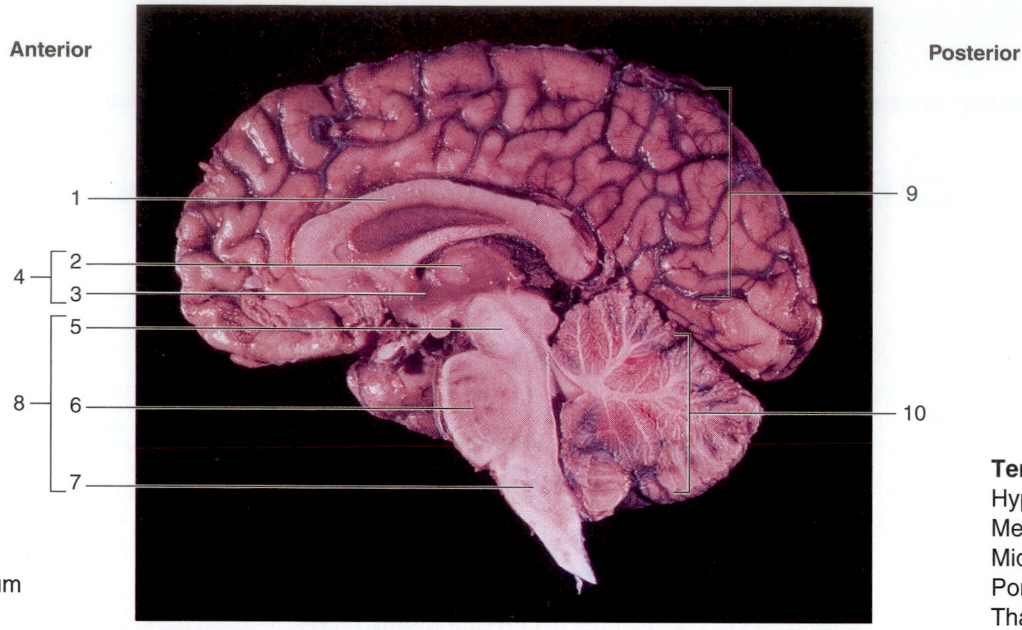

Terms:
Brain stem
Cerebellum
Cerebrum
Corpus callosum
Diencephalon

Terms:
Hypothalamus
Medulla oblongata
Midbrain
Pons
Thalamus

PART C

Indicate which cranial nerve (nerves) is (are) most closely associated with each of the following functions:

1. Sense of hearing ________________
2. Sense of taste ________________
3. Sense of sight ________________
4. Sense of smell ________________
5. Sense of equilibrium ________________
6. Conducting sensory impulses from upper teeth ________________
7. Conducting sensory impulses from lower teeth ________________
8. Raising eyelids ________________
9. Focusing lenses of eyes ________________
10. Adjusting amount of light entering eye ________________
11. Moving eyes ________________
12. Stimulating salivary secretions ________________
13. Movement of trapezius and sternocleidomastoid muscles ________________
14. Muscular movements associated with speech ________________
15. Muscular movements associated with swallowing ________________

DISSECTION OF THE SHEEP BRAIN

MATERIALS NEEDED

Dissectible model of human brain
Preserved sheep brain
Dissecting tray
Dissection instruments
Long knife

SAFETY

- Wear disposable gloves when handling the sheep brains.
- Save or dispose of the brains as instructed.
- Wash your hands before leaving the laboratory.

Mammalian brains have many features in common. Because human brains may not be available, sheep brains often are dissected as an aid to understanding mammalian brain structure. However, the adaptations of the sheep differ from the adaptations of the human, so comparisons of their structural features may not be precise.

PURPOSE OF THE EXERCISE

To observe the major features of the sheep brain and to compare these features with those of the human brain.

LEARNING OBJECTIVES

After completing this exercise, you should be able to

1. identify the major structures of the sheep brain;
2. locate the larger cranial nerves of the sheep brain;
3. list several differences and similarities between the sheep brain and the human brain.

PROCEDURE

1. Obtain a preserved sheep brain and rinse it thoroughly in water to remove as much of the preserving fluid as possible.
2. Examine the surface of the brain for the presence of meninges. (The outermost layers of these membranes may have been lost during removal of the brain from the cranial cavity.) If meninges are present, locate the following:

   dura mater—the thick, opaque outer layer

   arachnoid mater—the delicate, transparent middle layer that is attached to the undersurface of the dura mater

   pia mater—the thin, vascular layer that adheres to the surface of the brain (should be present)

3. Remove any remaining dura mater by pulling it gently from the surface of the brain.
4. Position the brain with its ventral surface down in the dissecting tray. Study figure 27.1, and locate the following structures on the specimen:

   cerebral hemispheres

   convolutions (gyri)

   sulci

   longitudinal fissure

   frontal lobe

   parietal lobe

   temporal lobe

   occipital lobe

   cerebellum

   medulla oblongata

Figure 27.1 Dorsal surface of the sheep brain.

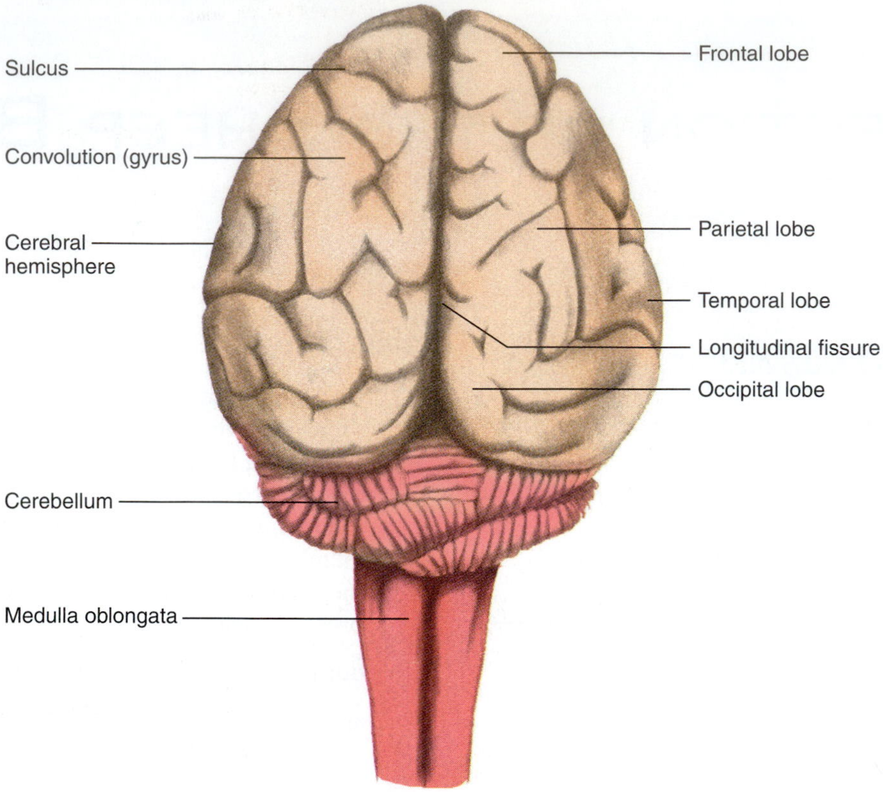

Figure 27.2 Gently bend the cerebellum and medulla oblongata away from the cerebrum to expose the pineal gland and the corpora quadrigemina.

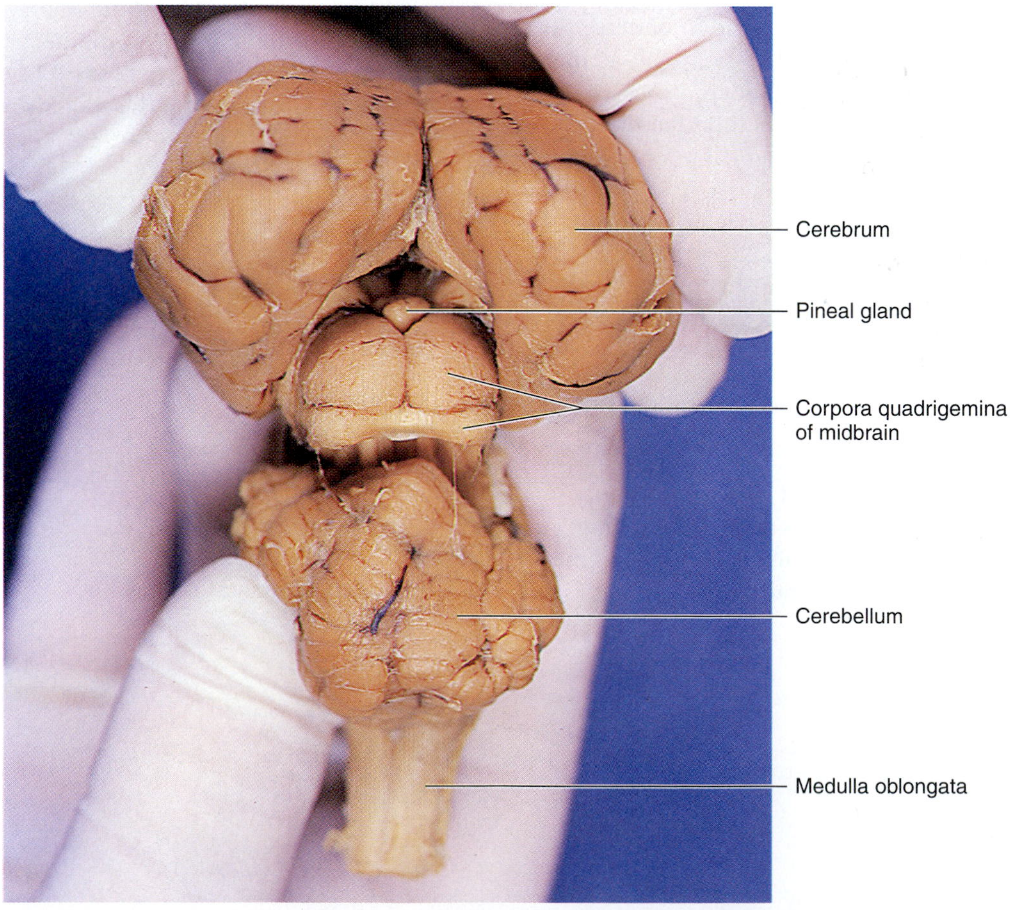

Figure 27.3 Lateral surface of the sheep brain.

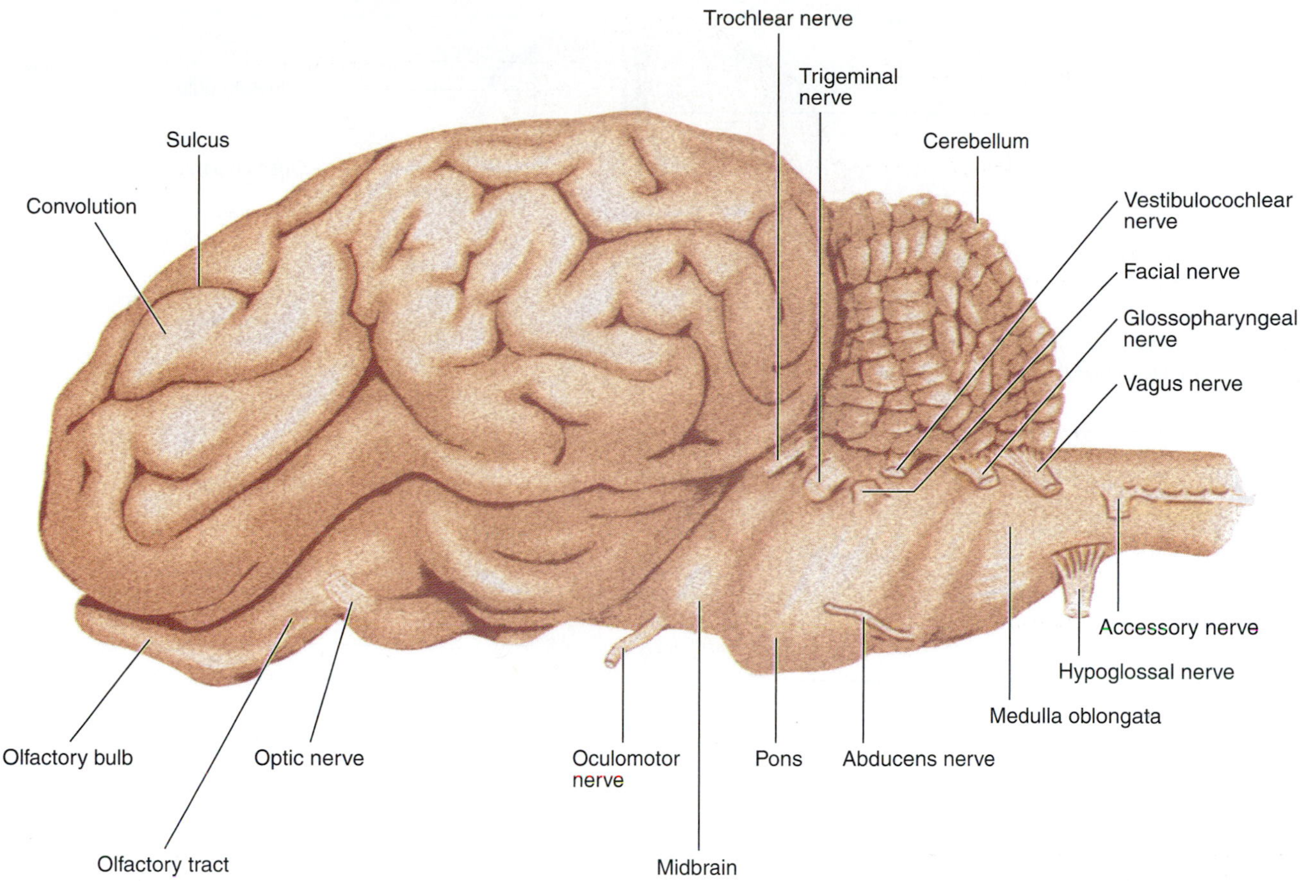

5. Gently separate the cerebral hemispheres along the longitudinal fissure and expose the transverse band of white fibers within the fissure that connects the hemispheres. This band is the *corpus callosum.*
6. Bend the cerebellum and medulla oblongata slightly downward and away from the cerebrum (figure 27.2). This will expose the *pineal gland* in the upper midline and the *corpora quadrigemina,* which consists of four rounded structures associated with the midbrain.
7. Position the brain with its ventral surface upward. Study figures 27.3 and 27.4, and locate the following structures on the specimen:

longitudinal fissure

olfactory bulbs

optic nerves

optic chiasma

optic tract

mammillary bodies

infundibulum (pituitary stalk)

midbrain

pons

Figure 27.4 Ventral surface of the sheep brain.

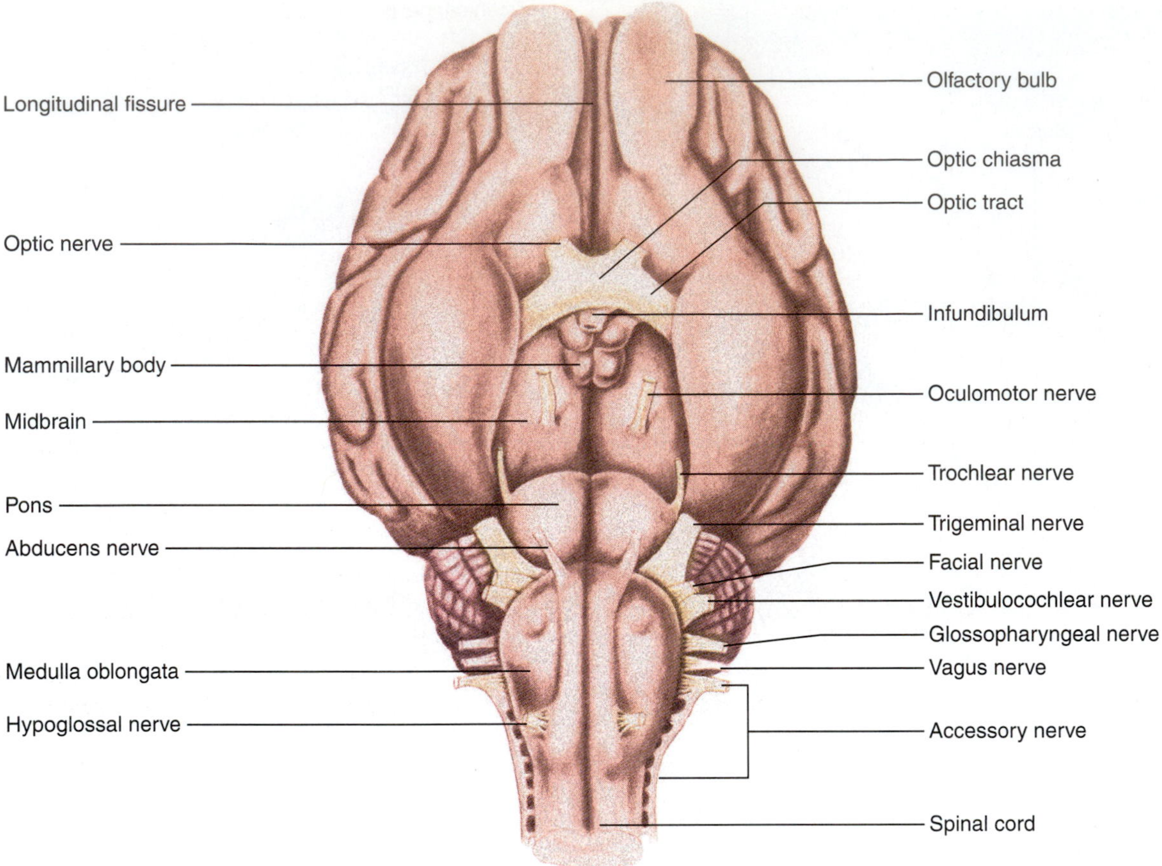

Figure 27.5 Midsagittal section of the left half of the sheep brain dissection.

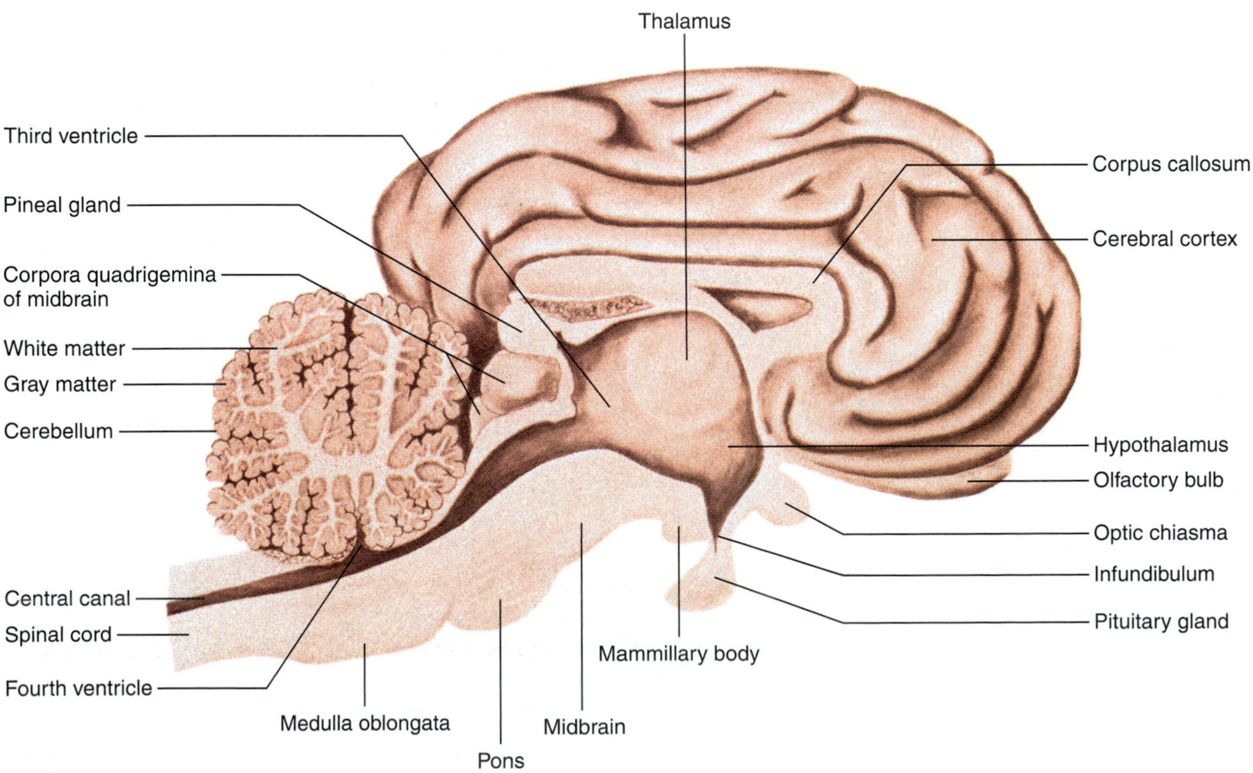

Figure 27.6 Midsagittal section of the right half of the sheep brain dissection.

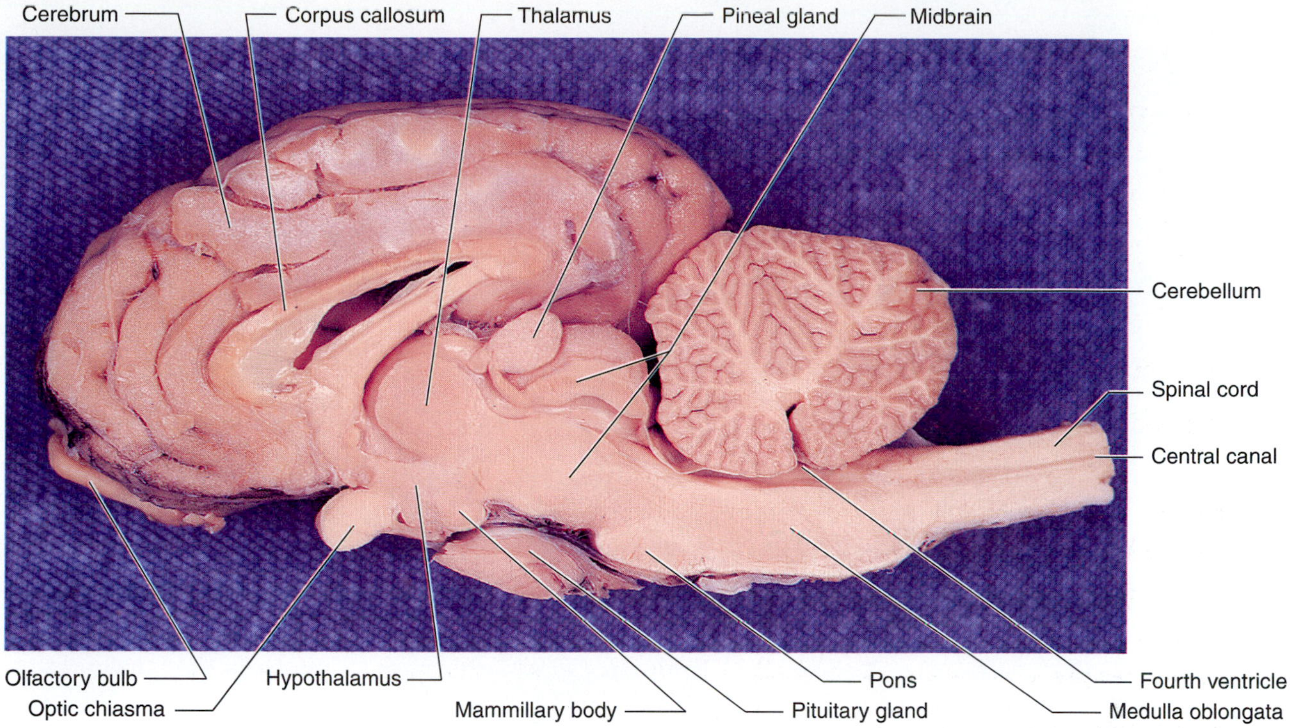

8. Although some of the cranial nerves may be missing or are quite small and difficult to find, locate as many of the following as possible, using figures 27.3 and 27.4 as references:

   oculomotor nerves

   trochlear nerves

   trigeminal nerves

   abducens nerves

   facial nerves

   vestibulocochlear nerves

   glossopharyngeal nerves

   vagus nerves

   accessory nerves

   hypoglossal nerves

9. Using a long, sharp knife, cut the sheep brain along the midline to produce a midsagittal section. Study figures 27.5 and 27.6, and locate the following structures on the specimen:

   cerebrum

   cerebral hemisphere

   cerebral cortex

   olfactory bulb

   corpus callosum

   cerebellum

   white matter

   gray matter

   third ventricle

   fourth ventricle

   diencephalon

   optic chiasma

   infundibulum

   pituitary gland (this structure may be missing)

   thalamus

   hypothalamus

   pineal gland

   midbrain

   pons

   medulla oblongata

10. Dispose of the sheep brain as directed by the laboratory instructor.
11. Complete Parts A and B of Laboratory Report 27.

Laboratory Report **27**

Name ______________________________

Date ______________________________

Section ______________________________

Dissection of the Sheep Brain

PART A

Answer the following questions:

1. How do the relative sizes of the sheep and human cerebral hemispheres differ? ______________________________

2. How do the convolutions and sulci of the sheep cerebrum compare with the human cerebrum in numbers?

3. What is the significance of the differences you noted in your answers for questions 1 and 2?

4. What difference did you note in the structures of the sheep cerebellum and the human cerebellum?

5. How do the sizes of the olfactory bulbs of the sheep brain compare with those of the human brain?

6. Based on their relative sizes, which of the cranial nerves seem to be most highly developed in the sheep brain?

7. What is the significance of the observations you noted in your answers for questions 5 and 6?

PART B

Critical Thinking Application

Prepare a list of at least five features to illustrate ways in which the brains of sheep and humans are similar.

1. ______________________________
2. ______________________________
3. ______________________________
4. ______________________________
5. ______________________________

THE EAR AND HEARING

MATERIALS NEEDED

Textbook
Dissectible ear model
Watch that ticks
Tuning fork (128 or 256 cps)
Rubber hammer
Cotton
Meterstick

For Demonstrations:

Compound microscope
Prepared microscope slide of cochlea (section)
Audiometer

The ear is composed of external, middle, and inner parts. The external structures gather sound waves and direct them inward to the tympanic membrane of the middle ear. The parts of the middle ear, in turn, transmit vibrations from the tympanic membrane (eardrum) to the inner ear, where the hearing receptors are located. As they are stimulated, these receptors initiate nerve impulses to pass over the vestibulocochlear nerve into the auditory cortex of the brain, where the impulses are interpreted and the sensations of hearing are created.

PURPOSE OF THE EXERCISE

To review the structural and functional characteristics of the ear and to conduct some simple hearing tests.

LEARNING OBJECTIVES

After completing this exercise, you should be able to

1. identify the major structures of the ear;
2. describe the functions of the structures of the ear;
3. trace the pathway of sound vibrations from the tympanic membrane to the hearing receptors;
4. conduct several simple hearing tests.

PROCEDURE A—STRUCTURE AND FUNCTION OF THE EAR

1. Review the section entitled "Sense of Hearing" in chapter 10 of the textbook.
2. As a review activity, label figures 28.1, 28.2, and 28.3.

DEMONSTRATION

Observe the section of the cochlea in the microscope set up by the laboratory instructor. Locate one of the turns of the cochlea and, using figures 28.3 and 28.4 as a guide, identify the scala vestibuli, scala media, scala tympani, vestibular membrane, basilar membrane, and the organ of Corti.

3. Examine the dissectible model of the ear and locate the following features:

external ear

- auricle
- external auditory meatus

middle ear

- tympanic cavity
- tympanic membrane (eardrum)
- auditory ossicles
  - malleus
  - incus
  - stapes
- oval window

auditory tube

Figure 28.1 Label the major structures of the ear.

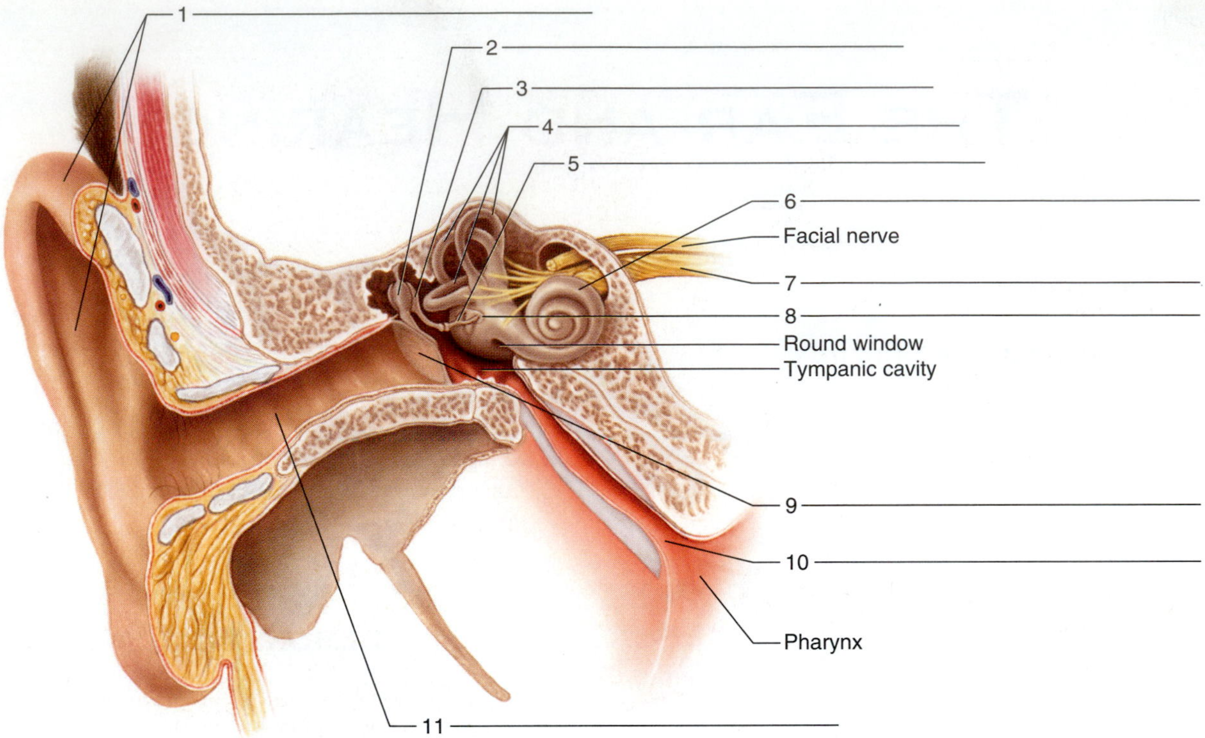

inner ear

osseous labyrinth

membranous labyrinth

cochlea

round window

semicircular canals

vestibule

vestibulocochlear nerve

vestibular nerve (balance branch)

cochlear nerve (hearing branch)

4. Complete Parts A and B of Laboratory Report 28.

PROCEDURE B—HEARING TESTS

Perform the following tests in a quiet room, using your laboratory partner as the test subject.

1. *Auditory acuity test.* To conduct this test, follow these steps:
   a. Have the test subject sit with eyes closed.
   b. Pack one of the subject's ears with cotton.
   c. Hold a ticking watch close to the open ear, and slowly move it straight out and away from the ear.
   d. Have the subject indicate when the sound of the ticking can no longer be heard.
   e. Use a meterstick to measure the distance in centimeters from the ear to the position of the watch.
   f. Repeat this procedure to test the acuity of the other ear.
   g. Record the test results in Part C of the laboratory report.
2. *Sound localization test.* To conduct this test, follow these steps:
   a. Have the subject sit with eyes closed.
   b. Hold the ticking watch somewhere within the audible range of the subject's ears, and ask the subject to point to the watch.
   c. Move the watch to another position and repeat the request. In this manner, determine how accurately the subject can locate the watch when it is in each of the following positions: in front of the head, behind the head, above the head, on the right side of the head, and on the left side of the head.
   d. Record the test results in Part C of the laboratory report.

Figure 28.2 Label the structures of the inner ear by placing the correct numbers in the spaces provided.

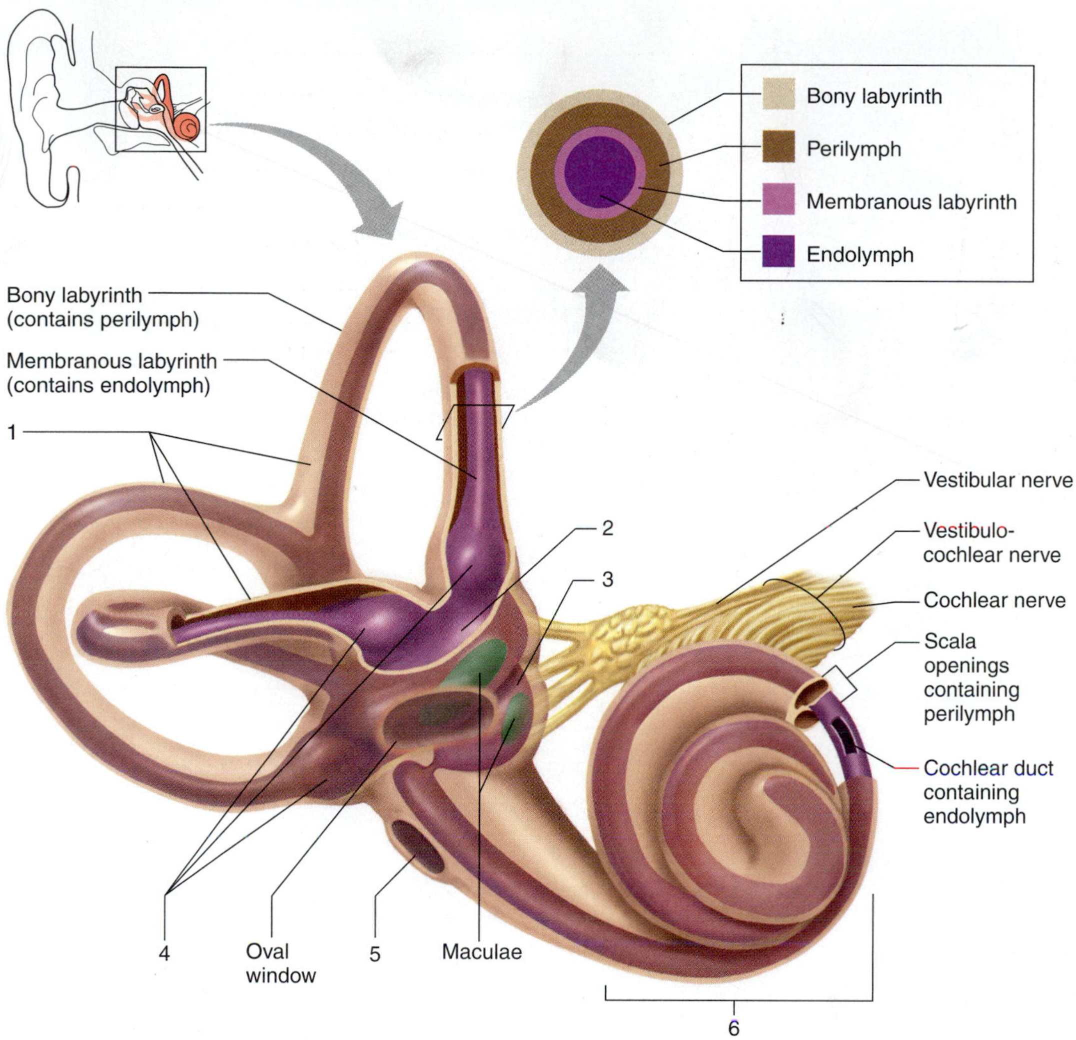

___ Ampullae ___ Round window ___ Semicircular canals

___ Cochlea ___ Saccule ___ Utricle

3. *Rinne test.* This test is done to assess possible conduction deafness by comparing bone and air conduction. To conduct this test, follow these steps:
   a. Obtain a tuning fork and strike it with a rubber hammer, or on the heel of your hand, causing it to vibrate.
   b. Place the end of the fork's handle against the subject's mastoid process behind one ear. Have the prongs of the fork pointed downward and away from the ear, and be sure nothing is touching them (fig. 28.5*a*). The sound sensation is that of bone conduction. If no sound is experienced, nerve deafness exists.
   c. Ask the subject to indicate when the sound is no longer heard.
   d. Then quickly remove the fork from the mastoid process and position it in the air close to the opening of the nearby external auditory meatus (fig. 28.5*b*).

      If hearing is normal, the sound (from air conduction) will be heard again; if there is conductive impairment, the sound will not be heard. Conductive impairment involves outer or middle ear defects. Hearing aids can improve hearing for conductive deafness because bone conduction transmits the sound into the inner ear. Surgery could possibly correct this type of defect.
   e. Record the test results in Part C of the laboratory report.

Figure 28.3 Label the organ of Corti structures of a cochlea.

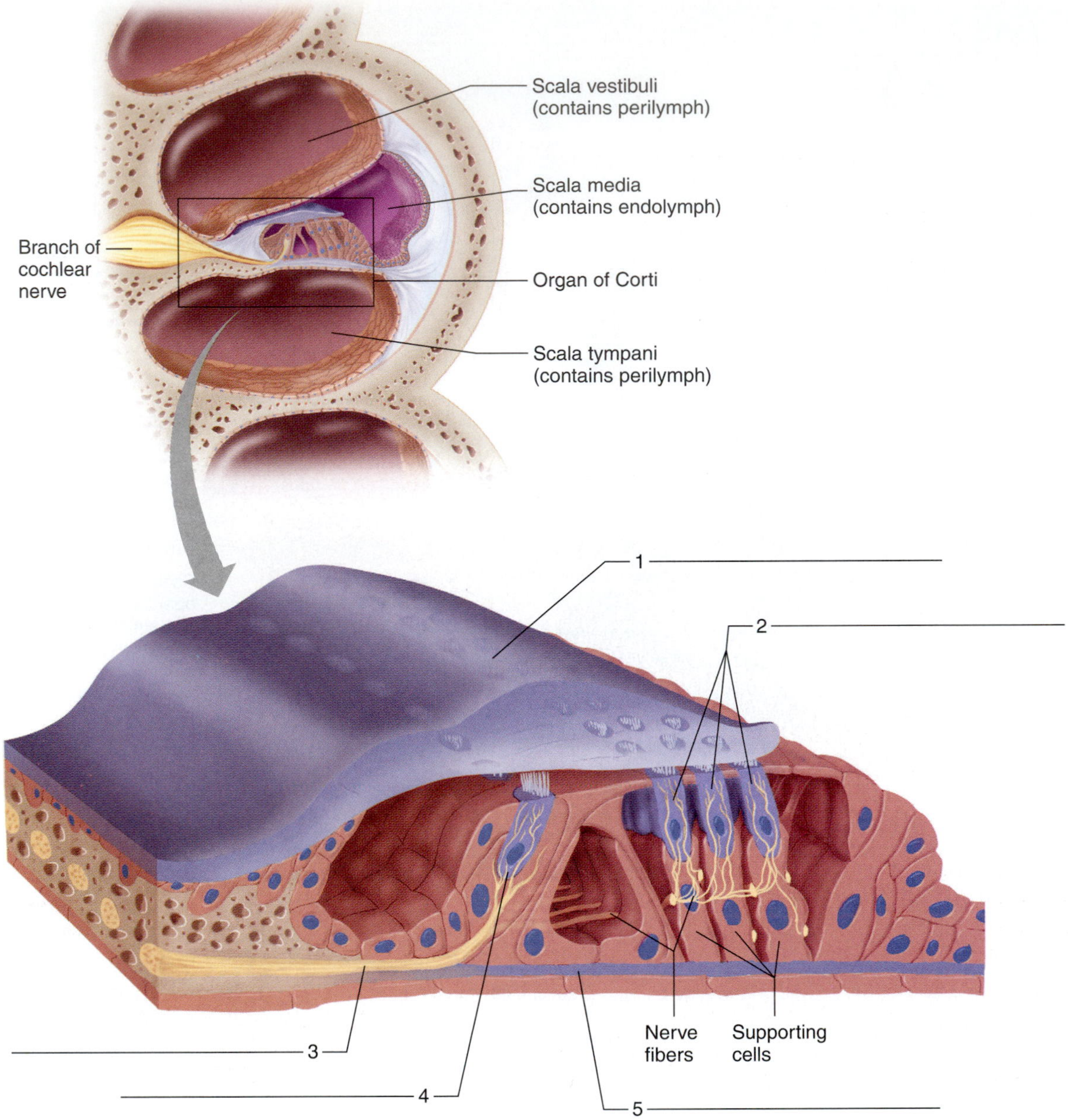

4. *Weber test.* This test is used to distinguish possible conduction or sensory deafness. To conduct this test, follow these steps:
   a. Strike the tuning fork with the rubber hammer.
   b. Place the handle of the fork against the subject's forehead in the midline (fig. 28.6).
   c. Ask the subject to indicate if the sound is louder in one ear than in the other or if it is equally loud in both ears.

      If hearing is normal, the sound will be equally loud in both ears. If there is conductive impairment, the sound will appear louder in the affected ear. If some degree of sensory (nerve) deafness exists, the sound will be louder in the normal ear. The impairment involves the organ of Corti or the cochlear nerve. Hearing aids will not improve sensory deafness.
   d. Have the subject experience the effects of conductive impairment by packing one ear with cotton and repeating the Weber test. Usually the sound appears louder in the plugged (or impaired) ear because extraneous sounds from the room are blocked out.
   e. Record the test results in Part C of the laboratory report.

5. Complete Part C of the laboratory report.

Figure 28.4 A section through the cochlea (22×).

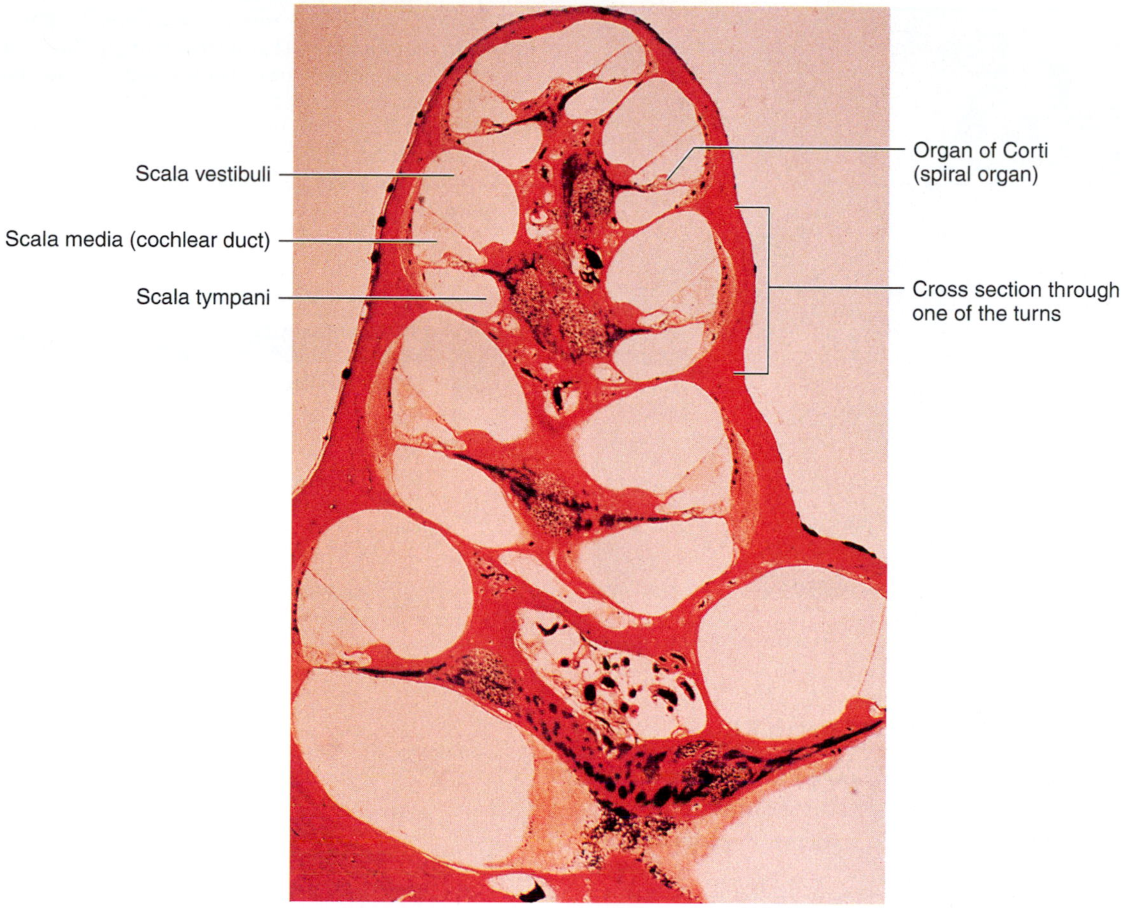

Figure 28.5 Rinne test: (*a*) first placement of vibrating tuning fork until sound is no longer heard; (*b*) second placement of tuning fork to assess air conduction.

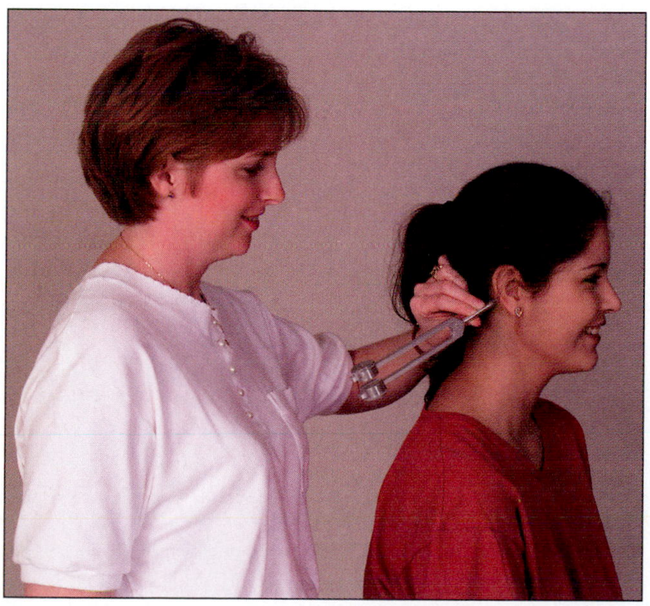

(a)

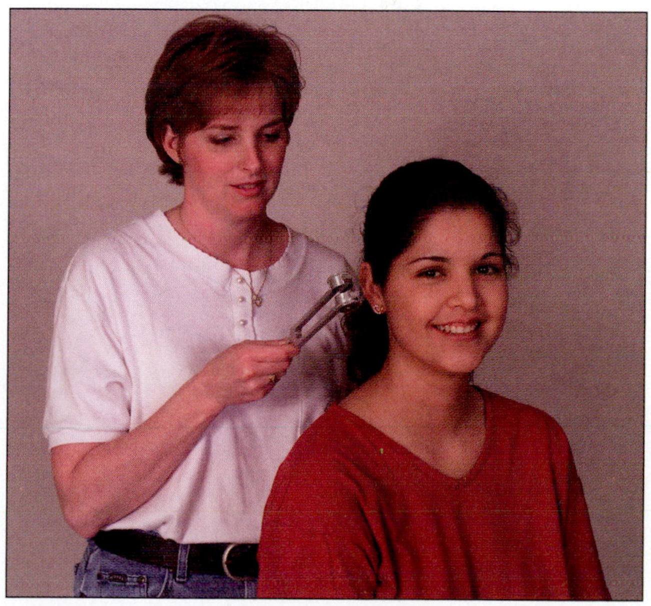

(b)

Figure 28.6 Weber test.

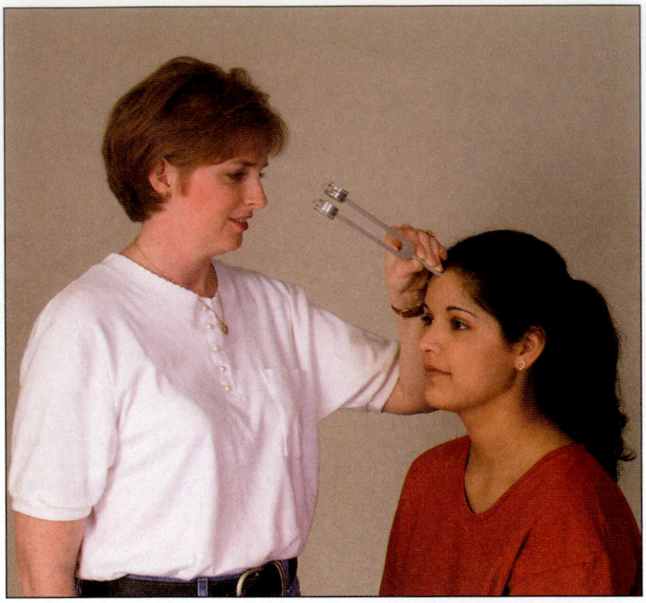

DEMONSTRATION

Ask the laboratory instructor to demonstrate the use of the audiometer. This instrument produces sound vibrations of known frequencies that are transmitted to one or both ears of a test subject through earphones. The audiometer can be used to determine the threshold of hearing for different sound frequencies, and, in the case of a hearing impairment, it can be used to determine the percentage of hearing loss for each frequency.

Web Quest

What are the causes of hearing impairment? Search this and review the anatomy and physiology of the ear at www.mhhe.com/shieress8

Critical Thinking Application

Ear structures from the outer ear into the inner ear are progressively smaller. Using results obtained from the hearing tests, explain this advantage.

Laboratory Report **28**

Name ______________________

Date ______________________

Section ______________________

THE EAR AND HEARING

PART A

Match the terms in column A with the descriptions in column B. Place the letter of your choice in the space provided.

Column A

a. Auditory tube
b. External auditory meatus
c. Malleus
d. Membranous labyrinth
e. Osseous labyrinth
f. Scala tympani
g. Scala vestibuli
h. Stapes
i. Tectorial membrane
j. Tympanic cavity
k. Tympanic membrane

Column B

_______ 1. Auditory ossicle attached to tympanic membrane
_______ 2. Air-filled space containing auditory ossicles
_______ 3. Contacts hairs of hearing receptors
_______ 4. Leads from oval window to apex of cochlea
_______ 5. S-shaped tube leading to tympanic membrane
_______ 6. Cone-shaped, semitransparent membrane attached to malleus
_______ 7. Auditory ossicle attached to oval window
_______ 8. Contains endolymph
_______ 9. Bony canal of inner ear in temporal bone
_______ 10. Connects middle ear and pharynx
_______ 11. Extends from apex of cochlea to round window

PART B

Label the structures indicated in the micrograph of the organ of Corti (spiral organ) in figure 28.7.

PART C

1. Results of auditory acuity test:

Ear Tested	**Audible Distance (cm)**
Right	
Left	

Figure 28.7 Label the structures of this organ of Corti (spiral organ) region of a cochlea, using the terms provided (75× micrograph enlarged to 300×).

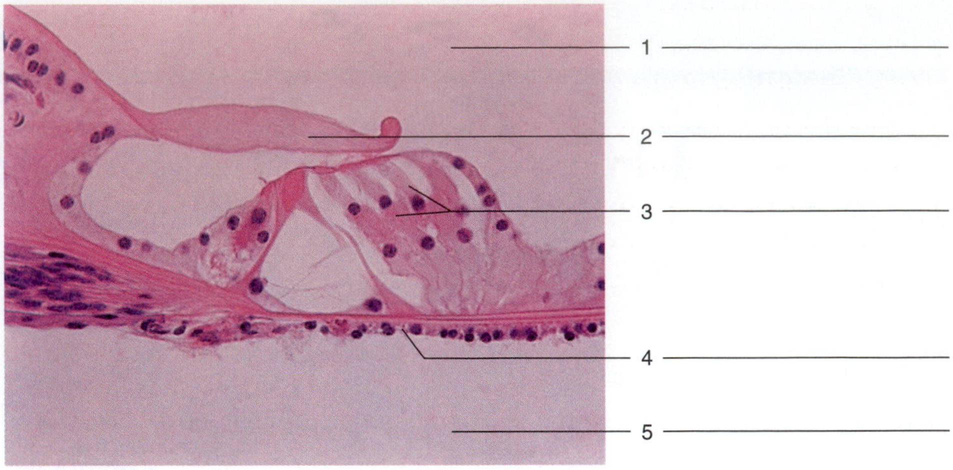

Terms:
Basilar membrane
Hair cells
Scala media (cochlear duct)
Scala tympani
Tectorial membrane

2. Results of sound localization test:

Actual Location	Reported Location
Front of the head	
Behind the head	
Above the head	
Right side of the head	
Left side of the head	

3. Results of experiments using tuning forks:

Test	Left Ear (normal or impaired)	Right Ear (normal or impaired)
Rinne		
Weber		

4. Summarize the results of the hearing tests you conducted on your laboratory partner.

Laboratory Exercise 29

The Eye

MATERIALS NEEDED

Textbook
Dissectible eye model
Sheep or beef eye (fresh or preserved)
Dissecting tray
Dissecting instruments—forceps, sharp scissors, and dissecting needle

For Optional Activity:

Ophthalmoscope

SAFETY

- Wear disposable gloves when working on the eye dissection.
- Dispose of tissue remnants and gloves as instructed.
- Wash the dissecting tray and instruments as instructed.
- Wash your laboratory table.
- Wash your hands before leaving the laboratory.

The eye contains photoreceptors, which are modified neurons located on its inner wall. Other parts of the eye provide protective functions or make it possible to move the eyeball. Still other structures serve to focus light entering the eye so that a sharp image is projected onto the receptor cells. Nerve impulses generated when the receptors are stimulated travel along the optic nerves to the brain, which interprets the impulses and creates the sensation of sight.

PURPOSE OF THE EXERCISE

To review the structure and function of the eye and to dissect a mammalian eye.

LEARNING OBJECTIVES

After completing this exercise, you should be able to

1. identify the major structures of an eye;
2. describe the functions of the structures of an eye;
3. list the structures through which light passes as it travels from the cornea to the retina;
4. dissect a mammalian eye and identify its major features.

PROCEDURE A—STRUCTURE AND FUNCTION OF THE EYE

1. Review the sections entitled "Visual Accessory Organs," "Structure of the Eye," and "Visual Receptors" in chapter 10 of the textbook.
2. As a review activity, label figures 29.1, 29.2, and 29.3.
3. Complete Part A of Laboratory Report 29.
4. Examine the dissectible model of the eye and locate the following features:

   eyelid

   conjunctiva

   orbicularis oculi

   levator palpebrae superioris

   lacrimal apparatus

   - lacrimal gland
   - canaliculi
   - lacrimal sac
   - nasolacrimal duct

   extrinsic muscles

   - superior rectus
   - inferior rectus
   - medial rectus
   - lateral rectus
   - superior oblique
   - inferior oblique

Figure 29.1 Label the structures of the lacrimal apparatus.

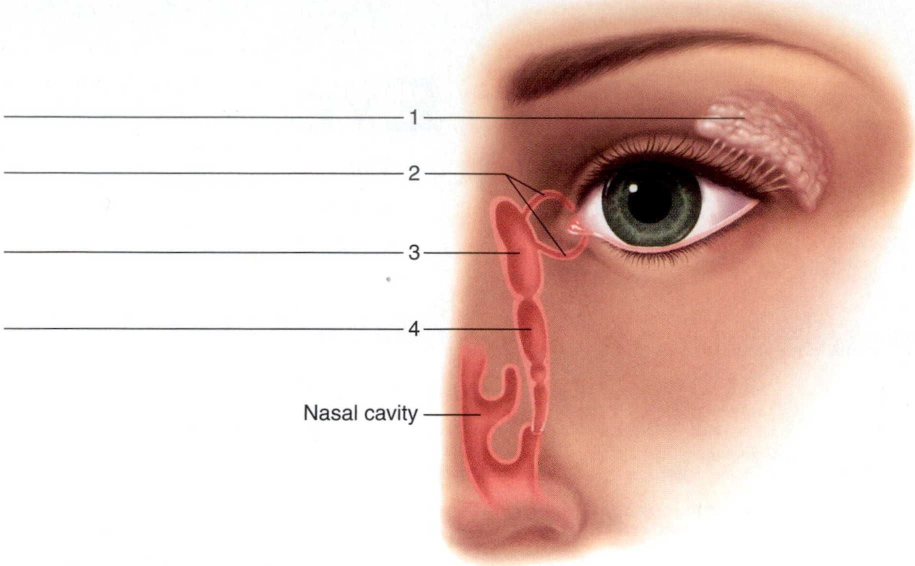

Figure 29.2 Label the extrinsic muscles of the right eye (lateral view).

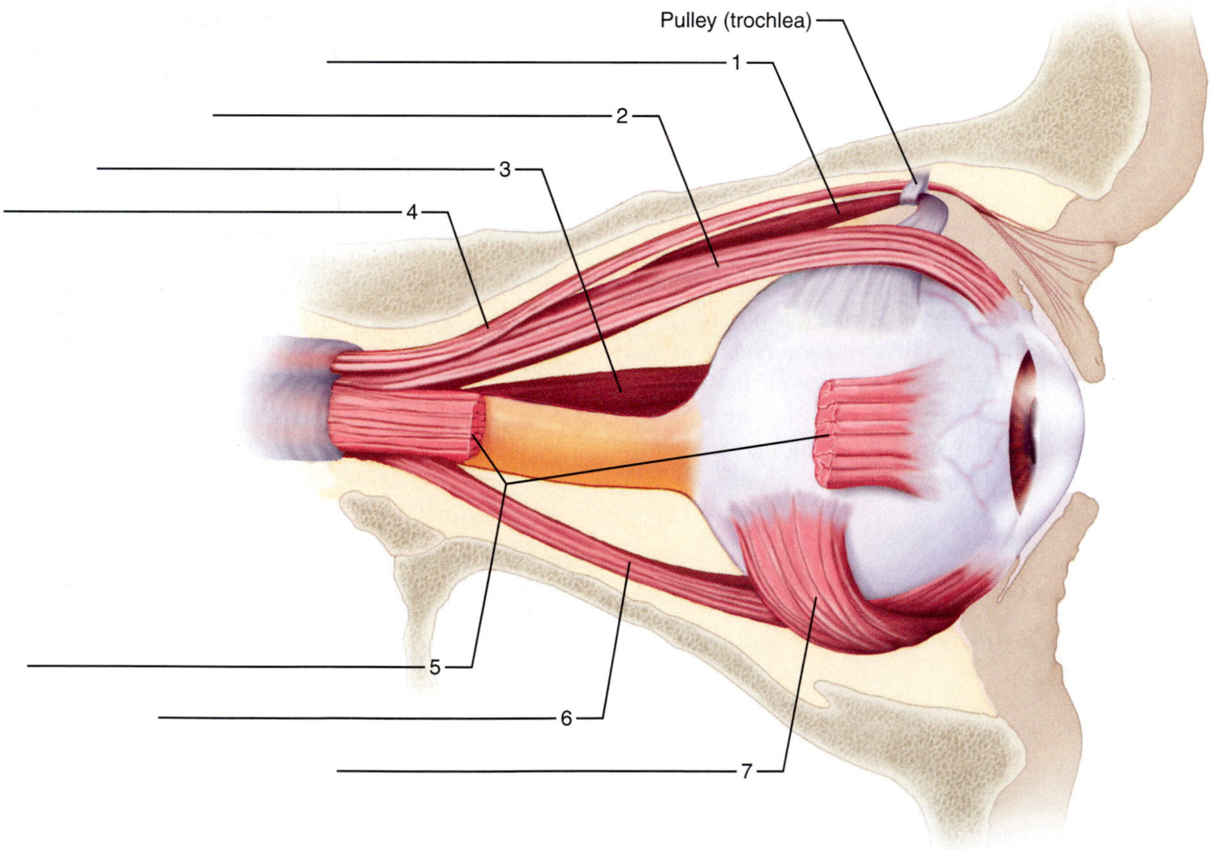

Figure 29.3 Label the structures indicated in this transverse section of the right eye (superior view).

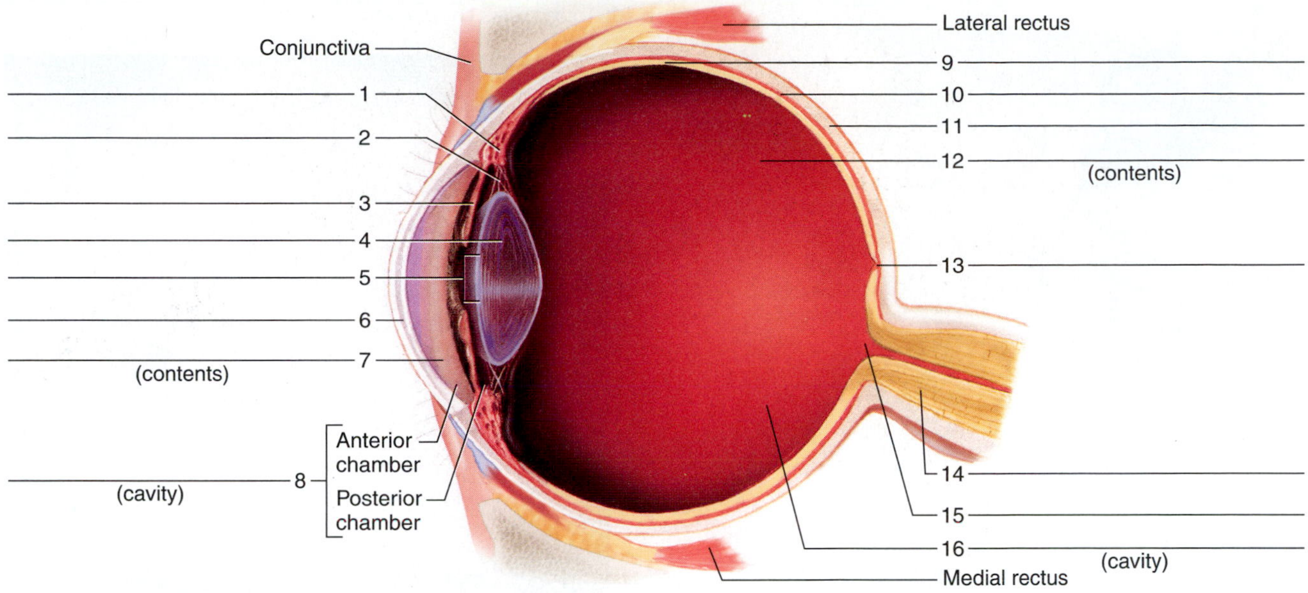

cornea

sclera

optic nerve

choroid coat

ciliary body

- ciliary processes
- ciliary muscles

lens

suspensory ligaments

iris

anterior cavity

- anterior chamber
- posterior chamber
- aqueous humor

pupil

retina

- macula lutea
- fovea centralis
- optic disc

posterior cavity

- vitreous humor

OPTIONAL ACTIVITY

Use an ophthalmoscope to examine the interior of your laboratory partner's eye. This instrument consists of a set of lenses held in a rotating disc, a light source, and some mirrors that reflect the light into the test subject's eye.

The examination should be conducted in a dimly lighted room. Have your partner seated and staring straight ahead at eye level. Move the rotating disc of the ophthalmoscope so that the *O* appears in the lens selection window. Hold the instrument in your right hand with the end of your index finger on the rotating disc (fig. 29.4). Direct the light at a slight angle from a distance of about 15 cm into the subject's right eye. The light beam should pass along the inner edge of the pupil. Look through the instrument, and you should see a reddish, circular area—the interior of the eye. Rotate the disc of lenses to higher values until sharp focus is achieved.

Move the ophthalmoscope to within about 5 cm of the eye being examined *being very careful that the instrument does not touch the eye,* and again rotate the lenses to sharpen the focus (fig. 29.5). Locate the optic disc and the blood vessels that pass through it. Also locate the yellowish macula lutea by having your partner stare directly into the light of the instrument (fig. 29.6).

Examine the subject's iris by viewing it from the side and by using a lens with a +15 or +20 value.

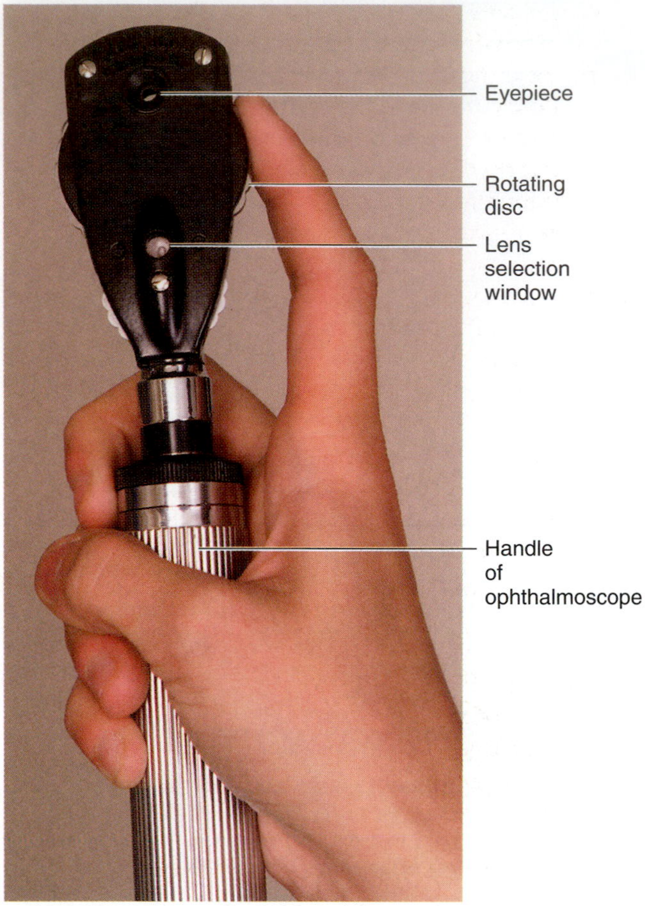

Figure 29.4 An ophthalmoscope is used to examine the interior of the eye.

Figure 29.5 Move the ophthalmoscope to within about 5 cm of the eye being examined and again rotate the lenses to sharpen the focus.

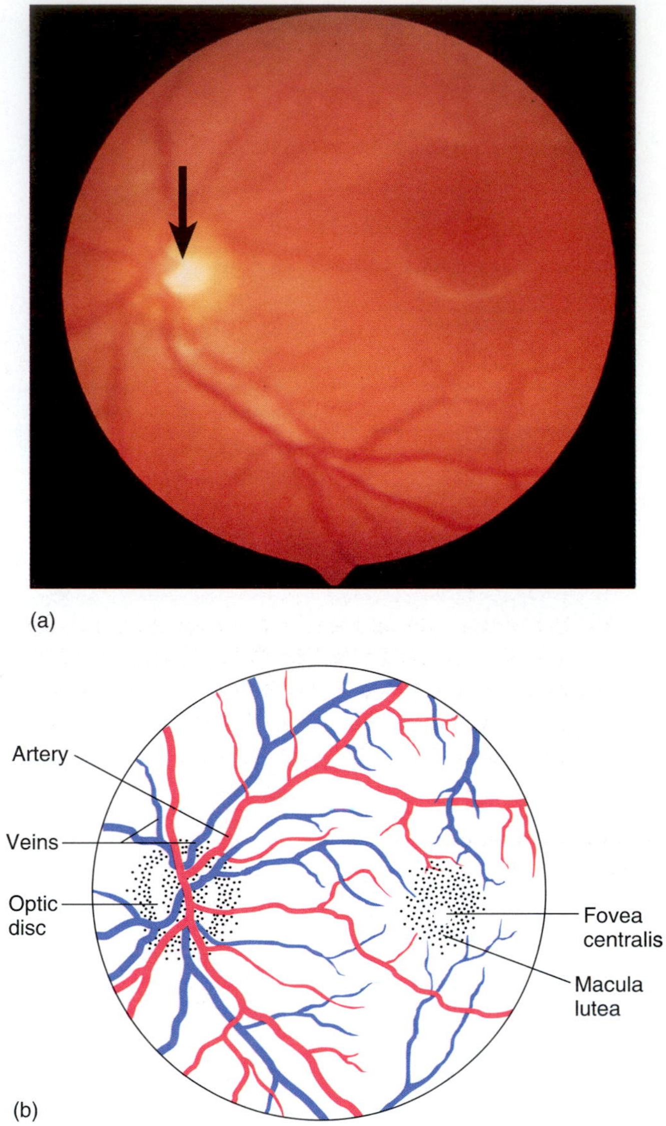

Figure 29.6 The interior of the eye as seen using an ophthalmoscope: (*a*) photograph with an arrow indicating the optic disc; (*b*) diagram.

PROCEDURE B—EYE DISSECTION

1. Obtain a mammalian eye, place it in a dissecting tray, and dissect it as follows:
   a. Trim away the fat and other connective tissues but leave the stubs of the *extrinsic muscles* and of the *optic nerve.* This nerve projects outward from the posterior region of the eyeball.
   b. Note the *conjunctiva,* which lines the eyelid and is reflected over the anterior surface (except cornea) of the eye. Lift some of this thin membrane away from the eye with forceps and examine it.

Figure 29.7 Prepare a coronal section of the eye.

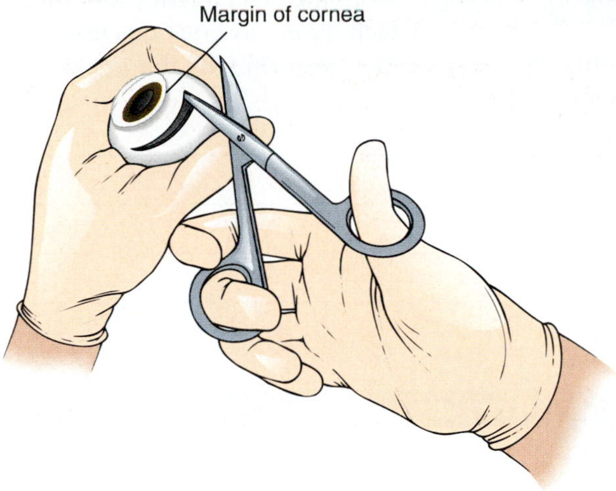

c. Locate and observe the *cornea, sclera,* and *iris.* Also note the *pupil* and its shape. The cornea from a fresh eye will be transparent; when preserved, it becomes opaque.

d. Use sharp scissors to make a coronal section of the eye. To do this, cut through the wall about 1 cm from the margin of the cornea and continue all the way around the eyeball. Try not to damage the internal structures of the eye (fig. 29.7).

e. Gently separate the eyeball into anterior and posterior portions. Usually the jellylike vitreous humor will remain in the posterior portion, and the lens may adhere to it. Place the parts in the dissecting tray with their contents facing upward.

f. Examine the anterior portion of the eye, and locate the *ciliary body,* which appears as a dark, circular structure. Also note the *iris* and the *lens* if it remained in the anterior portion. The lens is normally attached to the ciliary body by many *suspensory ligaments,* which appear as delicate, transparent threads.

g. Use a dissecting needle to gently remove the lens and examine it. If the lens is still transparent, hold it up and look through it at something in the distance and note that the lens inverts the image. The lens of a preserved eye is usually too opaque for this experience.

h. Examine the posterior portion of the eye. Note the *vitreous humor.* This jellylike mass helps to hold the lens in place anteriorly and helps to hold the *retina* against the choroid coat.

i. Carefully remove the vitreous humor and examine the retina. This layer will appear as a thin, nearly colorless to cream-colored membrane that detaches easily from the choroid coat. Compare the structures identified to figure 29.8.

Figure 29.8 Internal structures of the beef eye dissection.

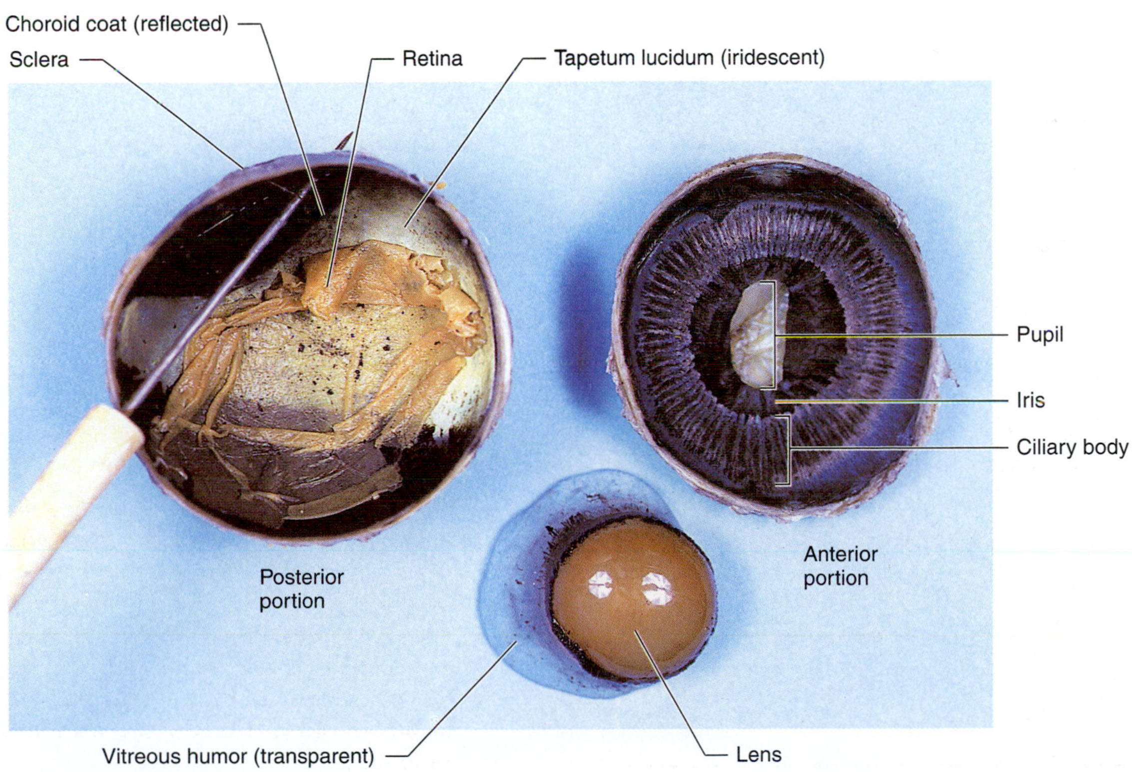

j. Locate the *optic disc*—the point where the retina is attached to the posterior wall of the eyeball and where the optic nerve originates. Because there are no receptor cells in the optic disc, this region is also called the "blind spot."

k. Note the iridescent area of the choroid coat beneath the retina. This colored surface is called the *tapetum lucidum.* It serves to reflect light back through the retina, an action that is thought to aid the night vision of some animals. The tapetum lucidum is lacking in the human eye.

l. Discard the tissues of the eye as directed by the laboratory instructor.

2. Complete Part B of the laboratory report.

Critical Thinking Application

A strong blow to the head might cause the retina to detach. From observations made during the eye dissection, explain why this could happen.

Web Quest

How do we see? Search this and review the anatomy and physiology of the eye at www.mhhe.com/shieress8

Laboratory Report 29

Name ______________________________

Date ______________________________

Section ______________________________

THE EYE

PART A

Match the terms in column A with the descriptions in column B. Place the letter of your choice in the space provided.

Column A		Column B
a. Aqueous humor	______	1. Posterior five-sixths of middle or vascular tunic loosely joined to sclera
b. Choroid coat	______	2. White part of outer tunic
c. Ciliary muscle	______	3. Transparent anterior portion of outer tunic
d. Conjunctiva	______	4. Inner lining of eyelid
e. Cornea	______	5. Secretes tears
f. Iris	______	6. Empties into nasal cavity
g. Lacrimal gland	______	7. Fills posterior cavity of eye
h. Lysozyme	______	8. Area where optic nerve originates
i. Nasolacrimal duct	______	9. Smooth muscle that controls light intensity
j. Optic disc	______	10. Fills anterior and posterior chambers of the anterior cavity of the eye
k. Retina	______	11. Contains visual receptors called rods and cones
l. Sclera	______	12. Connects lens to ciliary body
m. Suspensory ligament	______	13. Causes lens to change shape
n. Vitreous humor	______	14. Antibacterial agent in tears

Complete the following:

15. List the structures and fluids through which light passes as it travels from the cornea to the retina.

____

____

____

16. List three ways in which rods and cones differ in structure or function. ______________________

____

____

____

____

PART B

Complete the following:

1. Which tunic/layer of the eye was the most difficult to cut? ______________________
2. What kind of tissue do you think is responsible for this quality of toughness? ______________________
3. How do you compare the shape of the pupil in the dissected eye with your own pupil? ______________________
____
____
4. Where do you find aqueous humor in the dissected eye? ______________________
____
____
5. What is the function of the dark pigment in the choroid coat? ______________________
____
____
6. Describe the lens of the dissected eye. ______________________
____
____
____
7. Describe the vitreous humor of the dissected eye. ______________________
____
____
____

VISUAL TESTS AND DEMONSTRATIONS

MATERIALS NEEDED

Textbook
Snellen eye chart
3″ × 5″ card (plain)
3″ × 5″ card with word typed in center
Astigmatism chart
Meterstick
Metric ruler
Pen flashlight
Ichikawa's or Ishihara's color plates for color-blindness test

Normal vision results when light rays from objects in the external environment are refracted by the cornea and lens of the eye and focused onto the photoreceptors of the retina. Irregular curvatures in the surface of the cornea or lens, inability to change the shape of the lens, or defects in the shape of the eyeball can result in defective vision.

PURPOSE OF THE EXERCISE

To conduct tests for visual acuity, astigmatism, accommodation, color vision, the blind spot, and certain reflexes of the eye.

LEARNING OBJECTIVES

After completing this exercise, you should be able to

1. describe four conditions that can lead to defective vision;
2. conduct the tests used to evaluate visual acuity, astigmatism, the ability to accommodate for close vision, and color vision;
3. demonstrate the blind spot, photopupillary reflex, accommodation pupillary reflex, and convergence reflex.

PROCEDURE A—VISUAL TESTS

Perform the following visual tests, using your laboratory partner as a test subject. If your partner usually wears glasses, test each eye with and without the glasses.

1. *Visual acuity test.* Visual acuity (sharpness of vision) can be measured by using a Snellen eye chart (fig. 30.1). This chart consists of several sets of letters in different sizes printed on a white card. The letters near the top of the chart are relatively large, and those in each lower set become smaller. At one end of each set of letters is an acuity value in the form of a fraction. One of the sets near the bottom of the chart, for example, is marked 20/20. The normal eye can clearly see these letters from the standard distance of 20 feet and thus is said to have 20/20 vision. The letter at the top of the chart is marked 20/200. The normal eye can read letters of this size from a distance of 200 feet. Thus, an eye that is able to read only the top letter of the chart from a distance of 20 feet is said to have 20/200 vision. This person has less than normal vision. A line of letters near the bottom of the chart is marked 20/15. The normal eye can read letters of this size from a distance of 15 feet, but a person might be able to read it from 20 feet. This person has better than normal vision.

   To conduct the visual acuity test, follow these steps:

   a. Hang the Snellen eye chart on a well-illuminated wall at eye level.
   b. Have your partner stand 20 feet in front of the chart, gently cover the left eye with a 3″ × 5″ card, and read the smallest set of letters possible.
   c. Record the visual acuity value for that set of letters in Part A of Laboratory Report 30.
   d. Repeat the procedure, using the left eye.

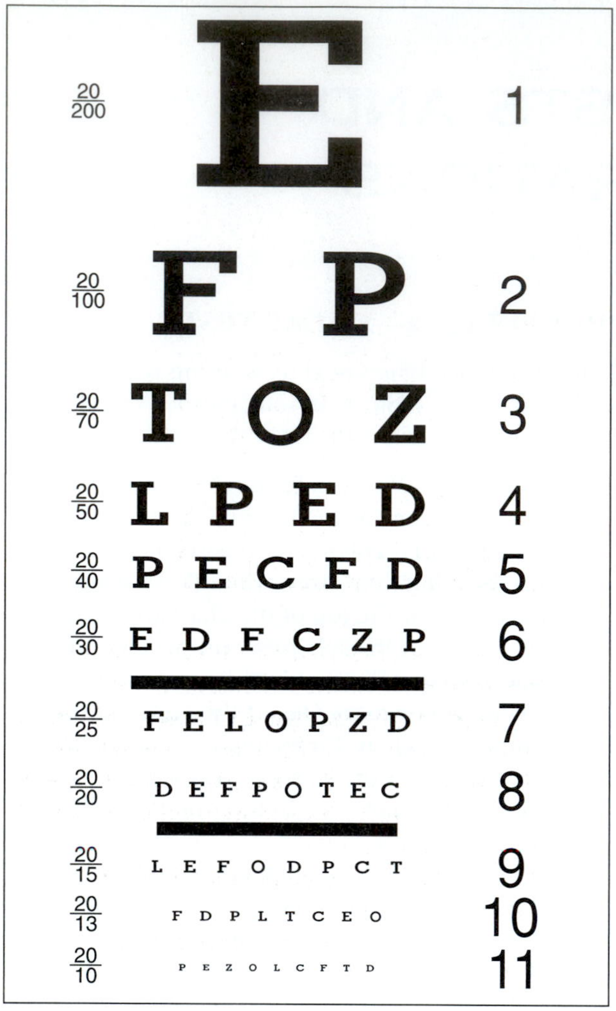

Figure 30.1 The Snellen eye chart looks similar to this but is somewhat larger.

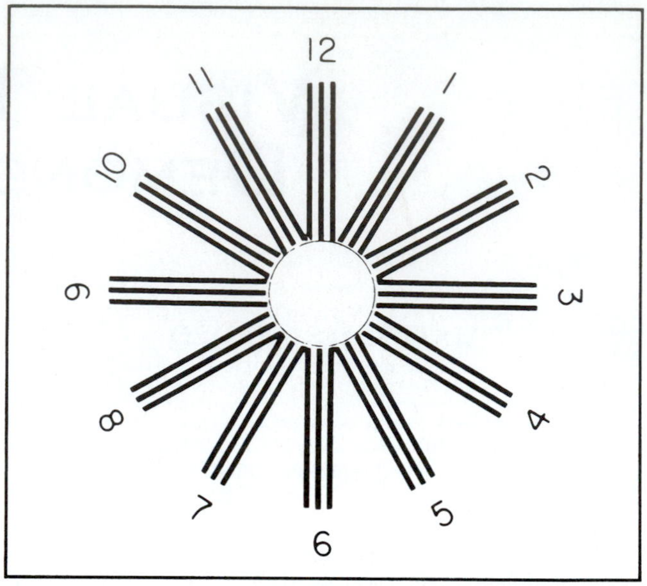

Figure 30.2 Astigmatism is evaluated using a chart such as this one.

2. *Astigmatism test.* Astigmatism is a condition that results from a defect in the curvature of the cornea or lens. As a consequence, some portions of the image projected on the retina are sharply focused, and other portions are blurred. Astigmatism can be evaluated by using an astigmatism chart (fig. 30.2). This chart consists of sets of black lines radiating from a central spot like the spokes of a wheel. To a normal eye, these lines appear sharply focused and equally dark; however, if the eye has an astigmatism, some sets of lines appear sharply focused and dark, while others are blurred and less dark.

   To conduct the astigmatism test, follow these steps:

   a. Hang the astigmatism chart on a well-illuminated wall at eye level.
   b. Have your partner stand 20 feet in front of the chart, gently cover the left eye with a 3″ × 5″ card, focus on the spot in the center of the radiating lines, and report which lines, if any, appear more sharply focused and darker.
   c. Repeat the procedure, using the left eye.
   d. Record the results in Part A of the laboratory report.

3. *Accommodation test.* Accommodation is the changing of the shape of the lens that occurs when the normal eye is focused for close vision. It involves a reflex in which muscles of the ciliary body are stimulated to contract, releasing tension on the suspensory ligaments that are fastened to the lens capsule. This allows the capsule to rebound elastically, causing the surface of the lens to become more convex. The ability to accommodate is likely to decrease with age because the tissues involved tend to lose their elasticity.

   To evaluate the ability to accommodate, follow these steps:

   a. Hold the end of a meterstick against your partner's chin so that the stick extends outward at a right angle to the plane of the face (fig. 30.3).
   b. Have your partner close the left eye.
   c. Hold a 3″ × 5″ card with a word typed in the center at the distal end of the meterstick.
   d. Slide the card along the stick toward your partner's open eye, and locate the *point closest to the eye* where your partner can still see the letters of the word sharply focused. This distance is called the *near point of accommodation,* and it tends to increase with age (table 30.1).
   e. Repeat the procedure with the right eye closed.
   f. Record the results in Part A of the laboratory report.

Figure 30.3 To determine the near point of accommodation, slide the 3″×5″ card along the meterstick toward your partner's open eye until the closest location where your partner can still see the word sharply focused.

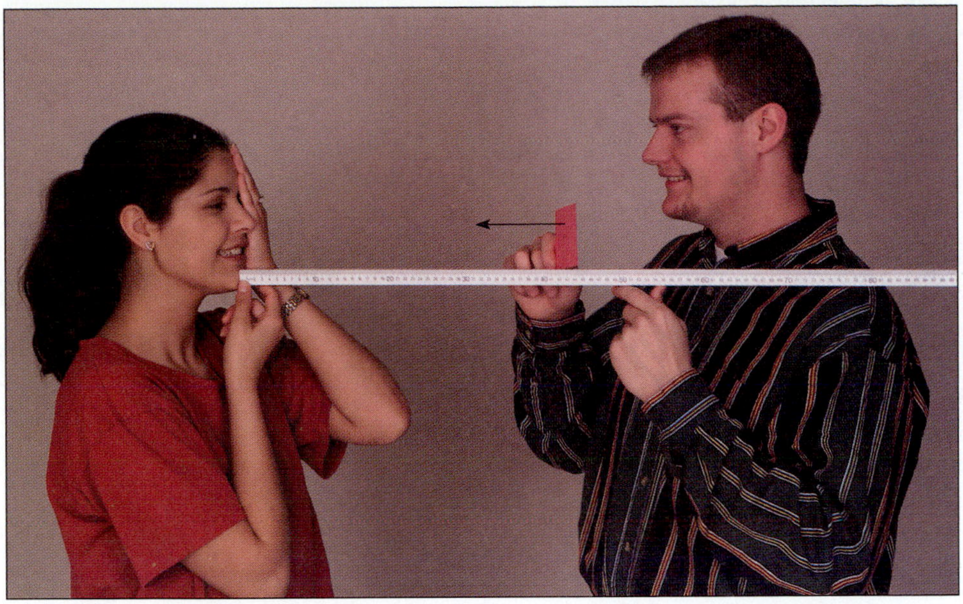

Table 30.1 Near Point of Accommodation

Age (years)	Average Near Point (cm)
10	7
20	10
30	13
40	20
50	45
60	90

4. *Color vision test.* Some individuals exhibit defective color vision because they lack certain cones, usually those sensitive to the reds or greens. Because this trait is an X-linked (sex-linked) inheritance, the condition is more prevalent in males (7%) than in females (0.4%). Individuals who lack or possess decreased sensitivity to the red-sensitive cones possess protanopia color blindness; those who lack or possess decreased sensitivity to green-sensitive cones possess deuteranopia color blindness. The color-blindness condition is often more of a deficiency or a weakness than one of blindness.

   To conduct the color vision test, follow these steps:

   a. Examine the color test plates in Ichikawa's or Ishihara's book to test for any red-green color vision deficiency. Also examine figure 30.4.
   b. Hold the test plates approximately 30 inches from the subject in bright light. All responses should occur within 3 seconds.
   c. Compare your responses with the correct answers in Ichikawa's or Ishihara's book. Determine the percentage of males and females in your class who exhibit any color-deficient vision. If an individual exhibits color-deficient vision, determine if the condition is protanopia or deuteranopia.
   d. Record the class results in Part A of the laboratory report.
5. Complete Part A of the laboratory report.

PROCEDURE B—VISUAL DEMONSTRATIONS

Perform the following demonstrations with the help of your laboratory partner.

1. *Blind-spot demonstration.* There are no photoreceptors in the optic disc, which is located where the nerve fibers of the retina leave the eye and enter the optic nerve. Consequently, this region of the retina is commonly called the *blind spot.*

   To demonstrate the blind spot, follow these steps:

   a. Close your left eye, hold figure 30.5 about 35 cm away from your face, and stare at the + sign in the figure with your right eye.
   b. Move the figure closer to your face as you continue to stare at the + until the dot on the figure suddenly disappears. This happens when the image of the dot is focused on the optic disc. Measure the distance using a metric ruler or a meterstick.

Figure 30.4 Samples of Ishihara's color plates. These plates are reproduced from *Ishihara's Tests for Colour Blindness* published by Kanehara & Co., Ltd., Tokyo, Japan, but tests for color blindness cannot be conducted with this material. For accurate testing, the original plates should be used.

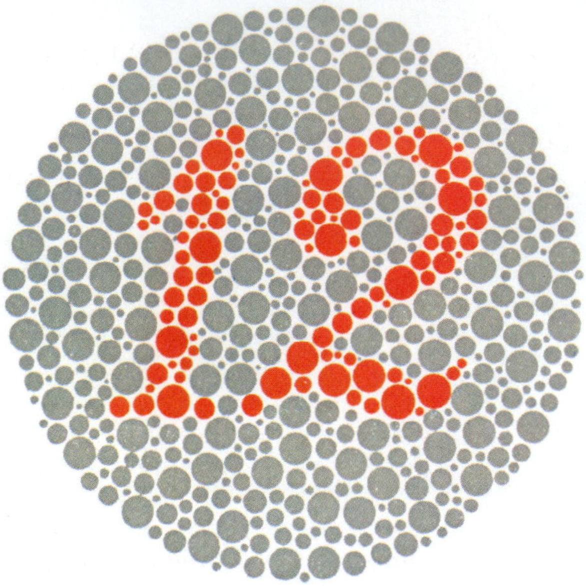

1 This color plate is nonselective. All individuals, normal or color-blind, will read this plate as 12.

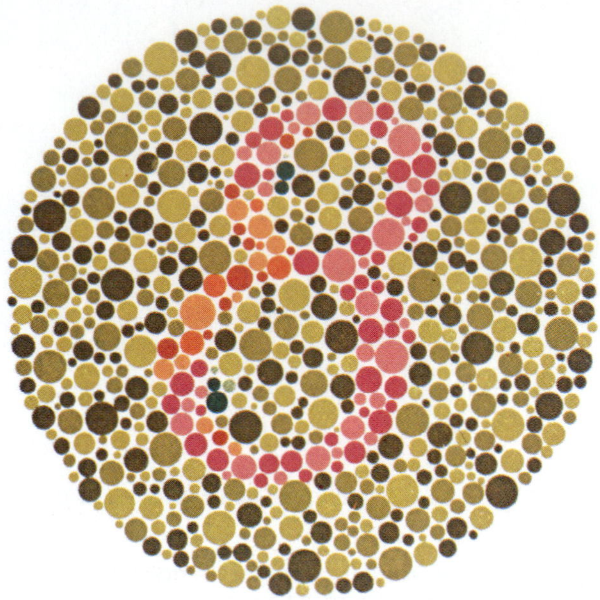

2 Normal individuals read this plate as 8. Red/green color-blind individuals see a 3 here.

3 A normal person sees a 16 here. Color-deficient individuals read this plate incorrectly or not at all.

4 Normal individuals read a 96 here. Protanopes see only a 6. Deuteranopes see only a 9.

Figure 30.5 Blind-spot demonstration.

c. Repeat the procedures with your right eye closed. This time stare at the dot, and the + will disappear when the image falls on the optic disc. Measure the distance.
d. Record the results in Part B of the laboratory report.

Critical Thinking Application

Under normal visual circumstances, explain why small objects are not lost from our vision.

2. *Photopupillary reflex.* The smooth muscles of the iris function to control the size of the pupil. For example, when the intensity of light entering the eye increases, a photopupillary reflex is triggered, and the circular muscles of the iris are stimulated to contract. As a result, the size of the pupil decreases, and less light enters the eye.

   To demonstrate this reflex, follow these steps:
   a. Ask your partner to sit with his or her hands thoroughly covering his or her eyes for 2 minutes.
   b. Position a pen flashlight close to one eye with the light shining on the hand that covers the eye.
   c. Ask your partner to remove the hand quickly.
   d. Observe the pupil and note any change in its size.
   e. Have your partner remove the other hand, but keep that uncovered eye shielded from extra light.
   f. Observe both pupils and note any difference in their sizes.
3. *Accommodation pupillary reflex.* The pupil constricts as a normal accommodation reflex to focusing on close objects. To demonstrate the accommodation reflex, follow these steps:
   a. Have your partner stare for several seconds at some dimly illuminated object in the room that is more than 20 feet away.
   b. Observe the size of the pupil of one eye. Then hold a pencil about 25 cm in front of your partner's face, and have your partner stare at it.
   c. Note any change in the size of the pupil.
4. *Convergence reflex.* The eyes converge as a normal convergence reflex to focusing on close objects. To demonstrate the convergence reflex, follow these steps:
   a. Repeat the procedure outlined for the accommodation pupillary reflex.
   b. Note any change in the position of the eyeballs as your partner changes focus from the distant object to the pencil.
5. Complete Part B of the laboratory report.

Laboratory Report **30**

Name ____________________

Date ____________________

Section ____________________

Visual Tests and Demonstrations

PART A

1. Visual acuity test results:

Eye Tested	Acuity Values
Right eye	
Right eye with glasses (if applicable)	
Left eye	
Left eye with glasses (if applicable)	

2. Astigmatism test results:

Eye Tested	Darker Lines
Right eye	
Right eye with glasses (if applicable)	
Left eye	
Left eye with glasses (if applicable)	

3. Accommodation test results:

Eye Tested	Near Point (cm)
Right eye	
Right eye with glasses (if applicable)	
Left eye	
Left eye with glasses (if applicable)	

4. Color vision test results:

	Males			Females		
Condition	**Class Number**	**Class Percentage**	**Expected Percentage**	**Class Number**	**Class Percentage**	**Expected Percentage**
Normal color vision			93			99.6
Deficient red-green color vision			7			0.4
Protanopia (lack red-sensitive cones)			less frequent type			less frequent type
Deuteranopia (lack green-sensitive cones)			more frequent type			more frequent type

5. Complete the following:

a. What is meant by 20/70 vision? ______________________________

b. What is meant by 20/10 vision? ______________________________

c. What visual problem is created by astigmatism? ______________________________

d. Why does the near point of accommodation often increase with age? ______________________________

e. Describe the eye defect that causes color-deficient vision. ______________________________

PART B

1. Blind-spot results:

a. Right eye distance _______

b. Left eye distance _______

Complete the following:

2. Explain why an eye has a blind spot. ______________________________

3. Describe the photopupillary reflex. ______________________________

4. What difference did you note in the size of the pupils when one eye was exposed to bright light and the other eye was shielded from the light? ______________________________

____

____

5. Describe the accommodation pupillary reflex. ______________________________

____

____

6. Describe the convergence reflex. ______________________________

____

____

LABORATORY EXERCISE 31

ENDOCRINE SYSTEM

MATERIALS NEEDED

Textbook
Human torso
Compound microscope
Prepared microscope slides of the following:
- Pituitary gland
- Thyroid gland
- Parathyroid gland
- Adrenal gland
- Pancreas

For Optional Activity:

Water bath equipped with temperature control mechanism set at 37°C (98.6°F)
Laboratory thermometer

The endocrine system consists of ductless glands that act together with parts of the nervous system to help control body activities. The endocrine glands secrete hormones that are transported in body fluids and affect cells possessing appropriate receptor molecules. In this way, hormones influence the rate of metabolic reactions, the transport of substances through cell membranes, and the regulation of water and electrolyte balances. By controlling cellular activities, endocrine glands play important roles in the maintenance of homeostasis.

PURPOSE OF THE EXERCISE

To review the structure and function of major endocrine glands and to examine microscopically the tissues of these glands.

LEARNING OBJECTIVES

After completing this exercise, you should be able to

1. name and locate the major endocrine glands;
2. name the hormones secreted by each of the major glands;
3. describe the principal functions of each hormone;
4. recognize tissue sections from the pituitary gland, thyroid gland, parathyroid glands, adrenal glands, and pancreas.

PROCEDURE

1. Review the sections entitled "Pituitary Gland," "Thyroid Gland," "Parathyroid Glands," "Adrenal Glands," "Pancreas," and "Other Endocrine Glands" in chapter 11 of the textbook.
2. As a review activity, label figures 31.1, 31.2, 31.3, 31.4, 31.5, and 31.6.
3. Complete Part A of Laboratory Report 31.
4. Examine the human torso and locate the following:

hypothalamus

pituitary stalk (infundibulum)

pituitary gland

anterior lobe

posterior lobe

thyroid gland

parathyroid glands

adrenal glands

adrenal medulla

adrenal cortex

pancreas

pineal gland

thymus gland

ovaries

testes

Figure 31.1 Label the major endocrine glands.

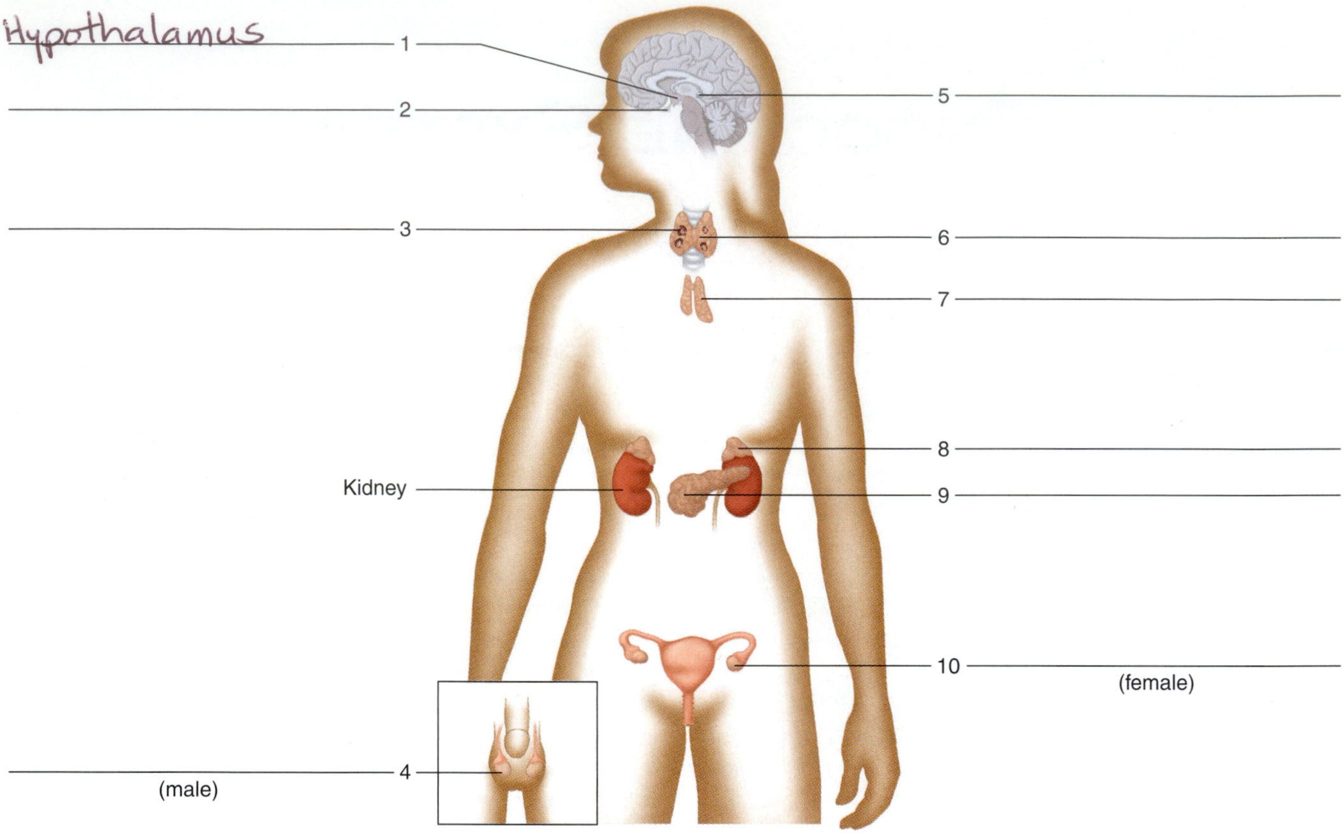

Figure 31.2 Label the features associated with the pituitary gland.

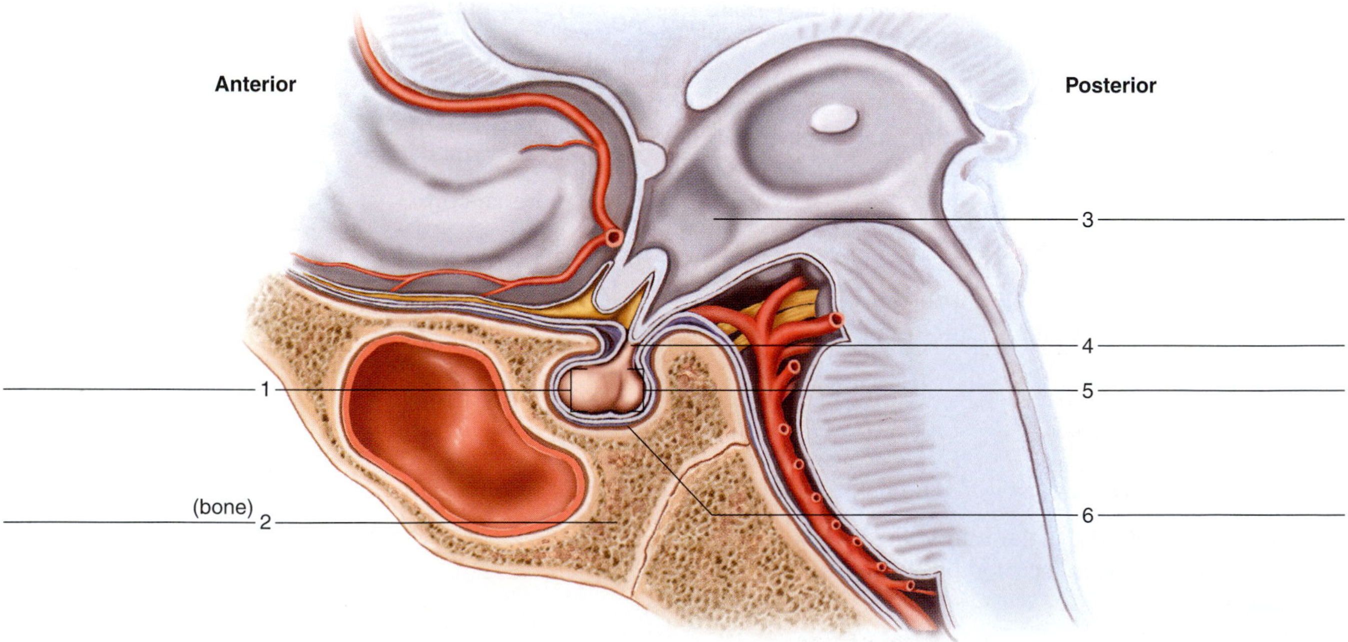

Figure 31.3 Label the features associated with the thyroid gland (anterior view).

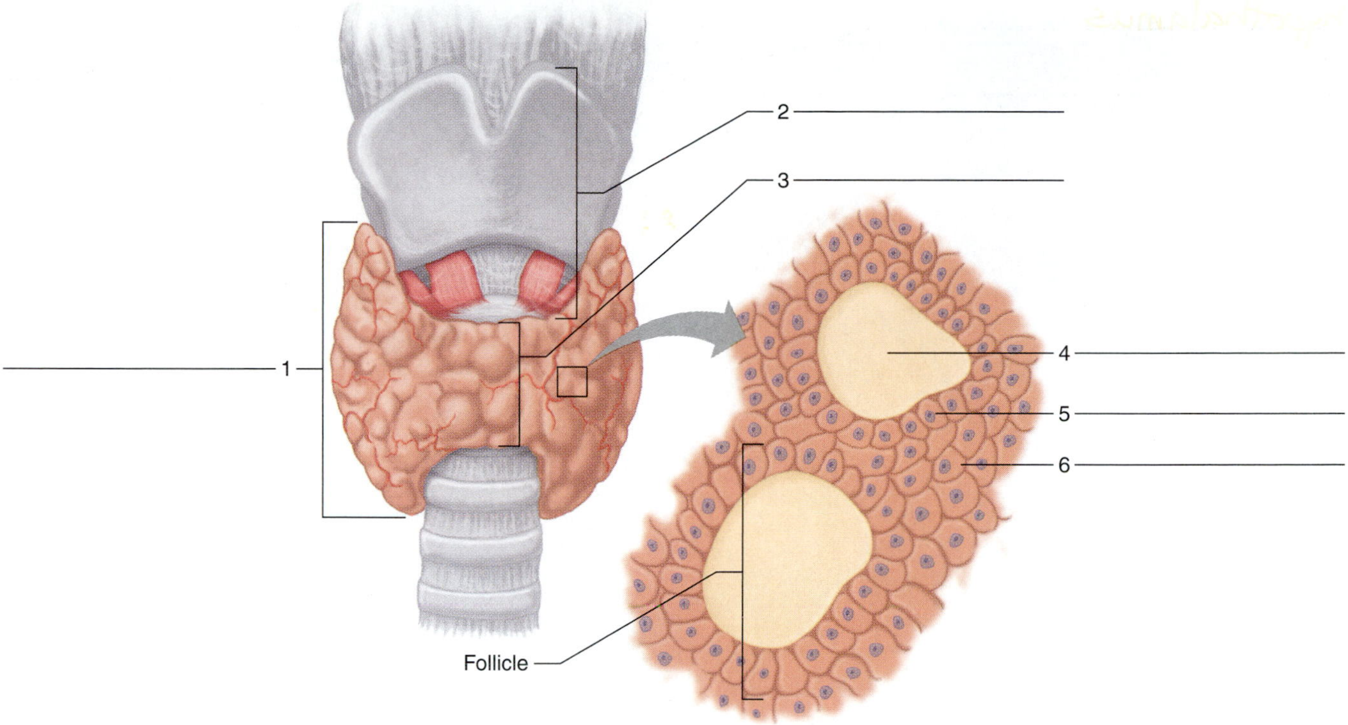

Figure 31.4 Label the features associated with the parathyroid glands (posterior view).

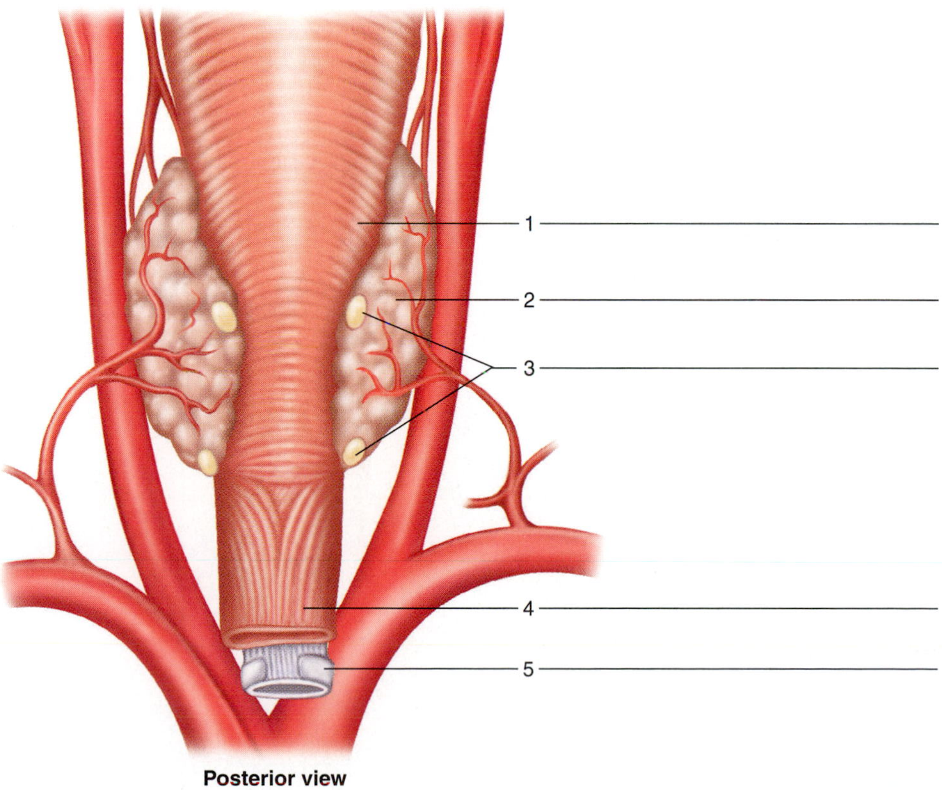

Figure 31.5 Label the features associated with the adrenal gland.

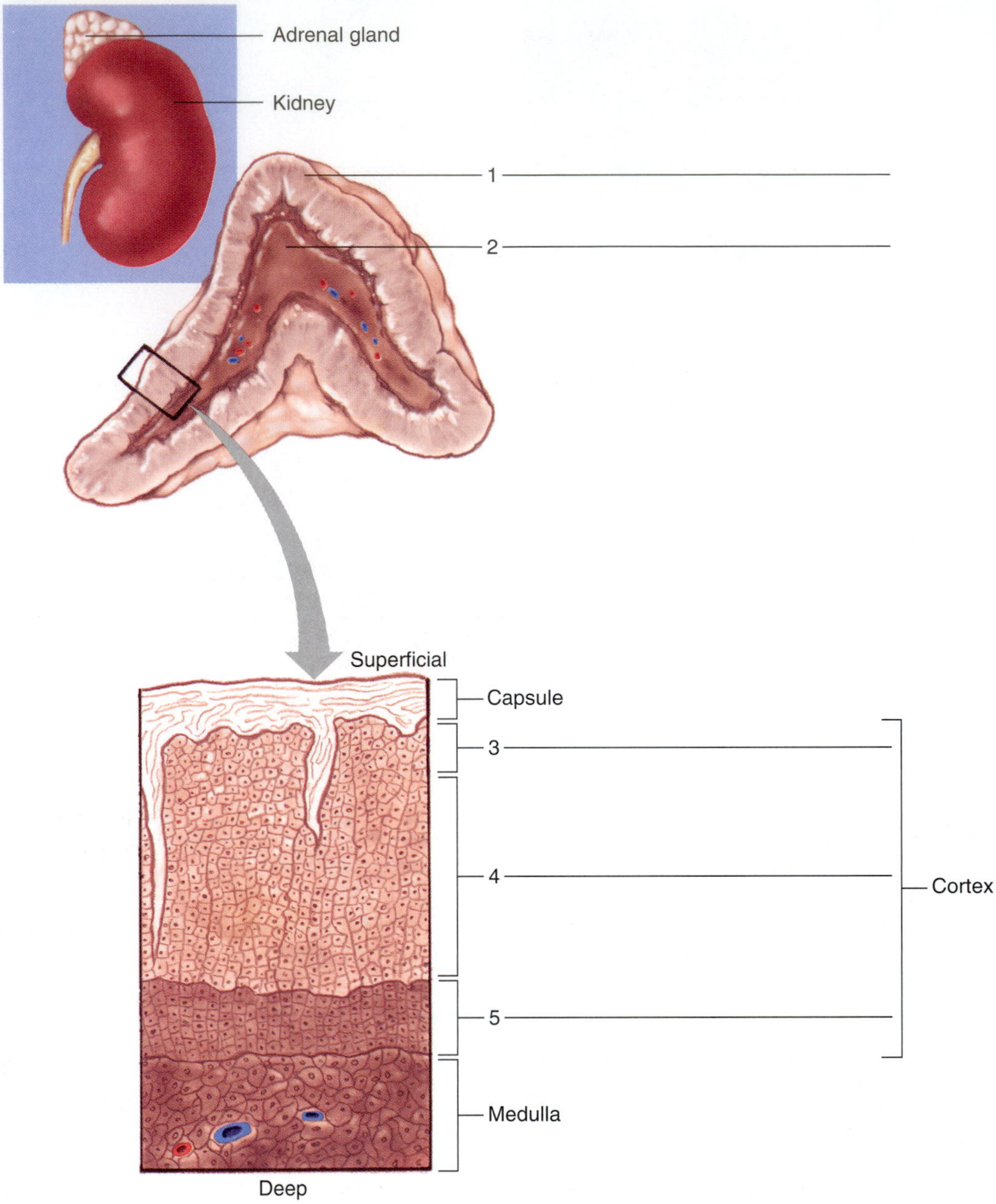

Figure 31.6 Label the features associated with the pancreas.

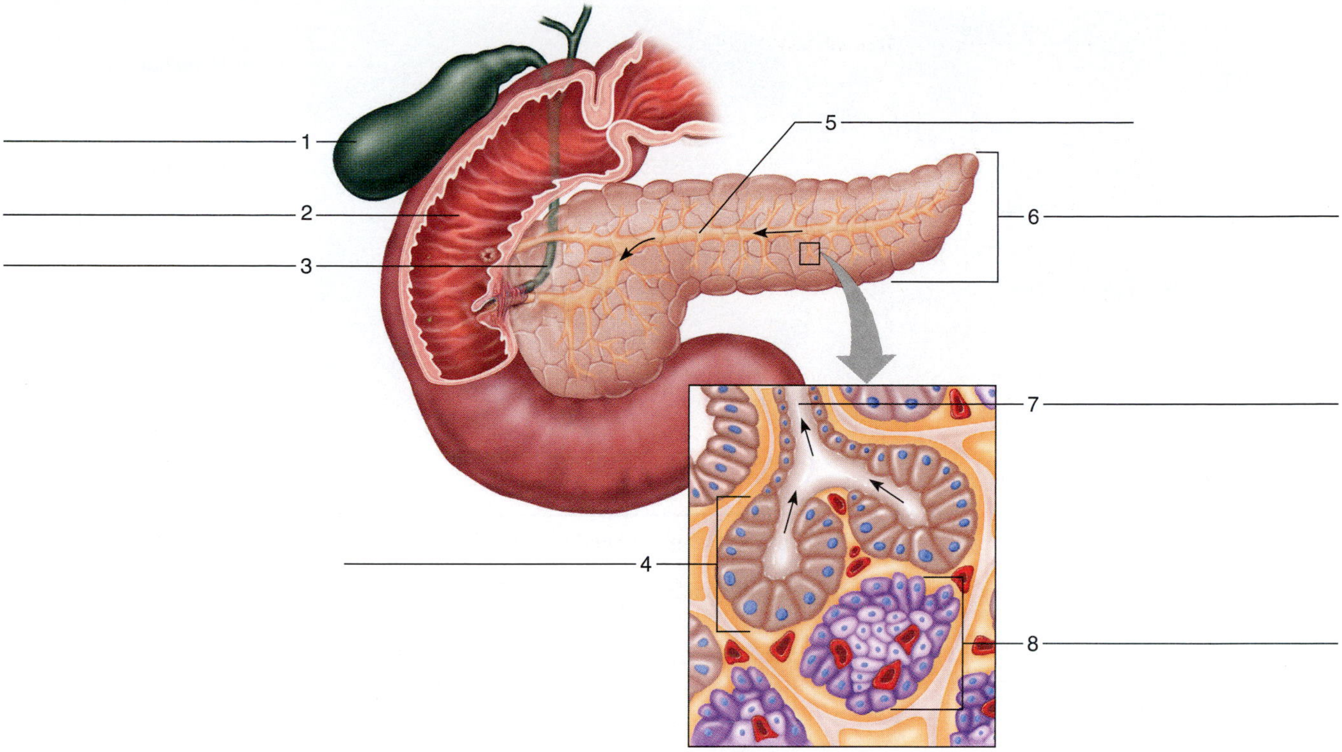

OPTIONAL ACTIVITY

The secretions of endocrine glands are usually controlled by negative feedback systems. As a result, the concentrations of hormones in body fluids remain relatively stable, although they will fluctuate slightly within a normal range. (See the section entitled "Control of Hormonal Secretions" in chapter 11 of the textbook.)

Similarly, the mechanism used to maintain the temperature of a laboratory water bath involves negative feedback. In this case, a temperature-sensitive thermostat in the water allows a water heater to operate whenever the water temperature drops below the thermostat's set point. Then, when the water temperature reaches the set point, the thermostat causes the water heater to turn off (a negative effect), and the water bath begins to cool again.

Use a laboratory thermometer to monitor the temperature of the water bath in the laboratory. Measure the temperature at regular intervals until you have recorded ten readings. What was the lowest temperature you recorded? ______________________
The highest temperature? ______________________
What was the average temperature of the water bath? ____________ How is the water bath temperature control mechanism similar to a hormonal control mechanism in the body? ______________________

____

____

____

____

____

5. Examine the microscopic tissue sections of the following glands, and identify the features described:

   Pituitary gland. To examine the pituitary tissue, follow these steps:

   a. Observe the tissues using low-power magnification (fig. 31.7).
   b. Locate the *pituitary stalk,* the *anterior lobe* (the largest part of the gland), and the *posterior lobe.*
   c. Observe an area of the anterior lobe with high-power magnification. Locate a cluster of relatively large cells and identify some *acidophil cells,* which contain pink-stained granules, and some *basophil cells,* which contain blue-stained granules. These acidophil and basophil cells are hormone-secreting cells.
   d. Observe an area of the posterior lobe with high-power magnification. Note the numerous unmyelinated nerve fibers present in this lobe. Also locate some *pituicytes,* a type of neuroglial cell, scattered among the nerve fibers.
   e. Prepare labeled sketches of representative portions of the anterior and posterior lobes of the pituitary gland in Part B of the laboratory report.

Figure 31.7 Micrograph of the pituitary gland (6×).

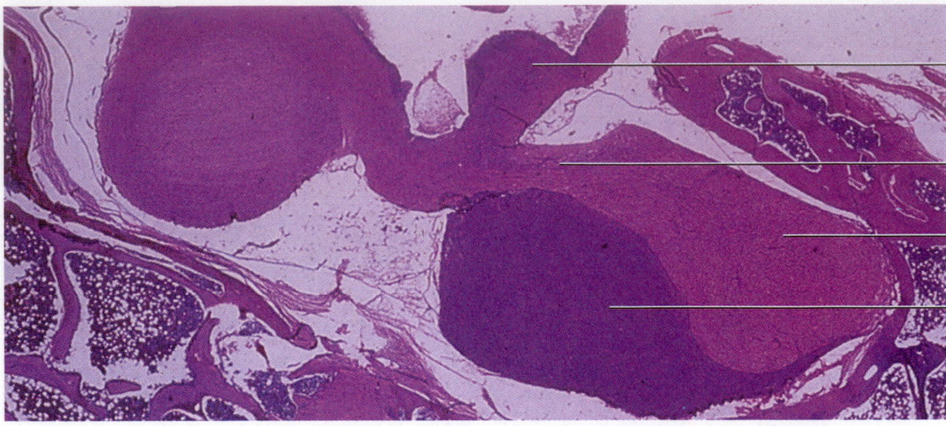

Figure 31.8 Micrograph of the thyroid gland (100× micrograph enlarged to 300×).

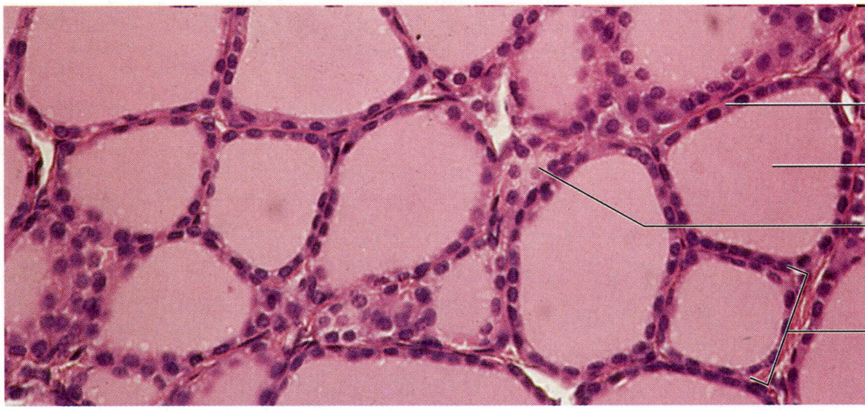

Thyroid gland. To examine the thyroid tissue, follow these steps:

a. Use low-power magnification to observe the tissue (fig. 31.8). Note the numerous *follicles,* each of which consists of a layer of cells surrounding a colloid-filled cavity.
b. Observe the tissue using high-power magnification. Note that the cells forming the wall of a follicle are simple cuboidal epithelial cells.
c. Prepare a labeled sketch of a representative portion of the thyroid gland in Part B of the laboratory report.

Parathyroid gland. To examine the parathyroid tissue, follow these steps:

a. Use low-power magnification to observe the tissue (fig. 31.9). Note that the gland consists of numerous tightly packed secretory cells.
b. Switch to high-power magnification and locate two types of cells—a smaller form (chief cells) that are arranged in cordlike patterns and a larger form (oxyphil cells) that have distinct cell boundaries and are present in clusters. The *chief cells* secrete parathyroid hormone. The function of the *oxyphil cells* is not well understood.
c. Prepare a labeled sketch of a representative portion of the parathyroid gland in Part B of the laboratory report.

Adrenal gland. To examine the adrenal tissue, follow these steps:

a. Use low-power magnification to observe the tissue (fig. 31.10). Note the thin capsule that covers the gland. Just beneath the capsule, there is a relatively thick *adrenal cortex.* The central portion of the gland is the *adrenal medulla.* The cells of the cortex are in three poorly

Figure 31.9 Micrograph of the parathyroid gland (65×).

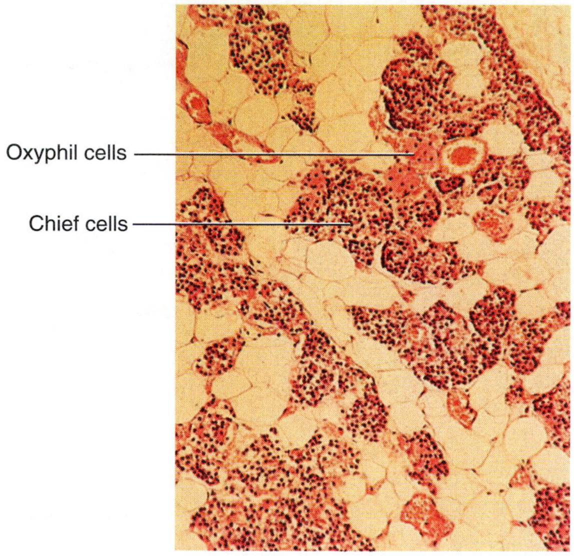

Figure 31.10 Micrograph of the adrenal cortex and the adrenal medulla (75×).

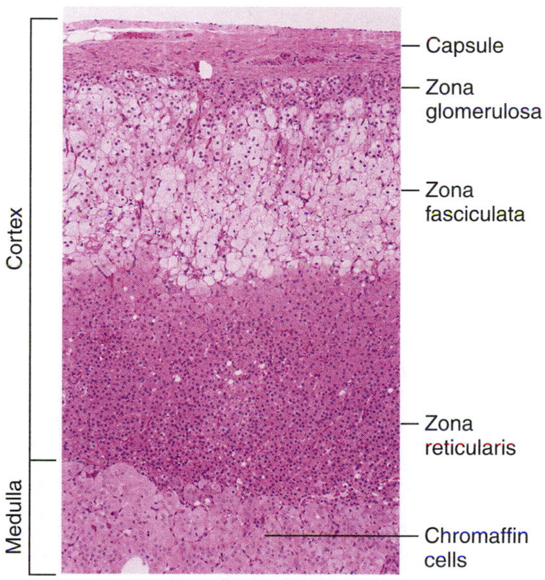

defined layers. Those of the outer layer (zona glomerulosa) are arranged irregularly; those of the middle layer (zona fasciculata) are in long cords; and those of the inner layer (zona reticularis) are arranged in an interconnected network of cords. The cells of the medulla are relatively large, irregularly shaped, and often occur in clusters.

b. Using high-power magnification, observe each of the layers of the cortex and the cells of the medulla.

Figure 31.11 Micrograph of the pancreas (100× micrograph enlarged to 400×).

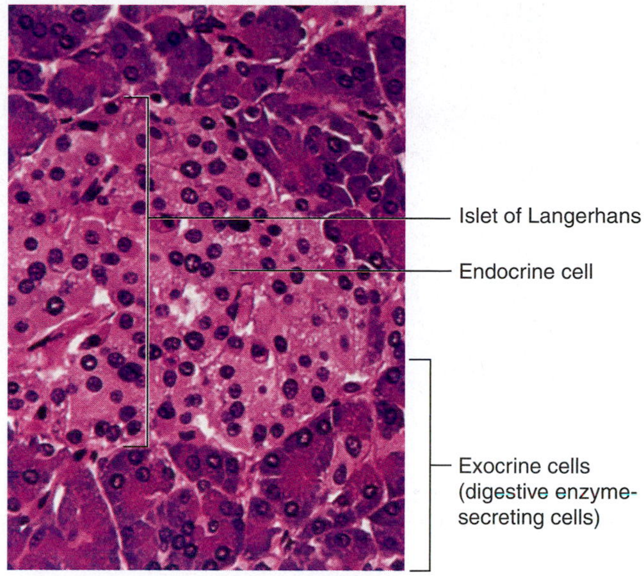

c. Prepare labeled sketches of representative portions of the adrenal cortex and medulla in Part B of the laboratory report.

Pancreas. To examine the pancreas tissue, follow these steps:

a. Use low-power magnification to observe the tissue (fig. 31.11). Note that the gland largely consists of deeply stained exocrine cells arranged in clusters around secretory ducts. These exocrine cells (acinar cells) secrete pancreatic juice rich in digestive enzymes. There are circular masses of lightly stained cells scattered throughout the gland. These clumps of cells constitute the *islets of Langerhans* (*pancreatic islets*), and they represent the endocrine portion of the pancreas.
b. Examine an islet, using high-power magnification.
c. Prepare a labeled sketch of a representative portion of the pancreas in Part B of the laboratory report.

Web Quest

What are common endocrine gland disorders? How can they be treated? Search these and review the endocrine system at www.mhhe.com/shieress8

Laboratory Report **31**

Name ______________________

Date ______________________

Section ______________________

Endocrine System

PART A

Complete the following:

1. Name six hormones secreted by the anterior lobe of the pituitary gland. ______________________
2. Name two hormones secreted by the posterior lobe of the pituitary gland. ______________________
3. Name the pituitary hormone responsible for the following actions:
   a. causes kidneys to conserve water ______________________
   b. stimulates cells to increase in size and divide more rapidly ______________________
   c. stimulates secretion from thyroid gland ______________________
   d. causes contraction of uterine wall muscles ______________________
   e. stimulates secretion from adrenal cortex ______________________
   f. stimulates milk production ______________________
4. Name two thyroid hormones that affect metabolic rate. ______________________
5. Name a hormone secreted by the thyroid gland that acts to lower blood calcium. ______________________
6. Name a hormone that acts to raise blood calcium. ______________________
7. Name three target organs of parathyroid hormone. ______________________
8. Name two hormones secreted by the adrenal medulla. ______________________
9. List five different effects produced by these medullary hormones. ______________________
10. Name the most important mineralocorticoid secreted by the adrenal cortex. ______________________
11. List two actions of this mineralocorticoid. ______________________
12. Name the most important glucocorticoid secreted by the adrenal cortex. ______________________

13. List three actions of this glucocorticoid. ______________________________

____

____

14. Distinguish the two hormones secreted by the alpha and beta cells of the islets of Langerhans. ______________________________

____

Critical Thinking Application

Briefly explain how the actions of pancreatic hormones complement one another.

____

____

____

____

PART B

Prepare labeled sketches to illustrate representative portions of the following endocrine glands:

Pituitary gland (_____×) (anterior lobe)	Pituitary gland (_____×) (posterior lobe)
Thyroid gland (_____×)	Parathyroid gland (_____×)

Adrenal gland (_____×)
(medulla)

Adrenal gland (_____×)
(cortex)

Pancreas (_____×)

BLOOD CELLS

MATERIALS NEEDED

Textbook
Compound microscope
Prepared microscope slides of human blood (Wright's stain)
Colored pencils

For Demonstration:

Microscope slides (precleaned)
Sterile disposable blood lancets
Sterile absorbent cotton
70% alcohol
Slide staining rack and tray
Wright's stain
Distilled water

For Optional Activity:

Prepared slides of pathological blood, such as eosinophilia, leukocytosis, leukopenia, and lymphocytosis

SAFETY

- It is important that students learn and practice correct procedures for handling body fluids. Consider using either mammal blood other than human or contaminant-free blood that has been tested and is available from various laboratory supply houses. Some of the procedures might be accomplished as demonstrations only. If student blood is utilized, it is important that students handle only their own blood.
- Use an appropriate disinfectant to wash the laboratory tables before and after the procedures.
- Wear disposable gloves when handling blood samples.
- Clean end of a finger with 70% alcohol before the puncture is performed.
- The sterile blood lancet should be used only once.
- Dispose of used lancets and blood-contaminated items into an appropriate container (never use the wastebasket).
- Wash your hands before leaving the laboratory.

Blood is a type of connective tissue whose cells are suspended in a liquid intercellular substance. These cells are mainly formed in red bone marrow, and they include *red blood cells, white blood cells,* and some cellular fragments called *platelets.*

Red blood cells function to transport gases between the body cells and the lungs, white blood cells serve to defend the body against infections, and platelets play an important role in stoppage of bleeding (hemostasis).

PURPOSE OF THE EXERCISE

To review the characteristics of blood cells, to examine them microscopically, and to perform a differential white blood cell count.

LEARNING OBJECTIVES

After completing this exercise, you should be able to

1. describe the structure and function of red blood cells, white blood cells, and platelets;
2. identify red blood cells, five types of white blood cells, and platelets on a stained blood slide;
3. perform a differential white blood cell count.

WARNING

Because of the possibility of blood infections being transmitted from one student to another if blood slides are prepared in the classroom, it is suggested that commercially prepared blood slides be used in this exercise. The instructor, however, may wish to demonstrate the procedure for preparing such a slide. Observe all safety procedures listed for this lab.

DEMONSTRATION—BLOOD SLIDE PREPARATION

To prepare a stained blood slide, follow these steps:

1. Obtain two precleaned microscope slides. Avoid touching their flat surfaces.
2. Thoroughly wash hands with soap and water and dry them with paper towels.
3. Cleanse the end of the middle finger with some sterile cotton moistened with 70% alcohol and let the finger dry in the air.
4. Remove a sterile disposable blood lancet from its package without touching the sharp end.
5. Puncture the skin on the side near the tip of the middle finger with the lancet and properly discard the lancet.
6. Wipe away the first drop of blood with the cotton ball. Place a drop of blood about 2 cm from the end of a clean microscope slide.
7. Use a second slide to spread the blood across the first slide, as illustrated in figure 32.1. Discard the slide used for spreading the blood in the appropriate container.
8. Place the blood slide on a slide staining rack and let it dry in the air.

Figure 32.1 To prepare a blood smear: (*a*) place a drop of blood about 2 cm from the end of a clean slide; (*b*) hold a second slide at about a 45° angle to the first one, allowing the blood to spread along its edge; (*c*) push the second slide over the surface of the first so that it pulls the blood with it; (*d*) observe the completed blood smear. The ideal smear should be 1½ inches in length, be evenly distributed, and contain a smooth, feathered edge.

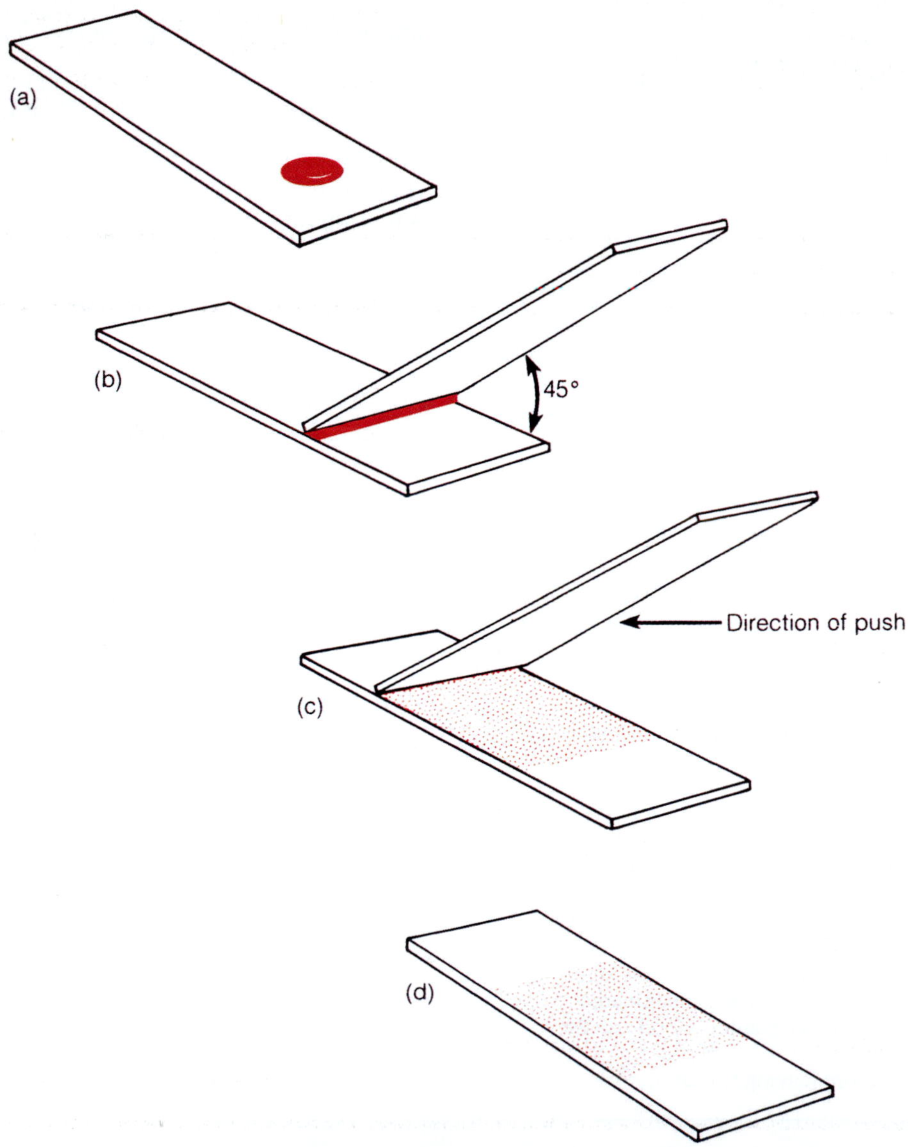

9. Put enough Wright's stain on the slide to cover the smear but not overflow the slide. Count the number of drops of stain that are used.
10. After 2–3 minutes, add an equal volume of distilled water to the stain and let the slide stand for 4 minutes. From time to time, gently blow on the liquid to mix the water and stain.
11. Flood the slide with distilled water until the blood smear appears light blue.
12. Tilt the slide to pour off the water, and let the slide dry in the air.
13. Examine the blood smear with low-power magnification, and locate an area where the blood cells are well distributed. Observe these cells, using high-power magnification and then with an oil immersion objective if one is available.

PROCEDURE A—TYPES OF BLOOD CELLS

1. Review the sections entitled "Red Blood Cells," "White Blood Cells," and "Blood Platelets" in chapter 12 of the textbook.
2. Complete Part A of Laboratory Report 32.
3. Refer to figures 12.6–12.10 in the textbook and figure 32.2 as an aid in identifying the various types of blood cells. Use the prepared slide of blood and locate each of the following:

   red blood cell (erythrocyte)

   white blood cell (leukocyte)

   - granulocytes
     - neutrophil
     - eosinophil
     - basophil
   - agranulocytes
     - lymphocyte
     - monocyte

   platelet (thrombocyte)

4. In Part B of the laboratory report, prepare sketches of single blood cells to illustrate each type. Pay particular attention to relative size, nuclear shape, and color of granules in the cytoplasm (if present).

Figure 32.2 Blood cells illustrating some of the numerous variations of each type.

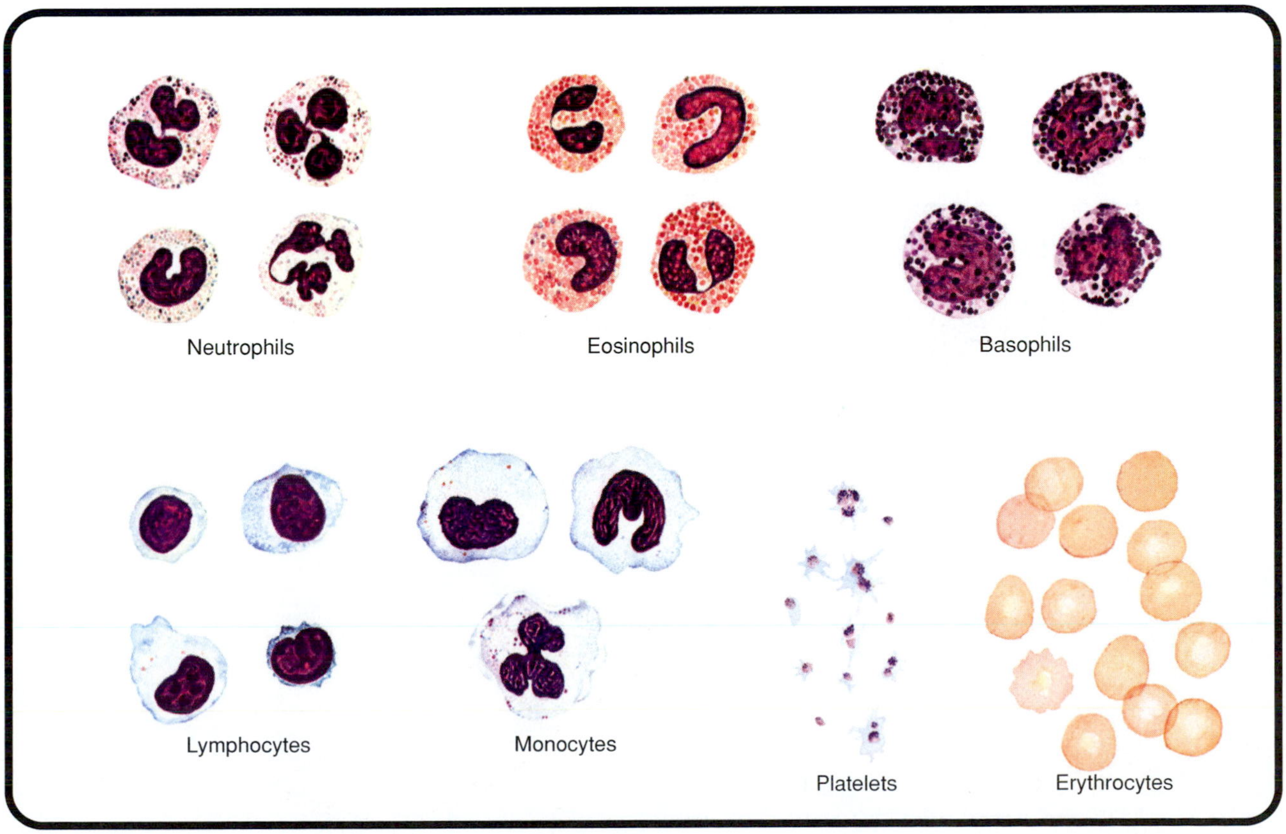

PROCEDURE B—DIFFERENTIAL WHITE BLOOD CELL COUNT

A differential white blood cell count is performed to determine the percentage of each of the various types of white blood cells present in a blood sample. The test is useful because the relative proportions of white blood cells may change in particular diseases. Neutrophils, for example, usually increase during bacterial infections, whereas eosinophils may increase during certain parasitic infections and allergic reactions.

1. To make a differential white blood cell count, follow these steps:
   a. Using high-power magnification or an oil immersion objective, focus on the cells at one end of a prepared blood slide where the cells are well distributed.
   b. Slowly move the blood slide back and forth, following a path that avoids passing over the same cells twice (fig. 32.3).
   c. Each time you encounter a white blood cell, identify its type and record it in Part C of the laboratory report.
   d. Continue searching for and identifying white blood cells until you have recorded 100 cells in the data table. Because *percent* means "parts of 100," for each type of white blood cell, the total number observed is equal to its percentage in the blood sample.
2. Complete Part C of the laboratory report.

Figure 32.3 Move the blood slide back and forth to avoid passing the same cells twice.

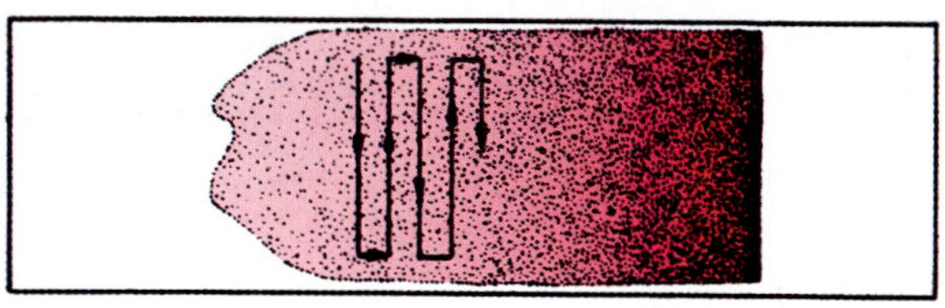

OPTIONAL ACTIVITY

Obtain a prepared slide of pathological blood that has been stained with Wright's stain. Perform a differential white blood cell count, using this slide, and compare the results with the values for normal blood listed in table 32.1. What differences do you note?

____

____

____

____

____

Table 32.1 Differential White Blood Cell Count

Cell Type	Normal Value (percent)
Neutrophil	54–62
Eosinophil	1–3
Basophil	<1
Lymphocyte	25–33
Monocyte	3–9

Web Quest

What are the functions of the various blood components? What do abnormal amounts indicate? Search these and review blood cell identification at www.mhhe.com/shieress8

Laboratory Report **32**

Name ______________________

Date ______________________

Section ______________________

Blood Cells

PART A

Complete the following:

1. Red blood cells are also called ______________________ .
2. The shape of a red blood cell can be described as a ______________________ disc.
3. The shape of a red blood cell is related to its function of ______________________ .
4. ______________________ is the oxygen-carrying substance in a red blood cell.
5. Red blood cells with high oxygen concentrations are bright red because of the presence of ______________________ .
6. Red blood cells cannot reproduce because they lack ______________________ when they are mature.
7. White blood cells are also called ______________________ .
8. White blood cells with granular cytoplasm are called ______________________ .
9. White blood cells lacking granular cytoplasm are called ______________________ .
10. Normally, the most numerous white blood cells are ______________________ .
11. White blood cells whose cytoplasmic granules stain red in acid stain are called ______________________ .
12. ______________________ are normally the least abundant of the white blood cells.
13. ______________________ are the largest of the white blood cells.
14. ______________________ are small agranulocytes that have relatively large, round nuclei with thin rims of cytoplasm.
15. In red bone marrow, platelets develop from cells called ______________________ .
16. A platelet, which is a fragment of a cell, lacks a ______________________ .

PART B

Sketch a single blood cell of each type in the following spaces. Use colored pencils to represent the stained colors of the cells. Label any features that can be identified.

Red blood cell (_______ ×)

Neutrophil (_______ ×)

Lymphocyte (_______ ×)

Monocyte (_______ ×)

Eosinophil (_______ ×)

Basophil (_______ ×)

Platelet (_______ ×)

PART C

1. Differential white blood cell count data table. As you identify white blood cells, record them on the table by using a tally system, such as ~~IIII~~ II. Place tally marks in the "Number Observed" column, and total each of the five WBCs when the differential count is completed. Obtain a total of all five WBCs counted to determine the percent of each WBC type.

Type of WBC	Number Observed	Total	Percent
Neutrophil			
Lymphocyte			
Monocyte			
Eosinophil			
Basophil			
		Total of column	

2. How do the results of your differential white blood cell count compare with the normal values listed in table 32.1? ______________________________

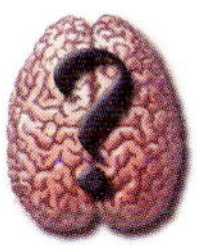

Critical Thinking Application

What is the difference between a differential white blood cell count and a total white blood cell count?

Blood Testing—A Demonstration

MATERIALS NEEDED

Textbook
Sterile disposable blood lancets
Sterile absorbent cotton
70% alcohol

For Demonstration A:

Heparinized microhematocrit capillary tube
Sealing clay (or Critocaps)
Microhematocrit centrifuge
Microhematocrit reader

For Demonstration B:

Hemoglobinometer
Lens paper
Hemolysis applicator

For Demonstration C:

Hemocytometer
Unopette system (a Becton Dickinson product) for counting red blood cells*
Hand counter (tally)

For Demonstration D:

Hemocytometer
Unopette system (a Becton Dickinson product) for counting white blood cells*
Hand counter (tally)

*See the Instructor's Manual for a supplier of Unopette systems.

SAFETY

- It is important that students learn and practice correct procedures for handling body fluids. Consider using either mammal blood other than human or contaminant-free blood that has been tested and is available from various laboratory supply houses. Some of the procedures might be accomplished as demonstrations only. If student blood is utilized, it is important that students handle only their own blood.
- Use an appropriate disinfectant to wash the laboratory tables before and after the procedures.
- Wear disposable gloves when handling blood samples.
- Clean the end of a finger with 70% alcohol before the puncture is performed.
- The sterile blood lancet should be used only once.
- Dispose of used lancets and blood-contaminated items into an appropriate container (never use the wastebasket).
- Wash your hands before leaving the laboratory.

As an aid in identifying various disease conditions, tests are often performed on blood to determine how its composition compares with normal values. These tests commonly include red blood cell percentage, hemoglobin content, red blood cell count, and total white blood cell count.

PURPOSE OF THE EXERCISE

To observe the blood tests used to determine red blood cell percentage, hemoglobin content, red blood cell count, and total white blood cell count.

LEARNING OBJECTIVES

After completing this exercise, you should be able to

1. determine the percentage of red blood cells in a blood sample;
2. determine the hemoglobin content of a blood sample;
3. describe how a red blood cell count is performed;
4. describe how a total white blood cell count is performed.

WARNING

Because of the possibility of blood infections being transmitted from one student to another during blood testing procedures, it is suggested that the following demonstrations be performed by the instructor. Observe all safety procedures listed for this lab.

DEMONSTRATION A—RED BLOOD CELL PERCENTAGE (HEMATOCRIT)

To determine the percentage of red blood cells in a whole blood sample, the cells must be separated from the liquid plasma. This separation can be rapidly accomplished by placing a tube of blood in a centrifuge. The force created by the spinning motion of the centrifuge causes the cells to be packed into the lower end of the tube. Then the quantities of cells and plasma can be measured, and the percentage of cells (hematocrit or packed cell volume) can be calculated.

1. To determine the percentage of red blood cells in a blood sample, follow these steps:
   a. Lance the end of a finger to obtain a drop of blood. *See the demonstration in Laboratory Exercise 32 for directions.*
   b. Touch the drop of blood with the colored end of a heparinized capillary tube. Hold the tube tilted slightly upward so that the blood will easily move into it by capillary action (fig. 33.1). To prevent an air bubble, keep the tip in the blood until filled.
   c. Allow the blood to fill about two-thirds of the length of the tube.
   d. Plug the blood end of the tube by pushing it with a rotating motion into sealing clay or by adding a plastic Critocap. By holding a finger over the tip of the dry end, blood will not drain out while sealing the blood end.
   e. Place the sealed tube into one of the numbered grooves of a microhematocrit centrifuge. The tube's sealed end should point outward from the center and should touch the rubber lining on the rim of the centrifuge (fig. 33.1).
   f. The centrifuge should be balanced by placing specimen tubes on opposite sides of the moving head, the inside cover should be tightened with the lock wrench, and the outside cover should be securely fastened.
   g. Run the centrifuge for 3–5 minutes.
   h. After the centrifuge has stopped, remove the specimen tube and note that the red blood cells have been packed into the bottom of the tube. The clear liquid on top of the cells is plasma.
   i. Use a microhematocrit reader to determine the percentage of red blood cells in the tube. If a microhematocrit reader is not available, measure the total length of the blood column in millimeters (red cells plus plasma) and the length of the red blood cell column alone in millimeters. Divide the red blood cell length by the total blood column length and multiply the answer by 100 to calculate the percentage of red blood cells.
   j. Record the test result in Part A of Laboratory Report 33.
2. Complete Part B of the laboratory report.

Figure 33.1 Steps in the hematocrit (red blood cell percentage) procedure: (*a*) load a heparinized capillary tube with blood; (*b*) plug the blood end of the tube with sealing clay; (*c*) place the tube in a microhematocrit centrifuge.

(a)
(b)
(c)

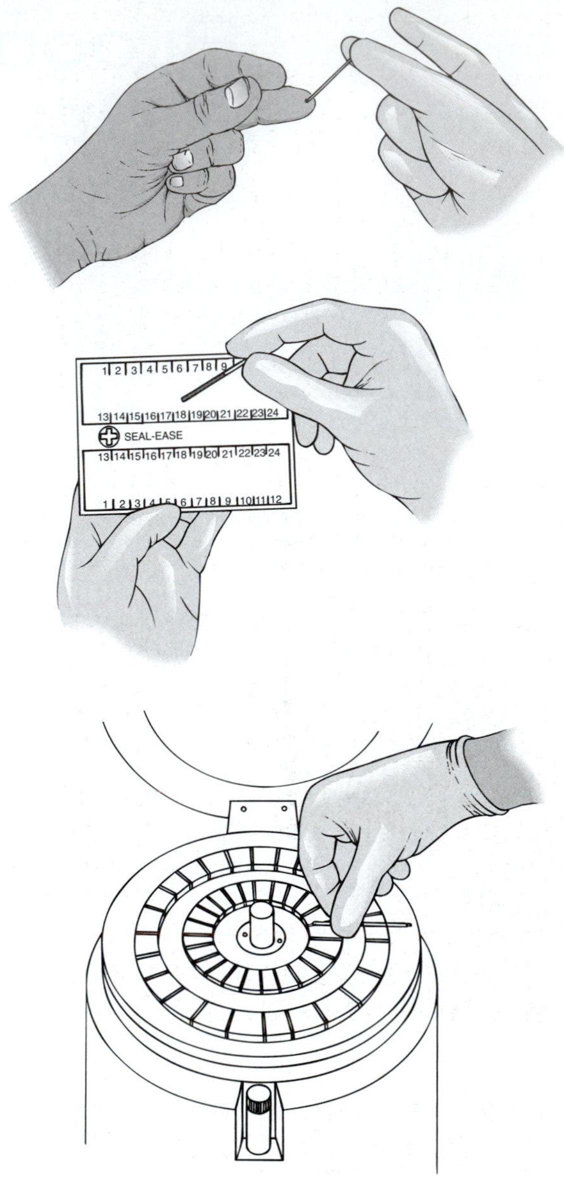

DEMONSTRATION B—HEMOGLOBIN CONTENT

Although the hemoglobin content of a blood sample can be measured in several ways, a common method uses a hemoglobinometer. This instrument is designed to compare the color of light passing through a hemolyzed blood sample with a standard color. The results of the test are expressed in grams of hemoglobin per 100 mL of blood or in percentage of normal values.

1. To measure the hemoglobin content of a blood sample, follow these steps:
   a. Obtain a hemoglobinometer and remove the blood chamber from the slot in its side.
   b. Separate the pieces of glass from the metal clip and clean them with 70% alcohol and lens paper. Note that one of the pieces of glass has two broad, U-shaped areas surrounded by depressions. The other piece is flat on both sides.
   c. Obtain a large drop of blood from a finger, as before.
   d. Place the drop of blood on one of the U-shaped areas of the blood chamber glass.
   e. Stir the blood with the tip of a hemolysis applicator until the blood appears clear rather than cloudy. This usually takes about 45 seconds.
   f. Place the flat piece of glass on top of the blood plate and slide both into the metal clip of the blood chamber.
   g. Push the blood chamber into the slot on the side of the hemoglobinometer, making sure that it is in all the way (fig. 33.2).

Figure 33.2 Steps in the hemoglobin content procedure: (*a*) load the blood chamber with blood; (*b*) stir the blood with a hemolysis applicator; (*c*) place the blood chamber in the slot of the hemoglobinometer; (*d*) match the colors in the green area by moving the slide on the side of the instrument.

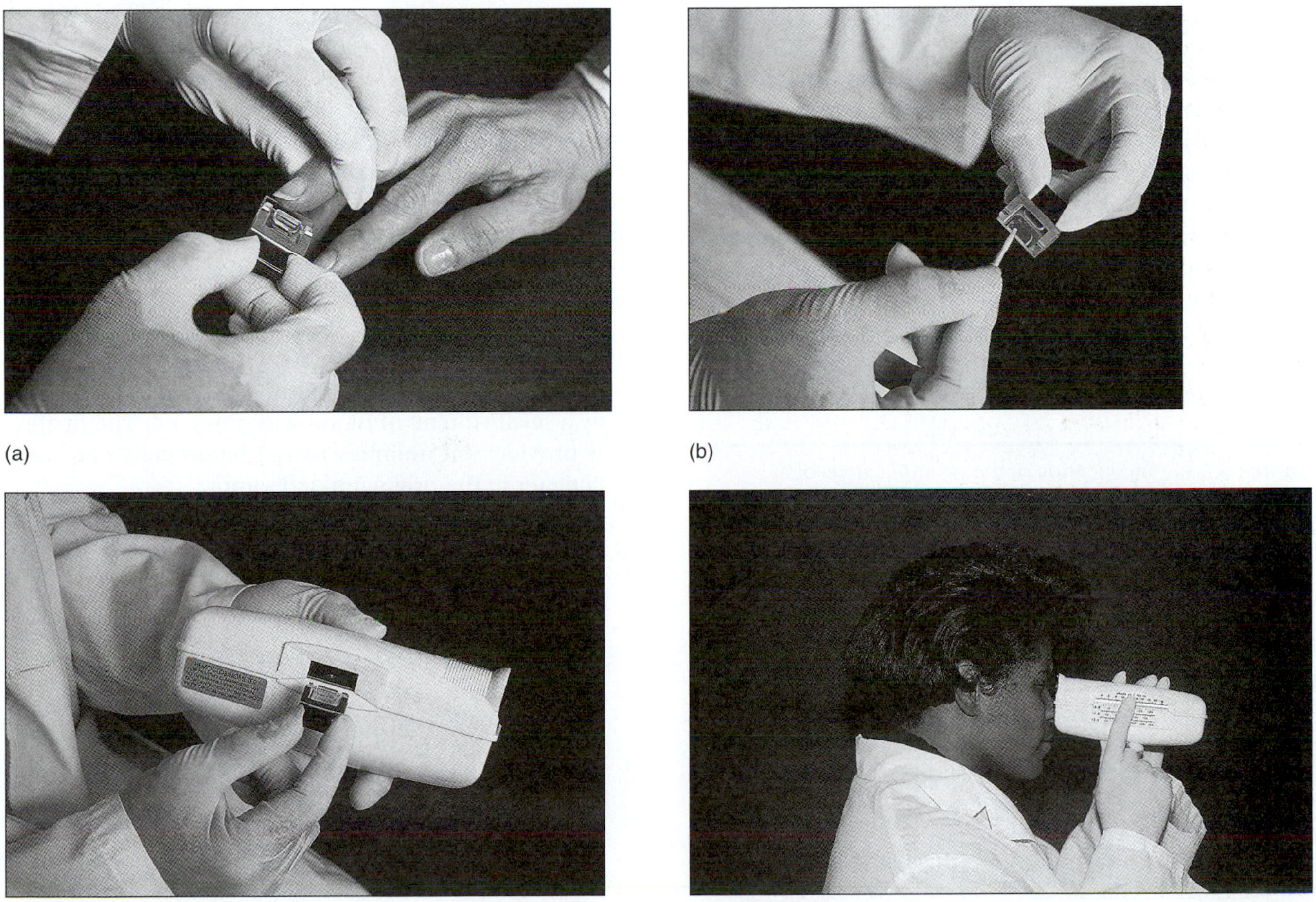

(a) (b) (c) (d)

h. Hold the hemoglobinometer in the left hand with the thumb on the light switch on the underside.
i. Look into the eyepiece and note the green area that is split in half.
j. Slowly move the slide on the side of the instrument back and forth with the right hand until the two halves of the green area look the same.
k. Note the value in the upper scale (grams of hemoglobin per 100 mL of blood), indicated by the mark in the center of the movable slide.
l. Record the test result in Part A of the laboratory report.

2. Complete Part C of the laboratory report.

DEMONSTRATION C—RED BLOOD CELL COUNT

Although modern clinical laboratories commonly use electronic instruments to obtain blood cell counts, a hemocytometer provides a convenient and inexpensive way to count both red and white blood cells. This instrument is a special microscope slide on which there are two counting areas (fig. 33.3). Each counting area contains a grid of tiny lines forming nine large squares that are further subdivided into smaller squares (fig. 33.4).

The squares on the counting area have known dimensions. When a coverslip is placed over the area, it is held a known distance above the grid by glass ridges of the hemocytometer. Thus, when a liquid is placed under the coverslip, the volume of liquid covering each part of the grid can be calculated. Also, if the number of blood cells in a tiny volume of blood can be counted, it is possible to calculate the number that must be present in any larger volume.

When red blood cells are counted using a hemocytometer, a small sample of blood is drawn into a special pipette, and the blood is diluted to reduce the number of cells that must be counted. Some of the diluted blood is spread over the counting area, and with the aid of a microscope, all of the cells in the grid areas marked *R* are counted (fig. 33.4). The total count is multiplied by 10,000 to correct for the dilution and for the fact that only a small volume of blood was observed. The final result provides the number of red blood cells per cubic millimeter in the original blood sample.

1. To perform a red blood cell count, follow these steps:
   a. Examine a hemocytometer, and use a microscope to locate the grid of a counting area. Focus on the large central square of the grid with low-power and then with high-power magnification. Adjust the light intensity so that the lines are clear and sharp.
   b. Wash the hemocytometer with soap and water, and dry it. Place the coverslip over the counting areas and set the instrument aside.
   c. Examine the Unopette system for counting red blood cells, which consists of a reservoir containing blood cell diluting fluid and a plastic capillary tube assembly within a plastic shield (fig. 33.5).

Figure 33.3 The location of the counting areas of a hemocytometer.

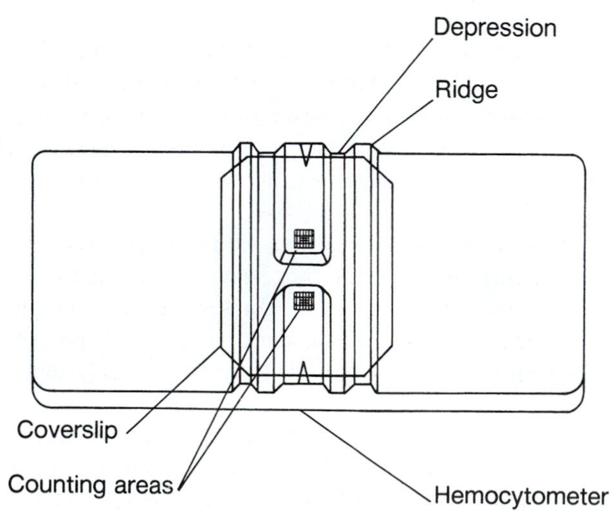

Figure 33.4 The pattern of lines in a counting area of a hemocytometer. The squares marked *R* are used to count red blood cells, whereas the squares marked *W* are used to count white blood cells.

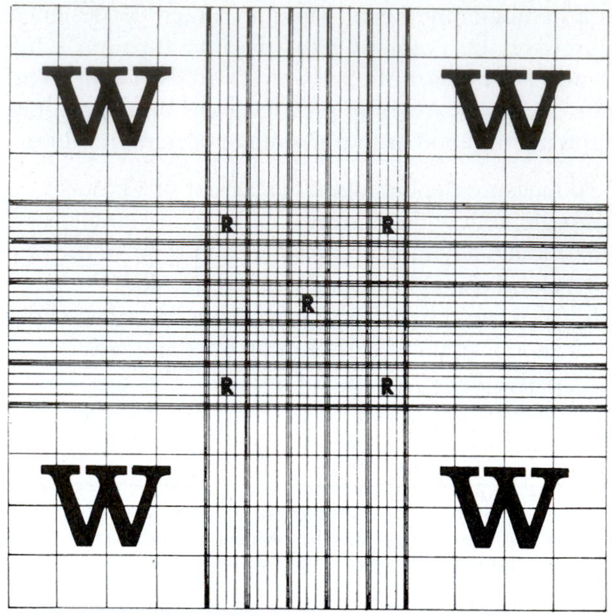

Figure 33.5 The Unopette system consists of a reservoir containing blood cell diluting fluid (*left*) and a plastic capillary tube within a plastic shield (*right*).

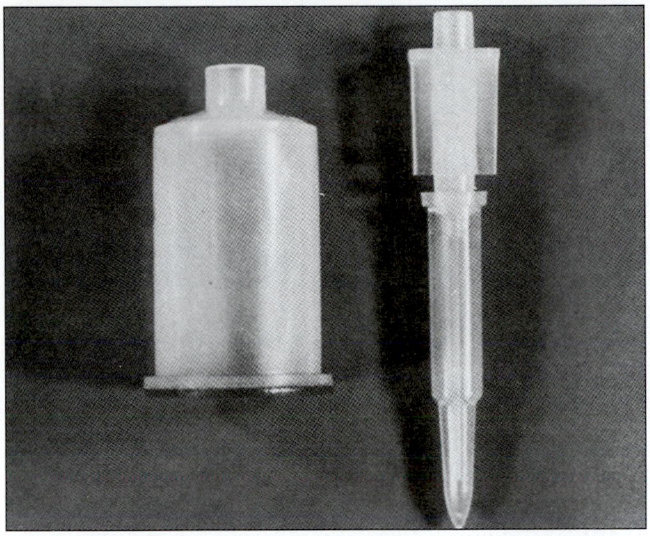

Figure 33.6 The Unopette capillary tube fills with blood by capillary action.

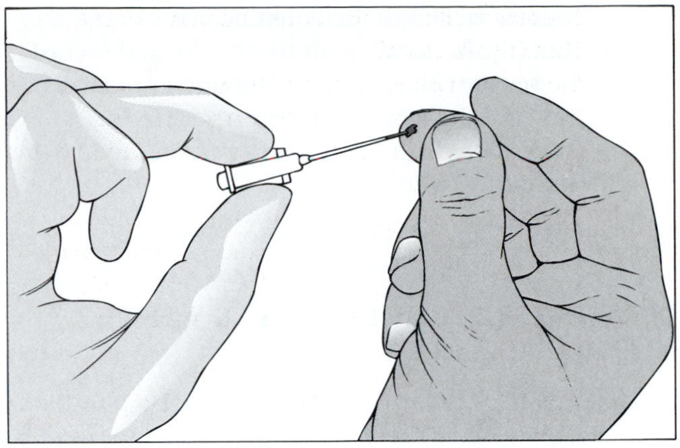

Figure 33.7 The Unopette dropper system being used to fill a hemocytometer with blood. By squeezing the reservoir, a drop of diluted blood enters the counting areas under the coverslip by capillary action.

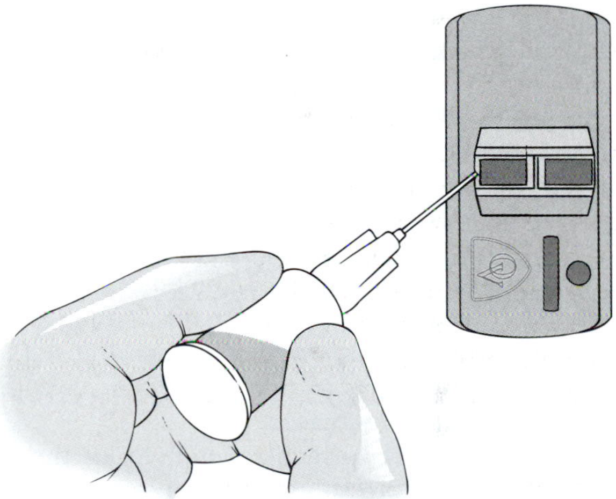

d. Place the Unopette reservoir on the table and gently force the pointed tip of the capillary tube shield through the thin diaphragm at the top of the reservoir.

e. Lance the tip of a finger, as before, to obtain a drop of blood.

f. Remove the Unopette capillary tube from its shield. Holding the tube horizontally, touch the tip of the tube to the drop of blood. Allow the tube to fill completely with blood by capillary action (fig. 33.6).

g. Gently squeeze the reservoir, taking care not to expel any of its fluid content, and while maintaining pressure on the sides of the reservoir, insert the capillary tube through the punctured diaphragm of the reservoir. Release the pressure on the sides of the reservoir and allow the blood to be drawn into its chamber.

h. Hold the capillary tube assembly in the reservoir and gently squeeze the sides of the reservoir several times to mix the blood with the diluting fluid inside. Also, invert the reservoir a few times to aid this mixing process.

i. Remove the capillary tube assembly from the reservoir and insert the opposite end of the assembly into the top of the reservoir, thus converting the parts into a dropper system (fig. 33.7).

j. Gently squeeze the sides of the reservoir to expel some of the diluted blood from the capillary tube. Discard the first four drops of this mixture, and place the next drop of diluted blood at the edge of the coverslip near a counting area of the hemocytometer (fig. 33.7). If the hemocytometer is properly charged with diluted blood, the space between the counting area and the coverslip will be filled and will lack air bubbles, but no fluid will spill over into the depression on either side.

k. Place the hemocytometer on the microscope stage and focus on the large, central square of the counting area with the low-power objective and then the high-power objective. Adjust the light so that the grid lines and blood cells are clearly visible.

l. Count all the cells in the five areas corresponding to those marked with *R* in figure 33.4. To obtain an accurate count, include cells that are touching the lines at the tops and left sides of the squares, but do not count those touching the bottoms and right sides of the squares. The use of a hand counter facilitates the counting task.

m. Multiply the total count by 10,000, and record the result (cells per cubic millimeter of blood) in Part A of the laboratory report.

n. Discard or clean the materials as directed by the laboratory instructor.

2. Complete Part D of the laboratory report.

DEMONSTRATION D—TOTAL WHITE BLOOD CELL COUNT

A hemocytometer is used to make a total white blood cell count in much the same way that it was used to count red blood cells. However, in the case of white cell counting, a diluting fluid is used that destroys red blood cells. Also, in making the white cell count, all of the cells in the four large squares marked *W* in figure 33.4 are included. The total count is multiplied by fifty to calculate the total number of white blood cells in a cubic millimeter of the blood sample.

1. To perform a total white blood cell count, follow these steps:
   a. Clean the hemocytometer as before.
   b. Lance the tip of a finger, as before, to obtain a drop of blood.
   c. Repeat steps 1*c*–1*g* of the procedure for counting red blood cells, but use a Unopette system for counting white blood cells.
   d. After mixing the blood with the diluting fluid in the reservoir, discard the first four drops of the mixture and charge the hemocytometer with diluted blood as before.
   e. Use low-power magnification to locate the areas of the grid corresponding to those marked with *W* in figure 33.4.
   f. Count all of the cells in the four large squares (remember that the red blood cells were destroyed by the white blood cell diluting fluid), following the same counting rules as before.
   g. Multiply the total by fifty, and record the result in Part A of the laboratory report.
   h. Discard or clean the materials as directed by the laboratory instructor.

2. Complete Part E of the laboratory report.

ALTERNATIVE PROCEDURE

Laboratories in modern hospitals and clinics use updated hematology analyzers (fig. 33.8) for evaluations of the blood factors that are described in Laboratory Exercise 33. The more traditional procedures performed in these laboratory activities will ensure the understanding of each blood characteristic.

Figure 33.8 Modern hematology analyzer being used in the laboratory of a clinic.

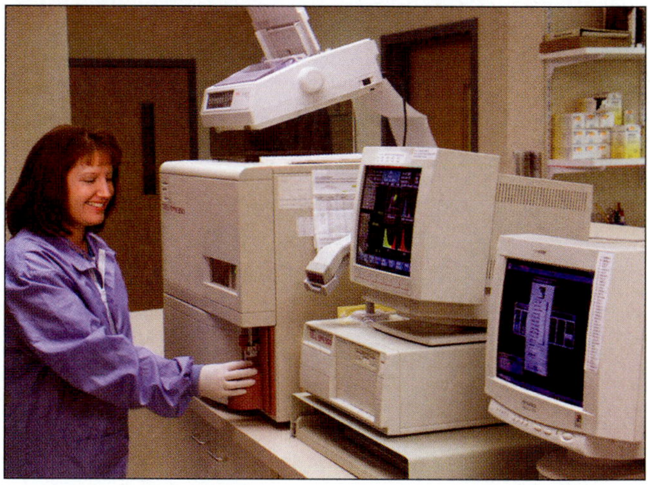

Laboratory Report 33

Name ______________________________

Date ______________________________

Section ______________________________

Blood Testing—A Demonstration

PART A

Blood test data:

Blood Test	Test Results	Normal Values
Hematocrit (mL per 100 mL blood)		Men: 40–54 Women: 37–47
Hemoglobin content (g per 100 mL blood)		Men: 14–18 Women: 12–16
Red blood cell count (cells per mm^3 blood)		Men: 4,600,000–6,200,000 Women: 4,200,000–5,400,000
White blood cell count (cells per mm^3 blood)		5,000–10,000

PART B

Complete the following:

1. How does the hematocrit from the demonstration blood test compare with the normal value?
____
____

2. What conditions might produce a decreased red blood cell percentage? ______________________________
____
____

3. What conditions might produce an increased red blood cell percentage? ______________________________
____
____

PART C

Complete the following:

1. How does the hemoglobin content from the demonstration blood test compare with the normal value? ______________________________

2. What conditions might produce a decreased hemoglobin content? ______________________________

3. What conditions might produce an increased hemoglobin content? ______________________________

PART D

Complete the following:

1. How does the red blood cell count from the demonstration blood test compare with the normal value? ______________________________

2. What conditions might produce a decreased red blood cell count? ______________________________

3. What conditions might produce an increased red blood cell count? ______________________________

PART E

Complete the following:

1. How does the white blood cell count from the demonstration blood test compare with the normal value? ______________________________

2. What conditions might produce a decreased white blood cell count? ______________________________

3. What conditions might produce an increased white blood cell count? ______________________________

Critical Thinking Application

Which blood tests performed in this lab could be used to determine possible anemia?

____

Laboratory Exercise 34

Blood Typing

MATERIALS NEEDED

Textbook
ABO blood typing kit

For Demonstration:

Microscope slide
70% alcohol
Sterile absorbent cotton
Sterile blood lancet
Toothpicks
Anti-D serum
Slide warming box (Rh blood-typing box or Rh view box)

SAFETY

- It is important that students learn and practice correct procedures for handling body fluids. Consider using contaminant-free blood that has been tested and is available from various laboratory supply houses. Some of the procedures might be accomplished as demonstrations only. If student blood is utilized, it is important that students handle only their own blood.
- Use an appropriate disinfectant to wash the laboratory tables before and after the procedures.
- Wear disposable gloves when handling blood samples.
- Clean the end of a finger with 70% alcohol before the puncture is performed.
- The sterile blood lancet should be used only once.
- Dispose of used lancets and blood-contaminated items into an appropriate container (never use the wastebasket).
- Wash your hands before leaving the laboratory.

Blood typing involves identifying protein substances called *antigens* that are present on red blood cell membranes. Although there are many different antigens associated with human red blood cells, only a few of them are of clinical importance. These include the antigens of the ABO group and those of the Rh group.

To determine which antigens are present, a blood sample is mixed with blood-typing sera that contain known types of antibodies. If a particular antibody contacts a corresponding antigen, a reaction occurs, and the red blood cells clump together (agglutination). Thus, if blood cells are mixed with serum containing antibodies that react with antigen A and the cells clump together, antigen A must be present in those cells.

PURPOSE OF THE EXERCISE

To determine the ABO blood type of a blood sample and to observe an Rh blood-typing test.

LEARNING OBJECTIVES

After completing this exercise, you should be able to

1. explain the basis of ABO blood typing;
2. determine the ABO type of a blood sample;
3. explain the basis of Rh blood typing;
4. describe how the Rh type of a blood sample is determined.

WARNING

Because of the possibility of blood infections being transmitted from one student to another if blood testing is performed in the classroom, it is suggested that commercially prepared blood-typing kits containing virus-free human blood be used for ABO blood typing. The instructor may wish to demonstrate Rh blood typing. Observe all of the safety procedures listed for this lab.

ABO BLOOD TYPING

1. Review the section entitled "ABO Blood Group" in chapter 12 of the textbook.
2. Complete Part A of Laboratory Report 34.

Table 34.1 Possible Reactions of ABO Blood-Typing Sera

Reactions		Blood Type
Anti-A serum	**Anti-B serum**	
Clumping	No clumping	Type A
No clumping	Clumping	Type B
Clumping	Clumping	Type AB
No clumping	No clumping	Type O

3. Perform the ABO blood type test using the blood-typing kit. To do this, follow these steps:
   a. Obtain a clean microscope slide and mark across its center with a wax pencil to divide it into right and left halves. Also write "anti-A" near the edge of the left half and "anti-B" near the edge of the right half (fig. 34.1).
   b. Place a small drop of blood on each half of the microscope slide. Work quickly so that the blood will not have time to clot.
   c. Add a drop of anti-A serum to the blood on the left half and a drop of anti-B serum to the blood on the right half. Note the color coding of the anti-A and anti-B typing sera. To prevent contaminating the serum, avoid touching the blood with the serum while it is in the dropper; instead allow the serum to fall from the dropper onto the blood.
   d. Use separate toothpicks to stir each sample of serum and blood together, and spread each over an area about as large as a quarter. Dispose of toothpicks in an appropriate container.
   e. Examine the samples for clumping of blood cells (agglutination) after 2 minutes.
   f. See table 34.1 and figure 34.2 for aid in interpreting the test results.
   g. Discard contaminated materials as instructed by the laboratory instructor.
4. Complete Part B of the laboratory report.

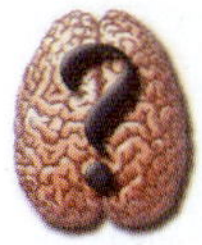

Critical Thinking Application

Judging from the observations of the blood-typing results, predict the components in the anti-A and anti-B sera that caused clumping.

___

___

___

___

DEMONSTRATION—Rh BLOOD TYPING

1. Review the section entitled "Rh Blood Group" in chapter 12 of the textbook.
2. Complete Part C of the laboratory report.
3. To determine the Rh blood type of a blood sample, follow these steps:
   a. Lance the tip of a finger. (See the demonstration procedures in Laboratory Exercise 32 for directions.) Place a small drop of blood in the center of a clean microscope slide.

Figure 34.1 Slide prepared for ABO blood typing.

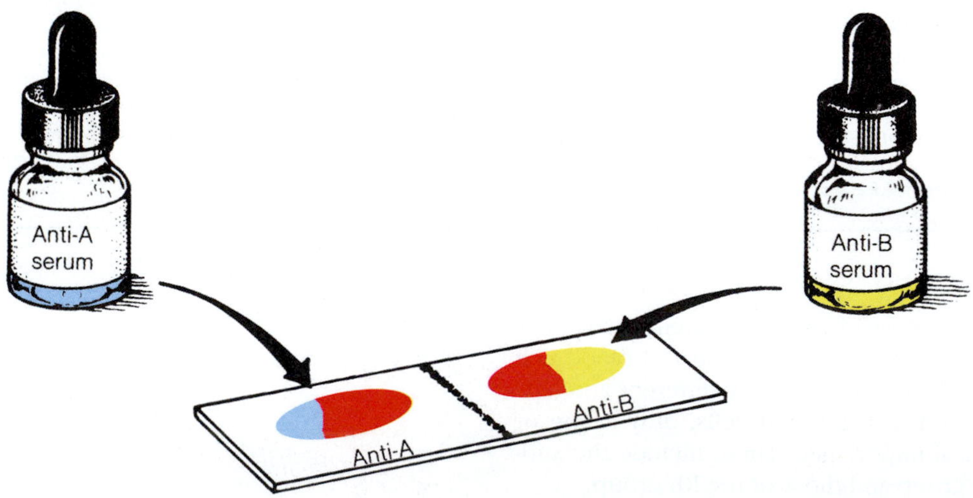

Figure 34.2 Four possible results of the ABO test.

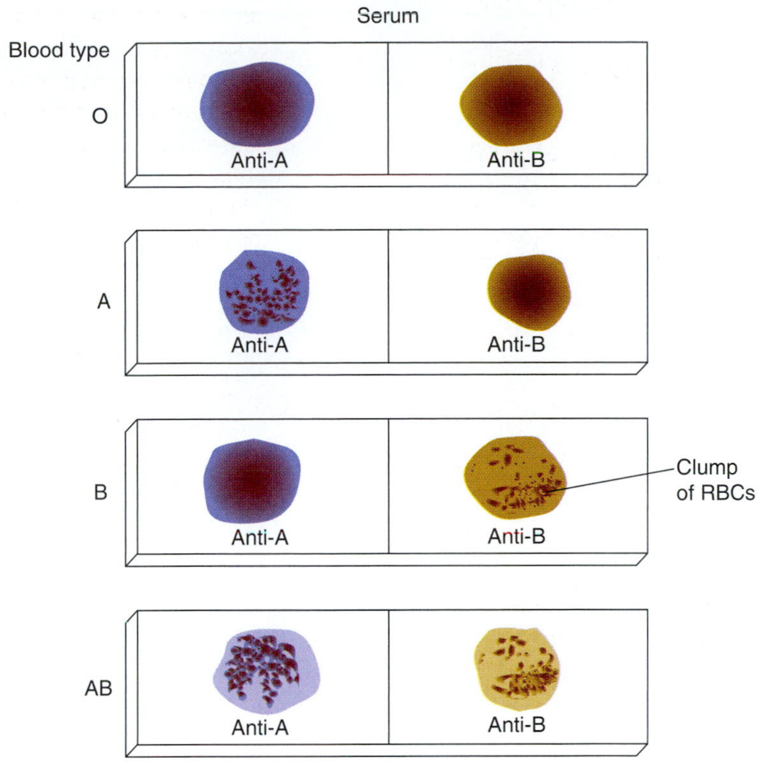

b. Add a drop of anti-D serum to the blood and mix them together with a clean toothpick.
c. Place the slide on the plate of a warming box (Rh blood-typing box or Rh view box) that has been prewarmed to 45°C (113°F) (fig. 34.3).
d. Slowly rock the box back and forth to keep the mixture moving, and watch for clumping of the blood cells. When clumping occurs in anti-D serum, the clumps usually are smaller than those that appear in anti-A or anti-B sera, so they may be less obvious. However, if clumping occurs, the blood is called Rh positive; if no clumping occurs *within 2 minutes,* the blood is called Rh negative.
e. Discard all contaminated materials in appropriate containers.

4. Complete Part D of the laboratory report.

Figure 34.3 Slide warming box used for Rh blood typing.

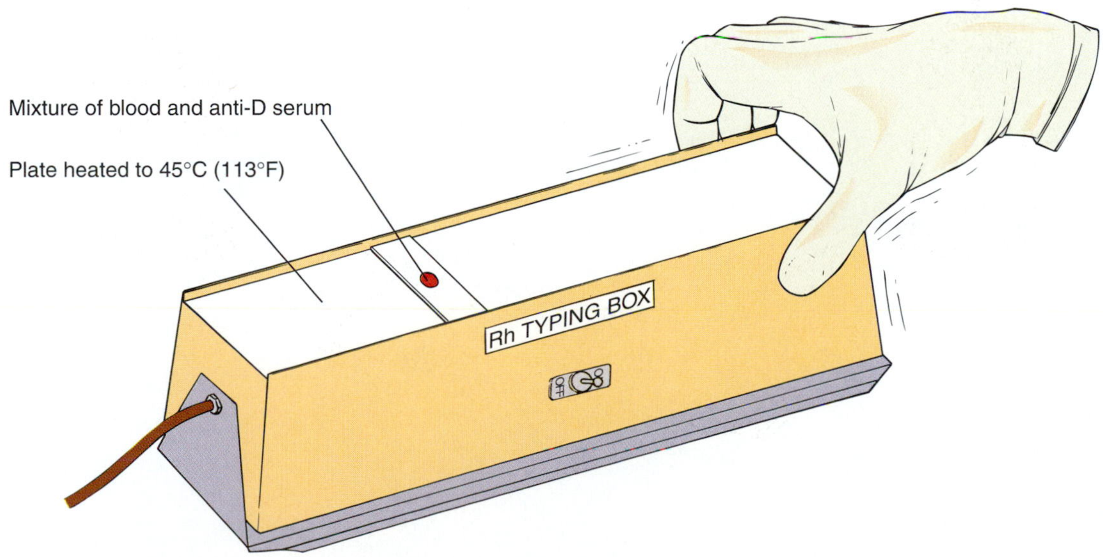

Laboratory Report **34**

Name ______________________

Date ______________________

Section ______________________

Blood Typing

PART A

Complete the following statements:

1. The antigens of the ABO blood group are located in the ______________________ .
2. The blood of every person contains one of (how many possible?) ______________________ combinations of antigens.
3. Type A blood contains antigen ______________________ .
4. Type B blood contains antigen ______________________ .
5. Type A blood contains antibody ______________________ in the plasma.
6. Type B blood contains antibody ______________________ in the plasma.
7. Persons with ABO blood type ______________________ are often called universal recipients.
8. Persons with ABO blood type ______________________ are often called universal donors.
9. Although antigens are present at the time of birth, antibodies usually do not appear in the blood until the age of ______________________ .

PART B

Complete the following:

1. What was the ABO type of the blood tested? ______________________
2. What ABO antigens are present in the red blood cells of this type of blood? ______________________
3. What ABO antibodies are present in the plasma of this type of blood? ______________________
4. If a person with this blood type needed a blood transfusion, what ABO type(s) of blood could be received safely?

   ____
5. If a person with this blood type was serving as a blood donor, what ABO blood type(s) could receive the blood safely? ______________________

PART C

Complete the following statements:

1. The Rh blood group was named after the ______________________________ .
2. Of the antigens in the Rh group, the most important is ______________________________ .
3. If red blood cells lack Rh antigens, the blood is called ______________________________ .
4. Rh antibodies form only in persons with ______________________________ type blood in response to special stimulation.
5. If an Rh-negative person who is sensitive to Rh-positive blood receives a transfusion of Rh-positive blood, the donor's cells are likely to ______________________________ .
6. An Rh-negative woman, who might be carrying an ______________________________ fetus, is given an injection of RhoGAM to prevent erythroblastosis fetalis.

PART D

Complete the following:

1. What was the Rh type of the blood tested in the demonstration? ______________________________
2. What Rh antigen is present in the red blood cells of this type of blood? ______________________________
3. What Rh antibody is normally present in the plasma of this type of blood? ______________________________
4. If a person with this blood type needed a blood transfusion, what type of blood could be received safely?

   ____
5. If a person with this blood type was serving as a blood donor, a person with what type of blood could receive the blood safely? ______________________________

Laboratory Exercise 35

Structure of the Heart

MATERIALS NEEDED

Textbook
Dissectible human heart model
Preserved sheep or other mammalian heart
Dissecting tray
Dissecting instruments

SAFETY

- Wear disposable gloves when working on the heart dissection.
- Save or dispose of the dissected heart as instructed.
- Wash the dissecting tray and instruments as instructed.
- Wash your laboratory table.
- Wash your hands before leaving the laboratory.

The heart is a muscular pump located within the mediastinum and resting upon the diaphragm. It is enclosed by the lungs, thoracic vertebrae, and sternum, and attached at its top (the base) are several large blood vessels. Its distal end extends downward to the left and terminates as a bluntly pointed apex.

The heart and the proximal ends of the attached blood vessels are enclosed by a double-layered pericardium. The inner layer of this membrane consists of a thin covering that is closely applied to the surface of the heart, whereas the outer layer forms a tough, protective sac surrounding the heart.

PURPOSE OF THE EXERCISE

To review the structural characteristics of the human heart and to examine the major features of a mammalian heart.

LEARNING OBJECTIVES

After completing this exercise, you should be able to

1. identify the major structural features of the human heart;
2. compare the features of the human heart with those of another mammal.

PROCEDURE A—THE HUMAN HEART

1. Review the section entitled "Structure of the Heart" in chapter 13 of the textbook.
2. As a review activity, label figures 35.1, 35.2, and 35.3.
3. Complete Part A of Laboratory Report 35.
4. Examine the human heart model, and locate the following features:

heart

- base
- apex

pericardial sac

- fibrous pericardium (outer layer)
- parietal pericardium (inner lining)

pericardial cavity

epicardium (visceral pericardium)

myocardium

endocardium

atria

- right atrium
- left atrium
- auricles

ventricles

- right ventricle
- left ventricle

Figure 35.1 Label this anterior view of the human heart.

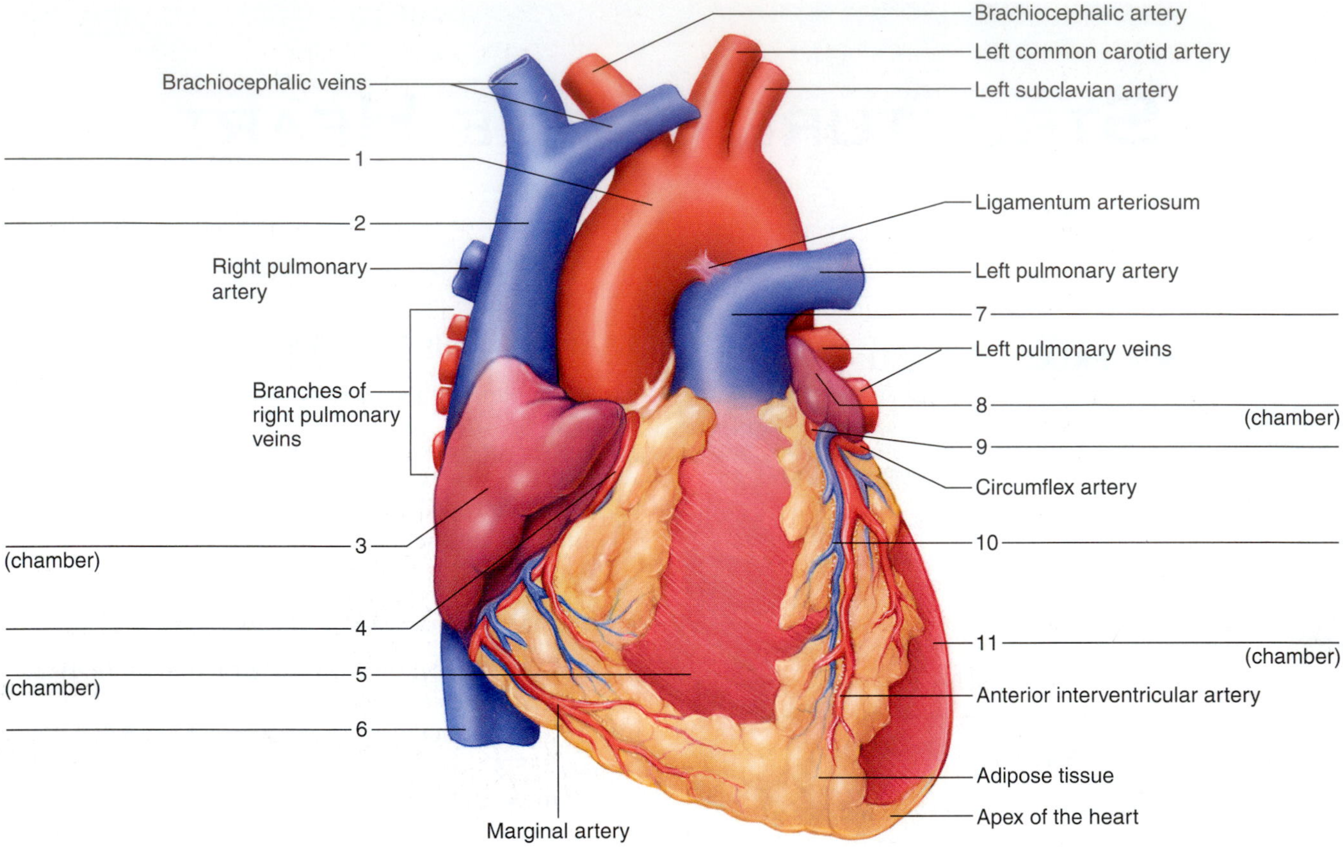

Figure 35.2 Label this posterior view of the human heart.

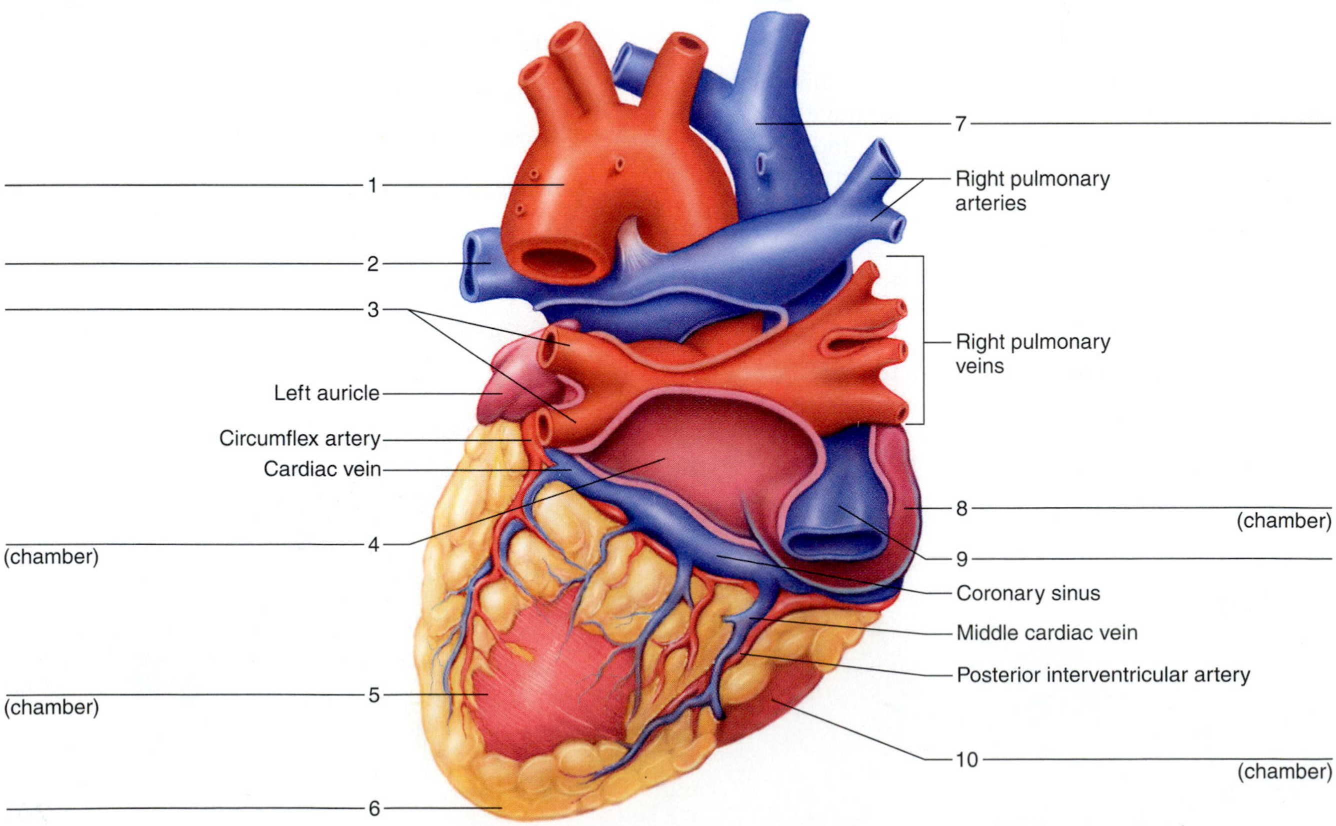

Figure 35.3 Label this coronal section of the human heart.

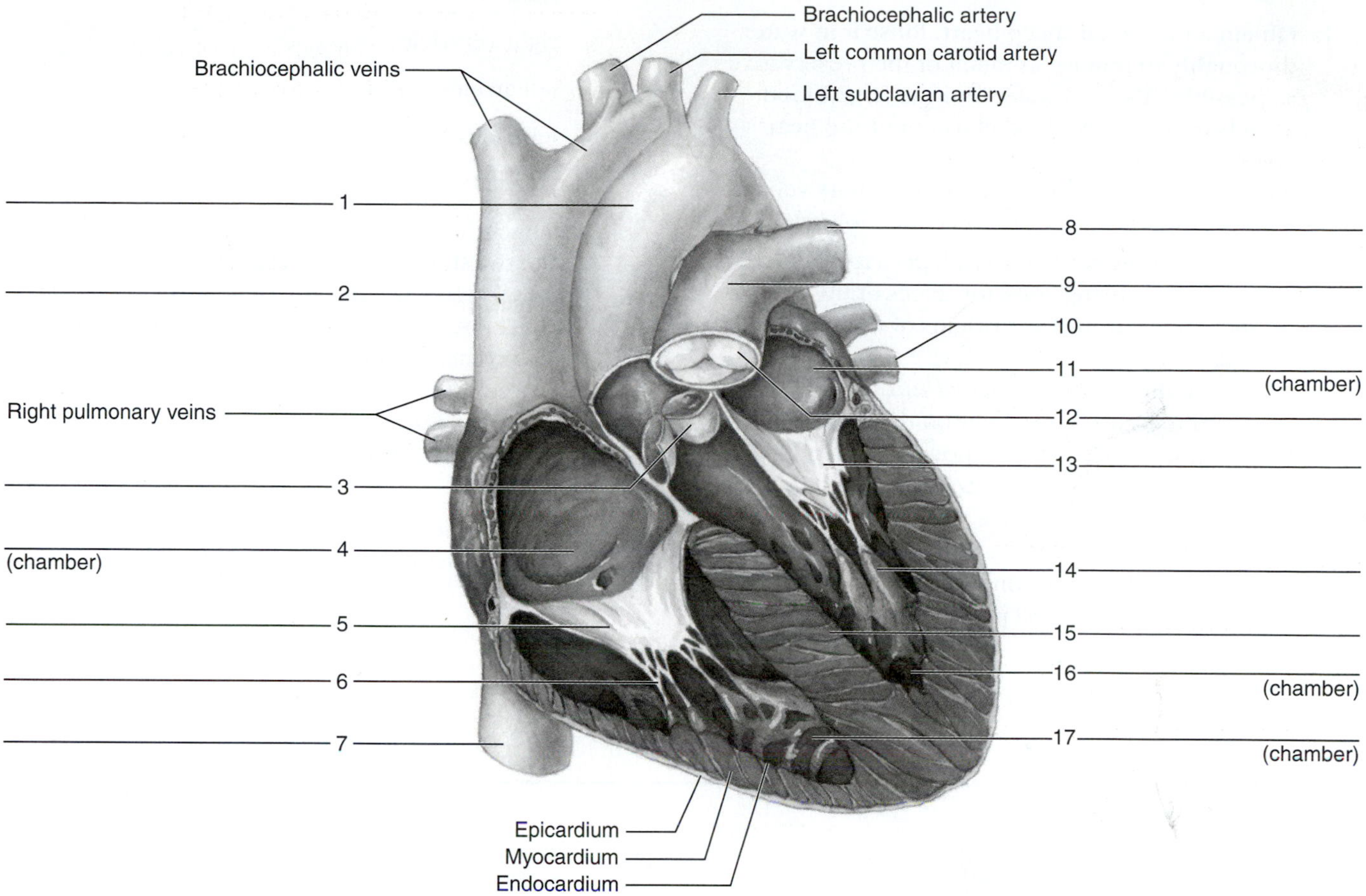

atrioventricular valves (A-V valves)

tricuspid valve

bicuspid (mitral) valve

chordae tendineae

papillary muscles

superior vena cava

inferior vena cava

pulmonary trunk

pulmonary arteries

pulmonary veins

aorta

semilunar valves

pulmonary valve

aortic valve

left coronary artery

right coronary artery

cardiac (coronary) veins

coronary sinus

OPTIONAL ACTIVITY

Use red and blue colored pencils to color the blood vessels in figure 35.3. Use red to illustrate a blood vessel high in oxygen, and use blue to illustrate a blood vessel low in oxygen. You can check your work by referring to the corresponding figures in the textbook, which are presented in full color.

PROCEDURE B—DISSECTION OF A SHEEP HEART

1. Obtain a preserved sheep heart. Rinse it in water thoroughly to remove as much of the preservative as possible. Also run water into the large blood vessels to force any blood clots out of the heart chambers.
2. Place the heart in a dissecting tray with its ventral surface up (fig. 35.4), and proceed as follows:
   a. Although the relatively thick *pericardial sac* is probably missing, look for traces of this membrane around the origins of the large blood vessels.
   b. Locate the *visceral pericardium,* which appears as a thin, transparent layer on the surface of the heart. Use a scalpel to remove a portion of this layer and expose the *myocardium* beneath. Also note the abundance of fat along the paths of various blood vessels. This adipose tissue occurs in the loose connective tissue that underlies the visceral pericardium.
   c. Identify the following:

      right atrium

      right ventricle

      left atrium

      left ventricle

      coronary arteries
3. Examine the dorsal surface of the heart (fig. 35.5). Locate the stumps of two relatively thin-walled blood vessels that enter the right atrium. Demonstrate this connection by passing a slender probe through them. The upper vessel is the *superior vena cava,* and the lower one is the *inferior vena cava.*
4. Open the right atrium. To do this, follow these steps:
   a. Insert a blade of the scissors into the superior vena cava and cut downward through the atrial wall (fig. 35.5).

Figure 35.4 Ventral side of sheep heart.

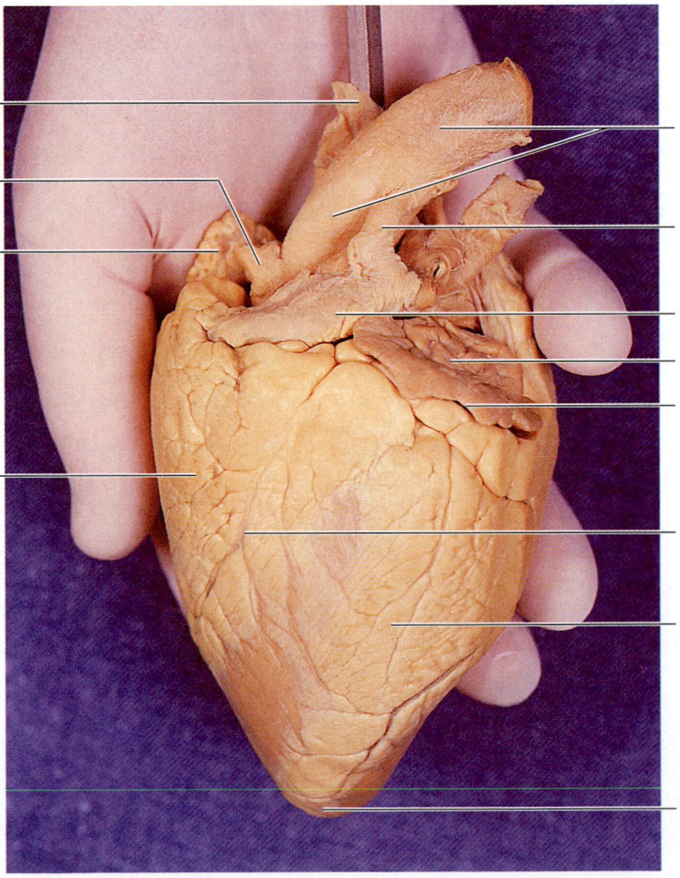

Figure 35.5 Dorsal side of sheep heart. To open the right atrium, insert a blade of the scissors into the superior (anterior in sheep) vena cava and cut downward.

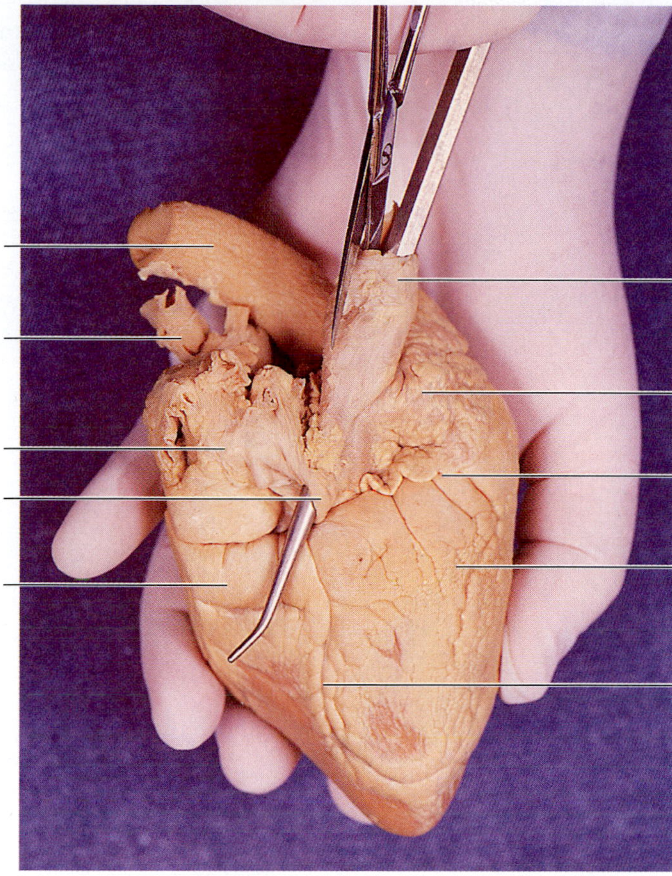

b. Open the chamber, locate the *tricuspid valve,* and examine its cusps.

c. Also locate the opening to the *coronary sinus* between the valve and the inferior vena cava.

d. Run some water through the tricuspid valve to fill the chamber of the right ventricle.

e. Gently squeeze the ventricles and watch the cusps of the valve as the water moves up against them.

5. Open the right ventricle as follows:

   a. Continue cutting downward through the tricuspid valve and the right ventricular wall until you reach the apex of the heart.

   b. Locate the *chordae tendineae* and the *papillary muscles.*

   c. Find the opening to the *pulmonary trunk* and use the scissors to cut upward through the wall of the right ventricle. Follow the pulmonary trunk until you have exposed the *pulmonary valve.*

   d. Examine the valve and its cusps.

6. Open the left side of the heart. To do this, follow these steps:

   a. Insert the blade of the scissors through the wall of the left atrium and cut downward to the apex of the heart.

   b. Open the left atrium and locate the four openings of the *pulmonary veins.* Pass a slender probe through each opening and locate the stump of its vessel.

   c. Examine the *bicuspid valve* and its cusps.

   d. Also examine the left ventricle and compare the thickness of its wall with that of the right ventricle.

7. Locate the aorta, which leads away from the left ventricle, and proceed as follows:

   a. Compare the thickness of the aortic wall with that of the pulmonary trunk.

   b. Use scissors to cut along the length of the aorta to expose the *aortic valve* at its base.

   c. Examine the cusps of the valve and locate the openings of the *coronary arteries* just distal to them.

8. As a review, locate and identify the stumps of each of the major blood vessels associated with the heart.
9. Discard or save the specimen as directed by the laboratory instructor.
10. Complete Part B of the laboratory report.

Web Quest

Trace blood flow through an animated heart at various rates. Identify heart structures and take an animated tour of the heart. Search these at www.mhhe.com/shieress8

Laboratory Report **35**

Name ______________________

Date ______________________

Section ______________________

Structure of the Heart

PART A

Match the terms in column A with the descriptions in column B. Place the letter of your choice in the space provided.

Column A		Column B
a. Aorta	______	1. Upper chamber of heart
b. Atrium	______	2. Structure from which chordae tendineae originate
c. Bicuspid valve	______	3. Prevents blood movement from right ventricle to right atrium
d. Cardiac vein	______	4. Double-layered membrane around heart
e. Coronary artery	______	5. Prevents blood movement from left ventricle to left atrium
f. Coronary sinus	______	6. Gives rise to left and right pulmonary arteries
g. Endocardium	______	7. Drains blood from myocardium into right atrium
h. Myocardium	______	8. Inner lining of heart chamber
i. Papillary muscle	______	9. Layer largely composed of cardiac muscle tissue
j. Pericardial cavity	______	10. Space containing serous fluid
k. Pericardial sac	______	11. Drains blood from myocardial capillaries
l. Pulmonary trunk	______	12. Supplies blood to heart muscle
m. Tricuspid valve	______	13. Distributes blood to body organs (systemic circuit) except lungs

PART B

Complete the following:

1. Compare the structure of the tricuspid valve with that of the pulmonary valve. ______________________

____

____

____

____

2. Describe the action of the tricuspid valve when you squeezed the water-filled right ventricle. ______________

____

____

____

____

3. Describe the function of the chordae tendineae and the papillary muscles. ______________________________

4. What is the significance of the difference in thickness between the wall of the aorta and the wall of the pulmonary trunk? ______________________________

5. List in order the major blood vessels, chambers, and valves through which blood must pass in traveling from a vena cava to the aorta. ______________________________

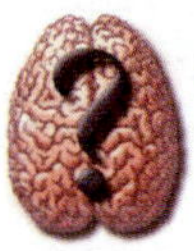

Critical Thinking Application

What is the significance of the difference in thickness of the ventricular walls?

____

Laboratory Exercise 36

The Cardiac Cycle

MATERIALS NEEDED

For Procedure A—Heart Sounds

Textbook
Stethoscope
70% alcohol
Absorbent cotton

For Procedure B—The Electrocardiogram

Electrocardiograph (or other instrument for recording an ECG)
Cot or table
70% alcohol
Absorbent cotton
Electrode cream (paste)
Plate electrodes and cables
Lead selector switch
Virtual Physiology Lab: 4, Effects of Drugs on the Frog Heart
Virtual Physiology Lab: 5, Electrocardiogram

A set of atrial contractions while the ventricular walls relax, followed by ventricular contractions while the atrial walls relax, constitutes a cardiac cycle. Such a cycle is accompanied by blood pressure changes within the heart chambers, opening and closing of heart valves, and movement of blood in and out of the chambers. These valves closing produce vibrations in the tissues and thus create the sounds associated with the heartbeat.

A number of electrical changes also occur in the myocardium as it contracts and relaxes. These changes can be detected by using metal electrodes and an instrument called an *electrocardiograph.* The recording produced by the instrument is an *electrocardiogram,* or *ECG (EKG).*

PURPOSE OF THE EXERCISE

To review the events of a cardiac cycle, to become acquainted with normal heart sounds, and to record an electrocardiogram.

LEARNING OBJECTIVES

After completing this exercise, you should be able to

1. describe the major events of a cardiac cycle;
2. identify the sounds produced during a cardiac cycle;
3. record an electrocardiogram;
4. identify the components of a normal ECG pattern;
5. describe the phases of a cardiac cycle represented by each part of a normal ECG pattern.

PROCEDURE A—HEART SOUNDS

1. Review the sections entitled "Cardiac Cycle" and "Heart Sounds" in chapter 13 of the textbook.
2. Complete Part A of Laboratory Report 36.
3. Listen to your own heart sounds. To do this, follow these steps:
   a. Obtain a stethoscope, and clean its earpieces and the diaphragm by using cotton moistened with alcohol.
   b. Fit the earpieces into your ear canals so that the angles are positioned in the forward direction.
   c. Firmly place the diaphragm (or bell) of the stethoscope on the chest over the apex of the heart (fig. 36.1) and listen to the sounds. This is a good location to hear the first sound (*lubb*) of a cardiac cycle when the A-V valves close.
   d. Move the diaphragm to the second intercostal space just to the left of the sternum, and listen to the sounds from this region. You should be able to hear the second sound (*dupp*) of the cardiac cycle clearly when the semilunar valves close.

Figure 36.1 The first sound of a cardiac cycle can be heard by placing the diaphragm of a stethoscope over the apex of the heart. The second sound can be heard over the second intercostal space, just left of the sternum.

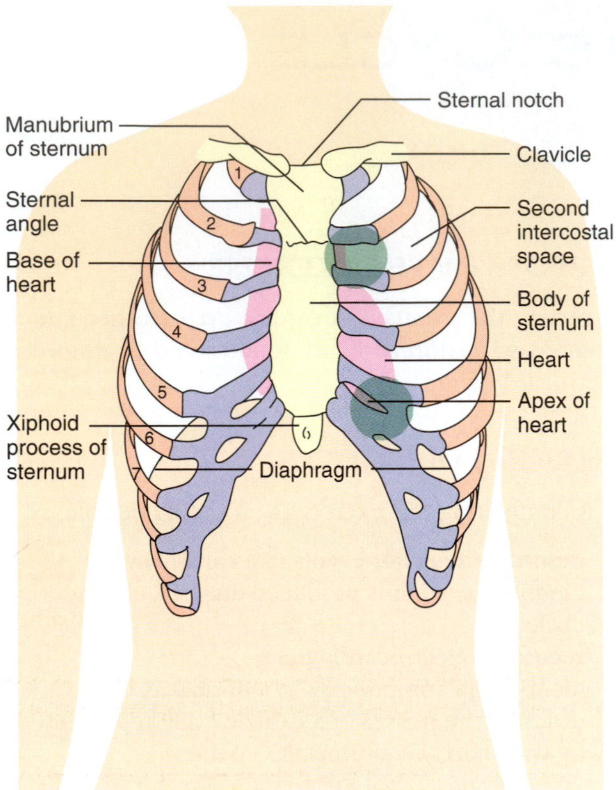

4. Inhale slowly and deeply, and exhale slowly while you listen to the heart sounds from each of the locations as before. Note any changes that have occurred in the sounds.
5. Exercise vigorously outside the laboratory for a few minutes so that other students listening to heart sounds will not be disturbed. After the exercise period, listen to the heart sounds and note any changes that have occurred in them.
6. Complete Part B of the laboratory report.

PROCEDURE B—THE ELECTROCARDIOGRAM

1. Review the sections entitled "Cardiac Conduction System" and "Electrocardiogram" in chapter 13 of the textbook.
2. Complete Part C of the laboratory report.
3. The laboratory instructor will demonstrate the proper adjustment and use of the instrument available to record an electrocardiogram.
4. Record your laboratory partner's ECG. To do this, follow these steps:
   a. Have your partner lie on a cot or table close to the electrocardiograph, remaining as relaxed and still as possible.
   b. Scrub the electrode placement locations with cotton moistened with alcohol (fig. 36.2). Apply a small quantity of electrode cream to the skin on the insides of the wrists and ankles. (Any jewelry on the wrists or ankles should be removed.)
   c. Spread some electrode cream over the inner surfaces of four plate electrodes and attach one to each of the prepared skin areas (fig. 36.2). Make sure there is good contact between the skin and the metal of the electrodes. The electrode plate on the right ankle is the grounding system.
   d. Attach the plate electrodes to the corresponding cables of a lead selector switch. When an ECG recording is made, only two electrodes are used at a time, and the selector switch allows various combinations of electrodes (leads) to be activated. Three standard limb leads placed on the two wrists and the left ankle are used for an ECG. This arrangement has become known as *Einthoven's triangle,*[1] which enables the recording of the potential difference between any two of the electrodes.

      The standard leads I, II, and III are called bipolar leads because they are the potential difference between two electrodes (a positive and a negative). Lead I measures the potential difference between the right wrist (negative) and the left wrist (positive). Lead II measures the potential difference between the right wrist and the left ankle, and lead III measures the potential difference between the left wrist and the left ankle. The right ankle is always the ground.
   e. Turn on the recording instrument and adjust it as previously demonstrated by the laboratory instructor. The paper speed should be set at 2.5 cm/second. This is the standard speed for ECG recordings.
   f. Set the lead selector switch to Lead I (right wrist, left wrist electrodes), and record the ECG for 1 minute.

[1] Willem Einthoven (1860–1927), a Dutch physiologist, received the Nobel prize for physiology or medicine for his work with electrocardiograms.

Figure 36.2 To record an ECG, attach electrodes to the wrists and ankles.

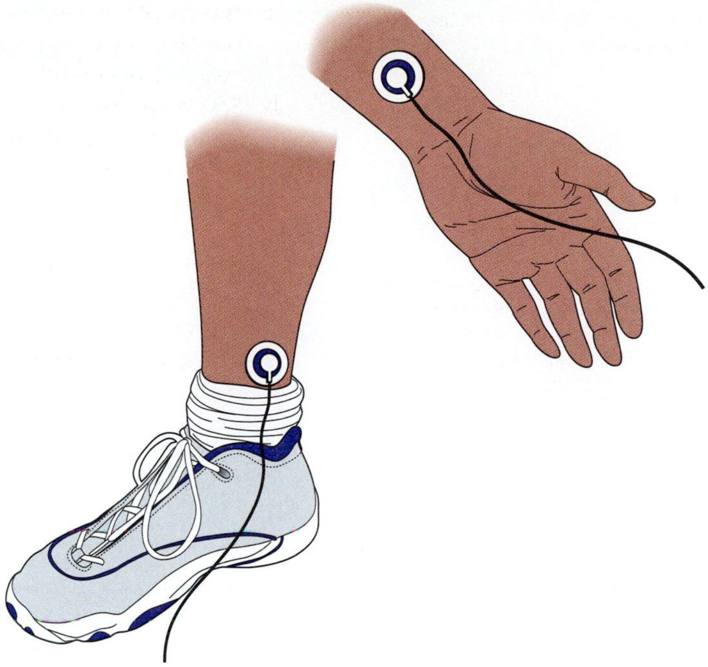

g. Set the lead selector switch to Lead II (right wrist, left ankle electrodes), and record the ECG for 1 minute.
h. Set the lead selector switch to Lead III (left wrist, left ankle electrodes), and record the ECG for 1 minute.
i. Remove the electrodes and clean the cream from the metal and skin.
j. Use figure 36.3 to label the ECG components of the results from leads I, II, and III. The P-Q interval is often called the P-R interval because the Q wave is frequently small or absent. The normal P-Q interval is 0.12–0.20 seconds. The normal QRS complex duration is less than 0.10 seconds.

5. Complete Part D of the laboratory report.

Figure 36.3 Components of a normal ECG pattern with a time scale.

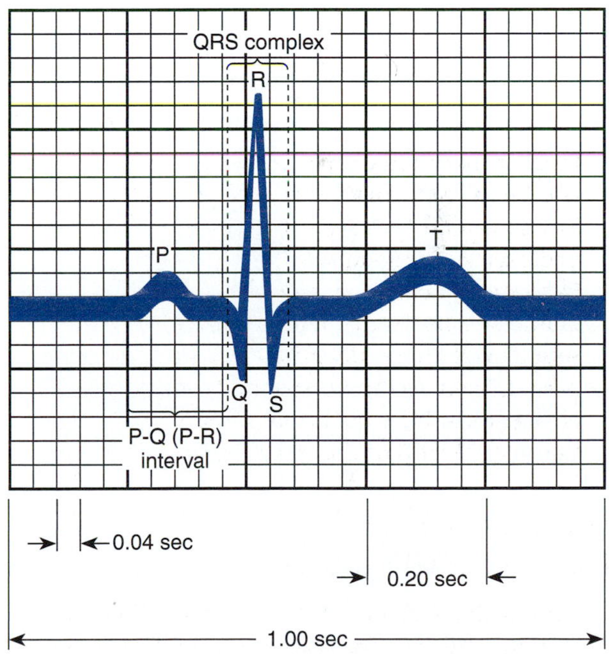

Web Quest

Identify normal heart sounds and a murmur. What is the purpose of stress electrocardiography? Compare normal and abnormal ECGs.

Search these at www.mhhe.com/shieress8

Laboratory Report **36**

Name ______________________

Date ______________________

Section ______________________

THE CARDIAC CYCLE

PART A

Complete the following statements:

1. The period during which a heart chamber is contracting is called ______________________ .
2. The period during which a heart chamber is relaxing is called ______________________ .
3. During ventricular contraction, the A-V valves (tricuspid and bicuspid valves) remain ______________________ .
4. During ventricular relaxation, the A-V valves remain ______________________ .
5. Heart sounds are due to ______________________ in the heart tissues created by changes in blood flow.
6. The first sound of the cardiac cycle occurs when the ______________________ valves are closing.
7. The second sound of the cardiac cycle occurs when the ______________________ valves are closing.

PART B

Complete the following:

1. What changes did you note in the heart sounds when you inhaled deeply? ______________________

____

____

2. What changes did you note in the heart sounds following the exercise period? ______________________

____

____

PART C

Complete the following statements:

1. The fibers of the cardiac conduction system are specialized ______________________ tissue.
2. Normally, the ______________________ node serves as the pacemaker of the heart.
3. The ______________________ node is located in the inferior portion of the interatrial septum.
4. The large fibers on the distal side of the A-V node make up the ______________________.
5. The fibers that carry cardiac impulses from the interventricular septum into the myocardium are called ______________________.
6. An electrocardiogram is a recording of electrical changes occurring in the myocardium during the ______________________.
7. Between cardiac cycles, cardiac muscle fibers remain ______________________ with no detectable electrical changes.
8. The P wave corresponds to the depolarization of the muscle fibers of the ______________________.
9. The QRS complex corresponds to the depolarization of the muscle fibers of the ______________________.
10. The T wave corresponds to the repolarization of the muscle fibers of the ______________________.

PART D

1. Attach a short segment of the ECG recording from each of the three leads you used, and label the waves of each.

Lead I

Lead II

Lead III

2. What differences do you find in the ECG patterns of these leads? ______

3. How much time passed from the beginning of the P wave to the beginning of the QRS complex (P-Q interval, or P-R interval) in the ECG from Lead I? ______

4. What is the significance of this P-Q (P-R) interval? ______

5. How can you determine the heart rate from an electrocardiogram? ______

6. What was your heart rate as determined from the ECG? ______

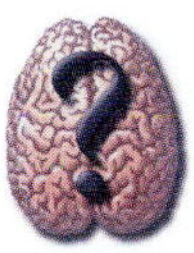

Critical Thinking Application

If a person's heart rate is 72 beats per minute, determine the number of QRS complexes that would have appeared on an ECG during the first 30 seconds.

Blood Vessels

MATERIALS NEEDED

Textbook
Compound microscope
Prepared microscope slides:
- Artery cross section
- Vein cross section

Live frog
Frog Ringer's solution
Paper towel
Rubber bands
Frog board or heavy cardboard (with a 1-inch hole cut in one corner)
Dissecting pins
Thread
Masking tape

For Optional Activity:

Ice
Hot plate
Thermometer

SAFETY

- Wear disposable gloves when handling the live frogs.
- Return the frogs to the location indicated after the experiment.
- Wash your hands before leaving the laboratory.

The blood vessels form a closed system of tubes that carry blood to and from the heart, lungs, and body cells. These tubes include arteries and arterioles that conduct blood away from the heart; capillaries in which exchanges of substances occur between the blood and surrounding tissues; and venules and veins that return blood to the heart.

PURPOSE OF THE EXERCISE

To review the structure and functions of blood vessels and to observe examples of blood vessels microscopically.

LEARNING OBJECTIVES

After completing this exercise, you should be able to

1. describe the structure and functions of arteries, capillaries, and veins;
2. distinguish cross sections of arteries and veins microscopically;
3. identify the three major layers in the wall of an artery or vein;
4. identify the types of blood vessels in the web of a frog's foot.

PROCEDURE

1. Review the section entitled "Blood Vessels" in chapter 13 of the textbook.
2. As a review activity, label figures 37.1 and 37.2.
3. Complete Part A of Laboratory Report 37.
4. Obtain a microscope slide of an artery cross section and examine it using low-power and high-power magnification. Identify the three distinct layers (tunics) of the arterial wall. The inner layer **(tunica interna)** is composed of an endothelium (simple squamous epithelium) and appears as a wavy line due to an abundance of elastic fibers that have recoiled just beneath it. The middle layer **(tunica media)** consists of numerous concentrically arranged smooth muscle cells with elastic fibers scattered among them. The outer layer **(tunica externa)** contains connective tissue that is rich in collagenous fibers (fig. 37.3).
5. Prepare a labeled sketch of the arterial wall in Part B of the laboratory report.

Figure 37.1 Label the tunics of the wall of this artery and vein.

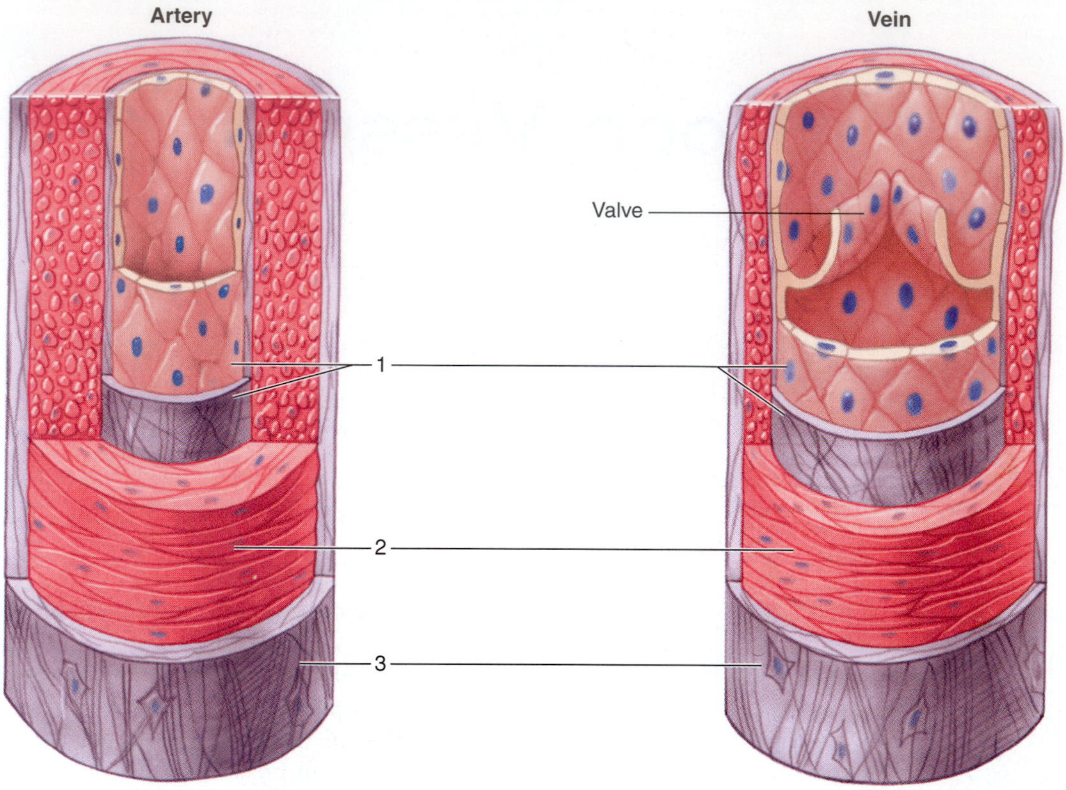

Figure 37.2 Label this arteriole by placing the correct numbers in the spaces provided.

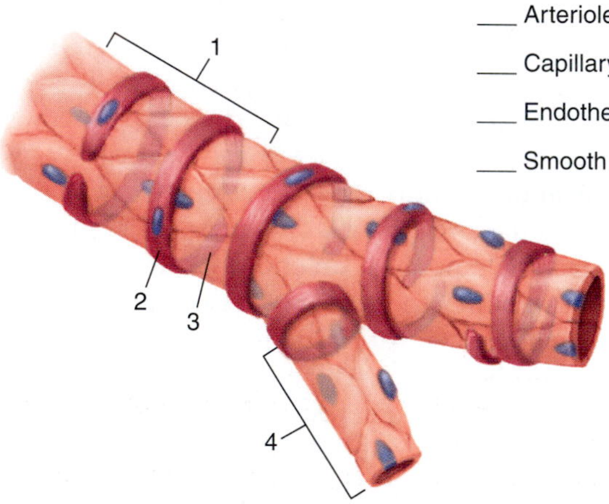

___ Arteriole

___ Capillary

___ Endothelium

___ Smooth muscle cell

6. Obtain a slide of a vein cross section and examine it as you did the artery cross section. Note the thinner wall and larger lumen relative to an artery of comparable size. Identify the three layers of the wall, and prepare a labeled sketch in Part B of the laboratory report.
7. Complete Part B of the laboratory report.
8. Observe the blood vessels in the webbing of a frog's foot. To do this, follow these steps:
    a. Obtain a live frog. Wrap its body in a moist paper towel, leaving one foot extending outward. Secure the towel with rubber bands, but be careful not to wrap the animal so tightly that it could be injured. Try to keep the nostrils exposed.
    b. Place the frog on a frog board or on a piece of heavy cardboard with the foot near the hole in one corner.
    c. Fasten the wrapped body to the board with masking tape.
    d. Carefully spread the web of the foot over the hole and secure it to the board with dissecting pins and thread (fig. 37.4). Keep the web moist with frog Ringer's solution.
    e. Secure the board on the stage of a microscope with heavy rubber bands and position it so that the web is beneath the objective lens.
    f. Focus on the web, using low-power magnification, and locate some blood vessels. Note the movement of the blood cells and the direction of the blood flow. You might notice that red blood cells of frogs are nucleated. Identify an arteriole, a capillary, and a venule.

Figure 37.3 Cross section of an artery and a vein (5×).

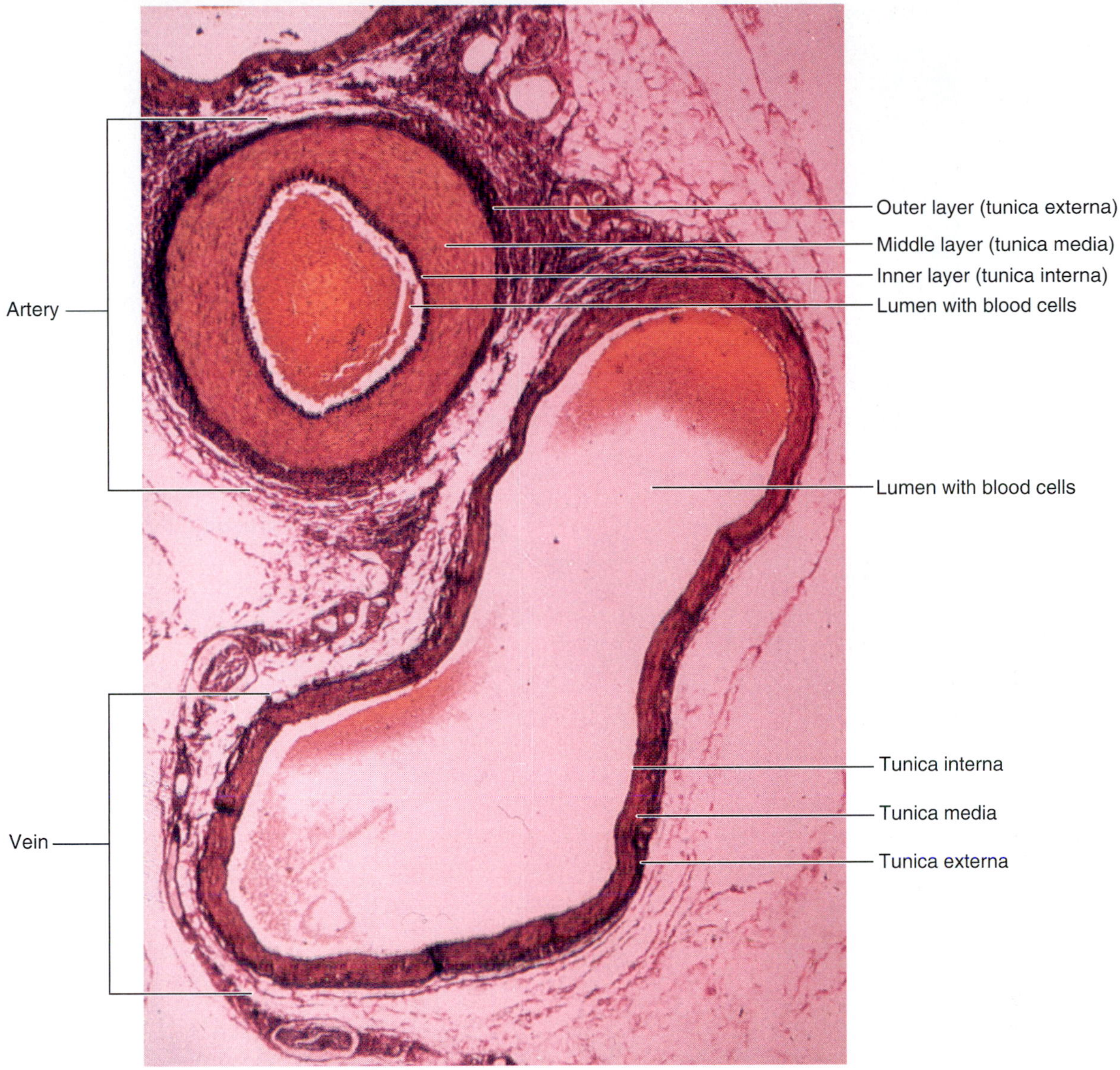

OPTIONAL ACTIVITY

Investigate the effect of temperature change on the blood vessels of the frog's foot by flooding the web with a small quantity of ice water. Observe the blood vessels with low-power magnification and note any changes in their diameters or the rate of blood flow. Remove the ice water and replace it with water heated to about 35°C (95°F). Repeat your observations. What do you conclude from this experiment?

____

____

____

____

g. Examine each of these vessels with high-power magnification.

h. When finished, return the frog to the location indicated by your instructor. The microscope lenses and stage will likely need cleaning after the experiment.

9. Complete Part C of the laboratory report.

Figure 37.4 Spread the web of the foot over the hole and secure it to the board with pins and thread.

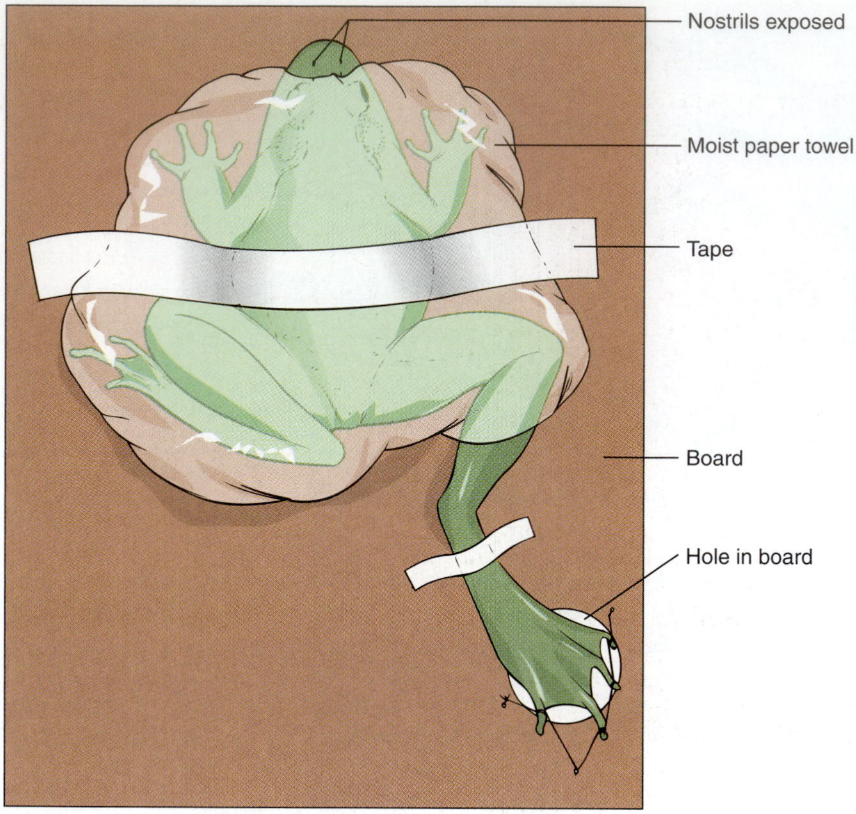

Laboratory Report **37**

Name ______________________________

Date ______________________________

Section ______________________________

Blood Vessels

PART A

Complete the following statements:

1. Simple squamous epithelial tissue called ______________________ forms the inner linings of the tunica interna of blood vessels.
2. The ______________________ of an arterial wall contains many smooth muscle cells.
3. The ______________________ of an arterial wall is largely composed of connective tissue.
4. When contraction of the smooth muscle in a blood vessel wall occurs, the vessel is referred to as being in a condition of ______________________ .
5. Relaxation of the smooth muscle in a blood vessel wall results in the vessel being in a condition of ____________ .
6. The smallest blood vessels are called ______________________ .
7. Tissues with high rates of metabolism tend to have ______________________ densities of capillaries.
8. ______________________ are composed of smooth muscles that encircle the entrances to capillaries and, thus, can control the distribution of the blood within the tissues.
9. The process called ______________________ provides the most important means of transfer of biochemicals through capillary walls.
10. Filtration results when substances are forced through capillary walls by ______________________ pressure.
11. The presence of plasma proteins in the blood increases its ______________________ pressure as compared to tissue fluids.
12. Excess tissue fluid is returned to the venous circulation by means of ______________________ vessels.
13. ______________________ in certain veins close if the blood begins to back up in that vein.
14. ______________________ are vessels that serve as blood reservoirs.

PART B

1. Sketch and label a section of an arterial wall.

2. Sketch and label a section of a venous wall.

3. Describe the differences you noted in the structures of the arterial and venous walls. Mention each of the three layers of the wall.

Critical Thinking Application

Explain the functional significance of the differences you noted in the structures of the arterial and venous walls.

PART C

1. How did you distinguish between arterioles and venules when you observed the vessels in the web of the frog's foot?

2. How did you recognize capillaries in the web?

3. What differences did you note in the rate of blood flow through the arterioles, capillaries, and venules?

PULSE RATE AND BLOOD PRESSURE

MATERIALS NEEDED

Textbook
Clock with second hand
Sphygmomanometer
Stethoscope
70% alcohol
Absorbent cotton

For Demonstration:

Pulse pickup transducer or plethysmogram
Physiological recording apparatus

The surge of blood that enters the arteries each time the ventricles of the heart contract causes the elastic walls of these vessels to swell. Then, as the ventricles relax, the walls recoil. This alternate expanding and recoiling of an arterial wall can be felt as a pulse in vessels that run close to the surface of the body.

The force exerted by the blood pressing against the inner walls of arteries also creates blood pressure. This pressure reaches a maximum during ventricular contraction and then drops to its lowest level while the ventricles are relaxed.

PURPOSE OF THE EXERCISE

To examine the pulse, determine the pulse rate, measure blood pressure, and investigate the effects of body position and exercise on pulse rate and blood pressure.

LEARNING OBJECTIVES

After completing this exercise, you should be able to

1. determine pulse rate;
2. test the effects of various factors on pulse rate;
3. measure blood pressure using a sphygmomanometer;
4. test the effects of various factors on blood pressure.

PROCEDURE A—PULSE RATE

1. Review the section entitled "Blood Pressure" in chapter 13 of the textbook.
2. Complete Part A of Laboratory Report 38.
3. Examine your laboratory partner's radial pulse. To do this, follow these steps:
   a. Have your partner sit quietly, remaining as relaxed as possible.
   b. Locate the pulse by placing your index and middle fingers over the radial artery on the anterior surface of the wrist. Do not use your thumb for sensing the pulse because you may feel a pulse coming from an artery in the thumb itself.
   c. Note the characteristics of the pulse. That is, could it be described as regular or irregular, strong or weak, hard or soft?
   d. To determine the pulse rate, count the number of pulses that occur in 1 minute. This can be accomplished by counting pulses in 30 seconds and multiplying that number by 2.
4. Repeat the procedure and determine the pulse rate in each of the following conditions:
   a. immediately after lying down;
   b. 5 minutes after lying down;
   c. immediately after standing;
   d. 5 minutes after standing quietly;
   e. immediately after 3 minutes of strenuous exercise (omit if the person has health problems);
   f. 5 minutes after exercise has ended.
5. Complete Part B of the laboratory report.

DEMONSTRATION

If the equipment is available, the laboratory instructor will demonstrate how a photoelectric pulse pickup transducer or plethysmogram can be used together with a physiological recording apparatus to record the pulse. Such a recording allows an investigator to analyze certain characteristics of the pulse more precisely than is possible using a finger to examine the pulse. For example, the pulse rate can be determined very accurately from a recording, and the heights of the pulse waves provide information concerning the blood pressure.

PROCEDURE B—BLOOD PRESSURE

1. Measure your laboratory partner's arterial blood pressure. To do this, follow these steps:
   a. Obtain a sphygmomanometer and a stethoscope.
   b. Clean the earpieces and the diaphragm of the stethoscope with cotton moistened with 70% alcohol.
   c. Have your partner sit quietly with bare upper limb resting on a table at heart level. Have the person remain as relaxed as possible.
   d. Locate the brachial artery at the antecubital space. Wrap the cuff of the sphygmomanometer around the arm so that its lower border is about 2.5 cm above the bend of the elbow. Center the bladder of the cuff in line with the *brachial pulse* (fig. 38.1).
   e. Palpate the *radial pulse.* Close the valve on the neck of the rubber bulb connected to the cuff, and pump air from the bulb into the cuff. Inflate the cuff while watching the sphygmomanometer, and note the pressure when the pulse disappears. (This is a rough estimate of the systolic pressure.) Immediately deflate the cuff.
   f. Position the stethoscope over the brachial artery. Reinflate the cuff to a level 30 mm Hg higher than the point where the pulse disappeared during palpation.
   g. Slowly open the valve of the bulb until the pressure in the cuff drops at a rate of about 2 or 3 mm Hg per second.
   h. Listen for sounds (Korotkoff sounds) from the brachial artery. When the first loud tapping sound is heard, record the reading as the systolic pressure. This indicates the pressure exerted against the arterial wall during systole.

Figure 38.1 Blood pressure is commonly measured by using a sphygmomanometer (blood pressure cuff).

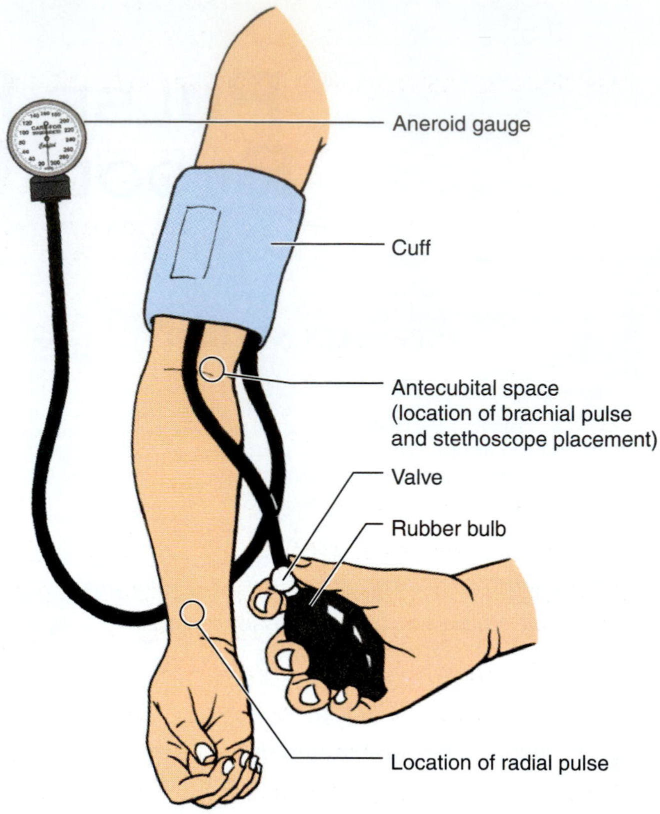

   i. Continue to listen to the sounds as the pressure drops, and note the level when the last sound is heard. Record this reading as the diastolic pressure, which measures the constant arterial resistance.
   j. Release all of the pressure from the cuff.
   k. Repeat the procedure until you have two blood pressure measurements from each arm, allowing 2–3 minutes of rest between readings.
   l. Average your readings and enter them in the table in Part C of the laboratory report.
2. Measure your partner's blood pressure in each of the following conditions:
   a. immediately after lying down;
   b. 5 minutes after lying down;
   c. immediately after standing;
   d. 5 minutes after standing quietly;
   e. immediately after 3 minutes of strenuous exercise (omit if the person has health problems);
   f. 5 minutes after exercise has ended.
3. Complete Part C of the laboratory report.

Laboratory Report **38**

Name ______________________

Date ______________________

Section ______________________

Pulse Rate and Blood Pressure

PART A

Complete the following statements:

1. Blood pressure is the force exerted against the ______________________ .
2. The term *blood pressure* is most commonly used to refer to systemic ______________ pressure.
3. The maximum pressure achieved during ventricular contraction is called ______________ pressure.
4. The lowest pressure that remains in the arterial system during ventricular relaxation is called ________ pressure.
5. Blood pressure rises and falls in a pattern corresponding to the phases of the ______________ .
6. A pulse is caused by the ______________ .
7. The ________ artery in the arm is the standard systemic artery in which blood pressure is measured.

PART B

1. Enter your observations of pulse characteristics and pulse rates in the table.

Test Subject	Pulse Characteristics	Pulse Rate
Sitting		
Lying down		
5 minutes later		
Standing		
5 minutes later		
After exercise		
5 minutes later		

2. Summarize the effects of body position and exercise on the characteristics and rates of the pulse. ______________
____
____
____

PART C

1. Enter the initial measurements of blood pressure in the table.

Reading	Blood Pressure in Right Arm	Blood Pressure in Left Arm
First		
Second		
Average		

2. Enter your test results in the table.

Test Subject	Blood Pressure
Lying down	
5 minutes later	
Standing	
5 minutes later	
After exercise	
5 minutes later	

3. Summarize the effects of body position and exercise on blood pressure. ______________________________

4. Summarize any correlations between pulse rate and blood pressure from any of the experimental conditions.

Critical Thinking Application

When a pulse is palpated and counted, which blood pressure (systolic or diastolic) would be characteristic at that moment? Explain your answer.

Laboratory Exercise 39

Major Arteries and Veins

MATERIALS NEEDED

Textbook
Human torso
Anatomical charts of the cardiovascular system

The blood vessels of the cardiovascular system can be divided into two major pathways—the pulmonary circuit and the systemic circuit. Within each circuit, arteries transport blood away from the heart. After exchanges of gases, nutrients, and wastes have occurred between the blood and the surrounding tissues, veins return the blood to the heart.

PURPOSE OF THE EXERCISE

To review the major circulatory pathways and to locate the major arteries and veins on anatomical charts and in the torso.

LEARNING OBJECTIVES

After completing this exercise, you should be able to

1. distinguish and trace the pulmonary and systemic circuits;
2. locate the major arteries in these circuits on a chart or model;
3. locate the major veins in these circuits on a chart or model.

PROCEDURE A—PATHS OF CIRCULATION

1. Review the sections entitled "Pulmonary Circuit" and "Systemic Circuit" in chapter 13 of the textbook.
2. As a review activity, label figure 39.1.
3. Locate the following blood vessels on the available anatomic charts and the human torso:

   pulmonary trunk 1

   pulmonary arteries 2

   pulmonary veins 4

   aorta 1

   superior vena cava 1

   inferior vena cava 1

Critical Thinking Application

Why is the left ventricle wall thicker than the right ventricle wall?

____

____

____

____

PROCEDURE B—THE ARTERIAL SYSTEM

1. Review the section entitled "Arterial System" in chapter 13 of the textbook.
2. As a review activity, label figures 39.2, 39.3, 39.4, and 39.5.
3. Locate the following arteries of the systemic circuit on the charts and torso:

   aorta

   ascending aorta

   aortic sinus

   aortic arch (arch of the aorta)

   thoracic aorta

   abdominal aorta

Label the major blood vessels associated with the pulmonary and systemic circuits, using the terms provided.

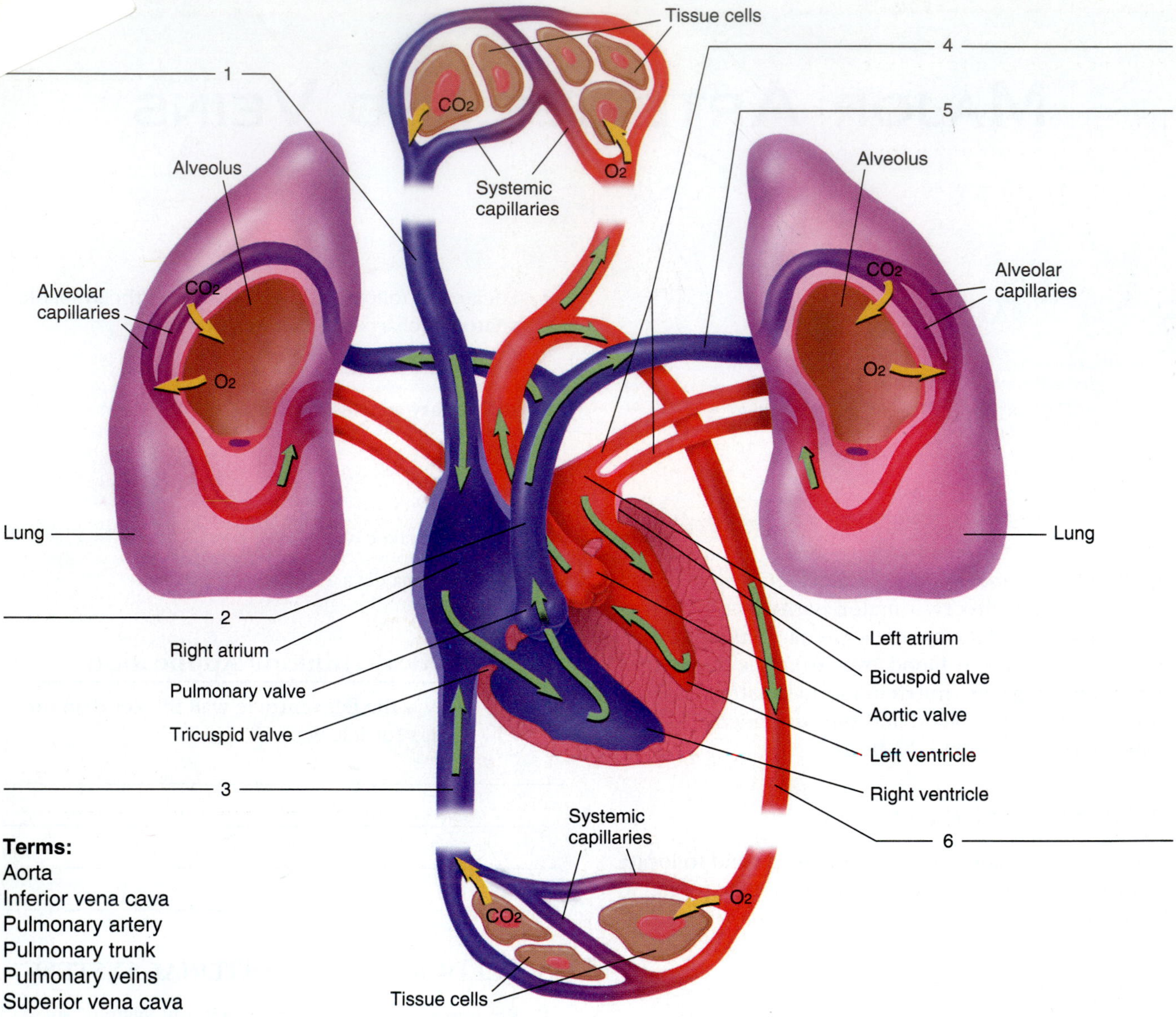

Terms:
Aorta
Inferior vena cava
Pulmonary artery
Pulmonary trunk
Pulmonary veins
Superior vena cava

branches of the aorta

- coronary arteries
- brachiocephalic artery
- left common carotid artery
- left subclavian artery
- celiac artery
- superior mesenteric artery
- suprarenal arteries

renal arteries

gonadal arteries

- ovarian arteries
- testicular arteries

inferior mesenteric artery

middle sacral artery

common iliac artery

Figure 39.2 Label the portions of the aorta and its principal branches.

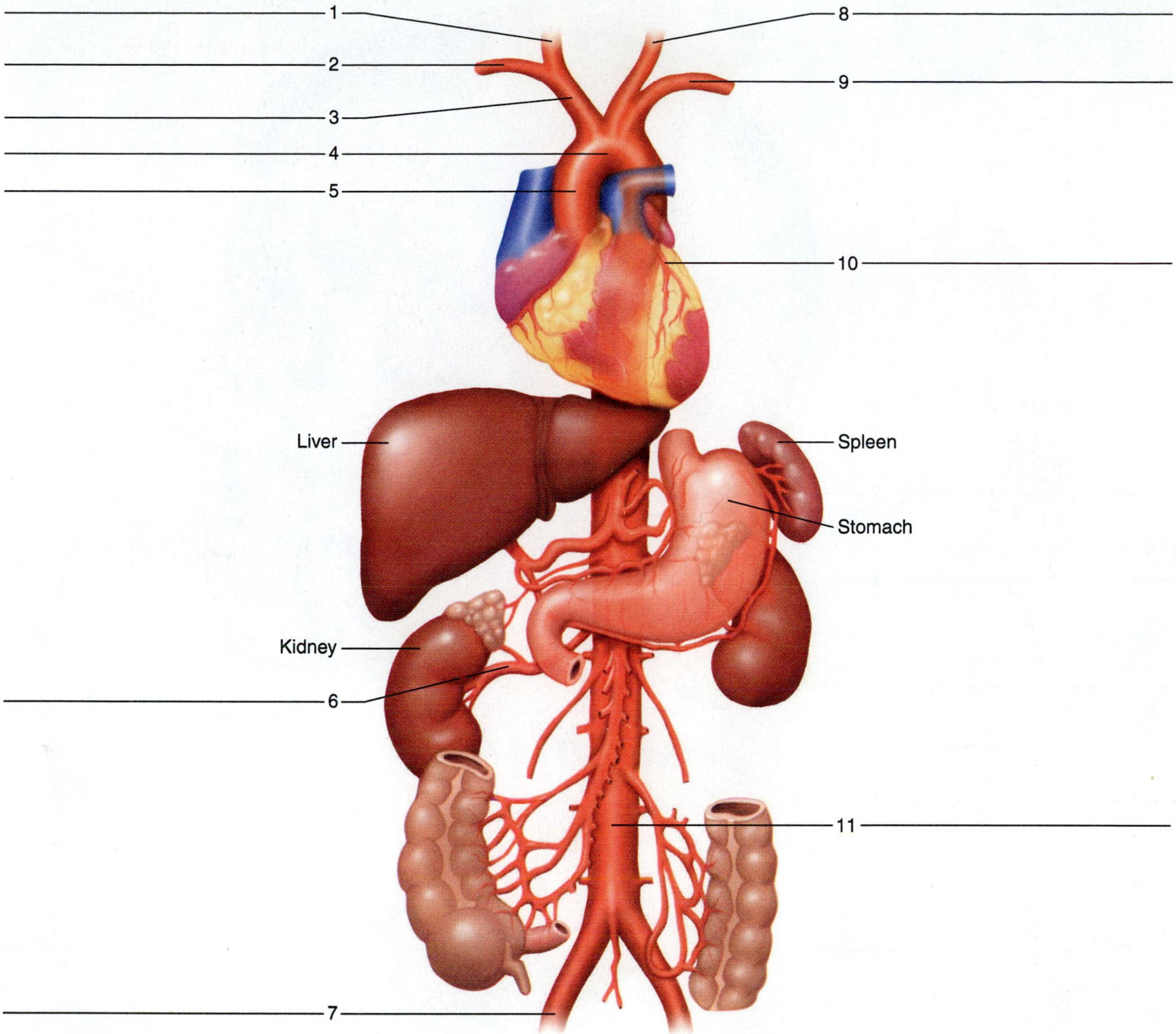

arteries to neck, head, and brain

vertebral arteries

thyrocervical arteries

common carotid arteries

external carotid arteries

internal carotid arteries

superior thyroid artery

facial artery

arteries to shoulder and upper limb

axillary artery

brachial artery

deep brachial artery

ulnar artery

radial artery

Figure 39.3 Label the arteries supplying the right side of the neck and head. (Note that the clavicle has been removed.)

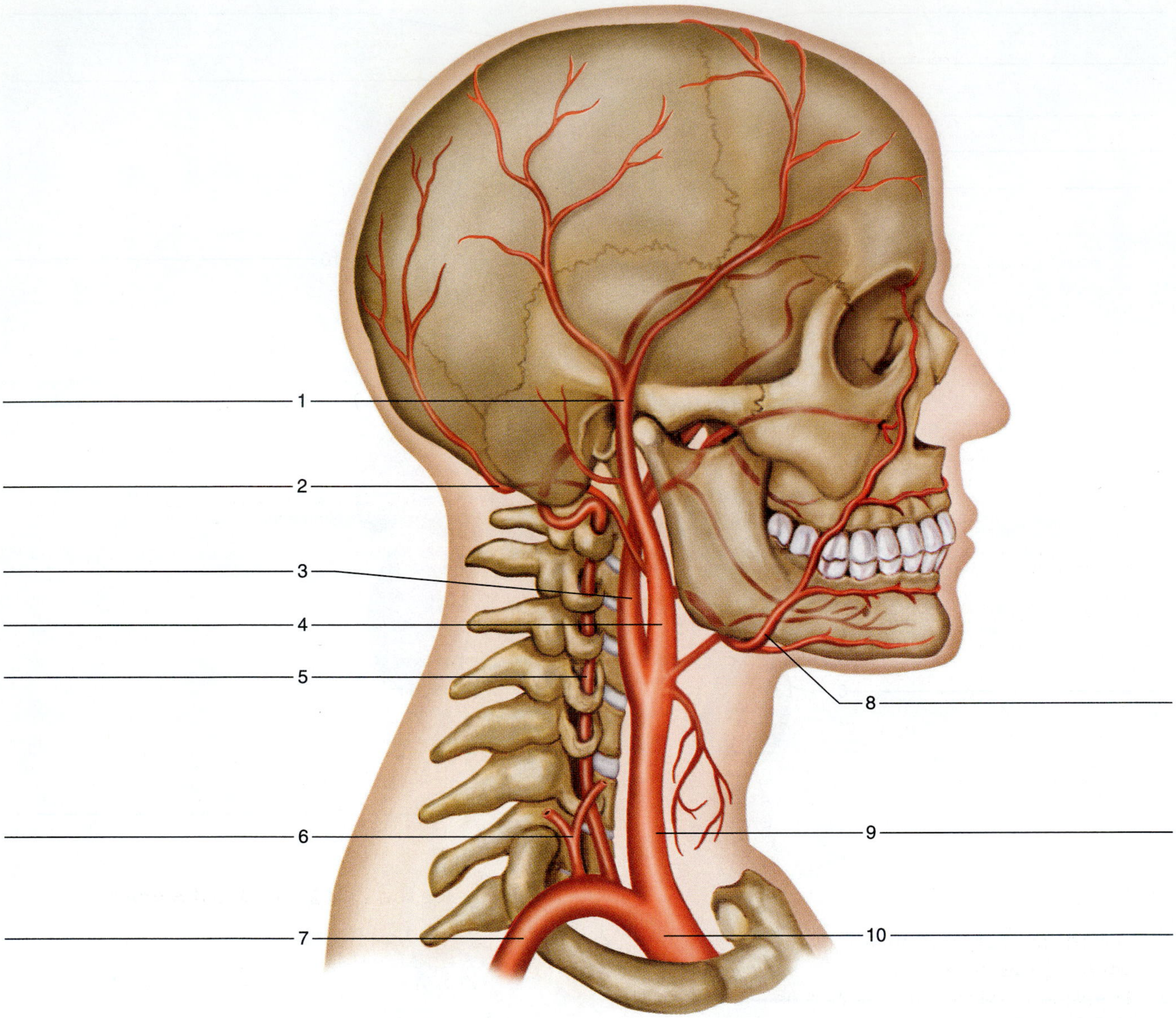

arteries to the thoracic and abdominal walls

internal thoracic artery

anterior intercostal artery

posterior intercostal artery

arteries to pelvis and lower limb

common iliac artery

internal iliac artery

external iliac artery

femoral artery

popliteal artery

anterior tibial artery

dorsalis pedis artery (dorsal pedis artery)

posterior tibial artery

4. Complete Parts A and B of Laboratory Report 39.

Figure 39.4 Label the major arteries of the shoulder and upper limb.

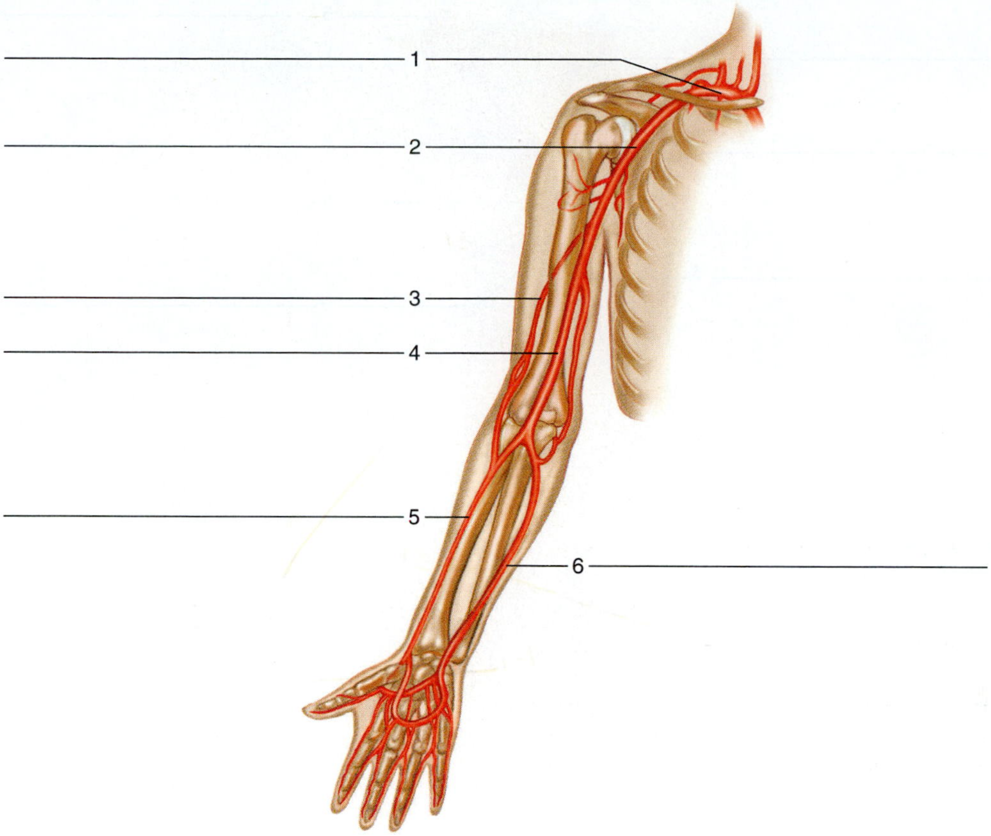

PROCEDURE C—THE VENOUS SYSTEM

1. Review the section entitled "Venous System" in chapter 13 of the textbook.
2. As a review activity, label figures 39.6, 39.7, 39.8, 39.9, and 39.10.
3. Locate the following veins of the systemic circuit on the charts and torso:

veins from brain, head, and neck

external jugular veins

internal jugular veins

brachiocephalic veins

superior vena cava

veins from upper limb and shoulder

radial vein

ulnar vein

brachial vein

basilic vein

cephalic vein

median cubital vein

axillary vein

subclavian vein

Figure 39.5 Label the major arteries supplying the pelvis and lower limb, using the terms provided.

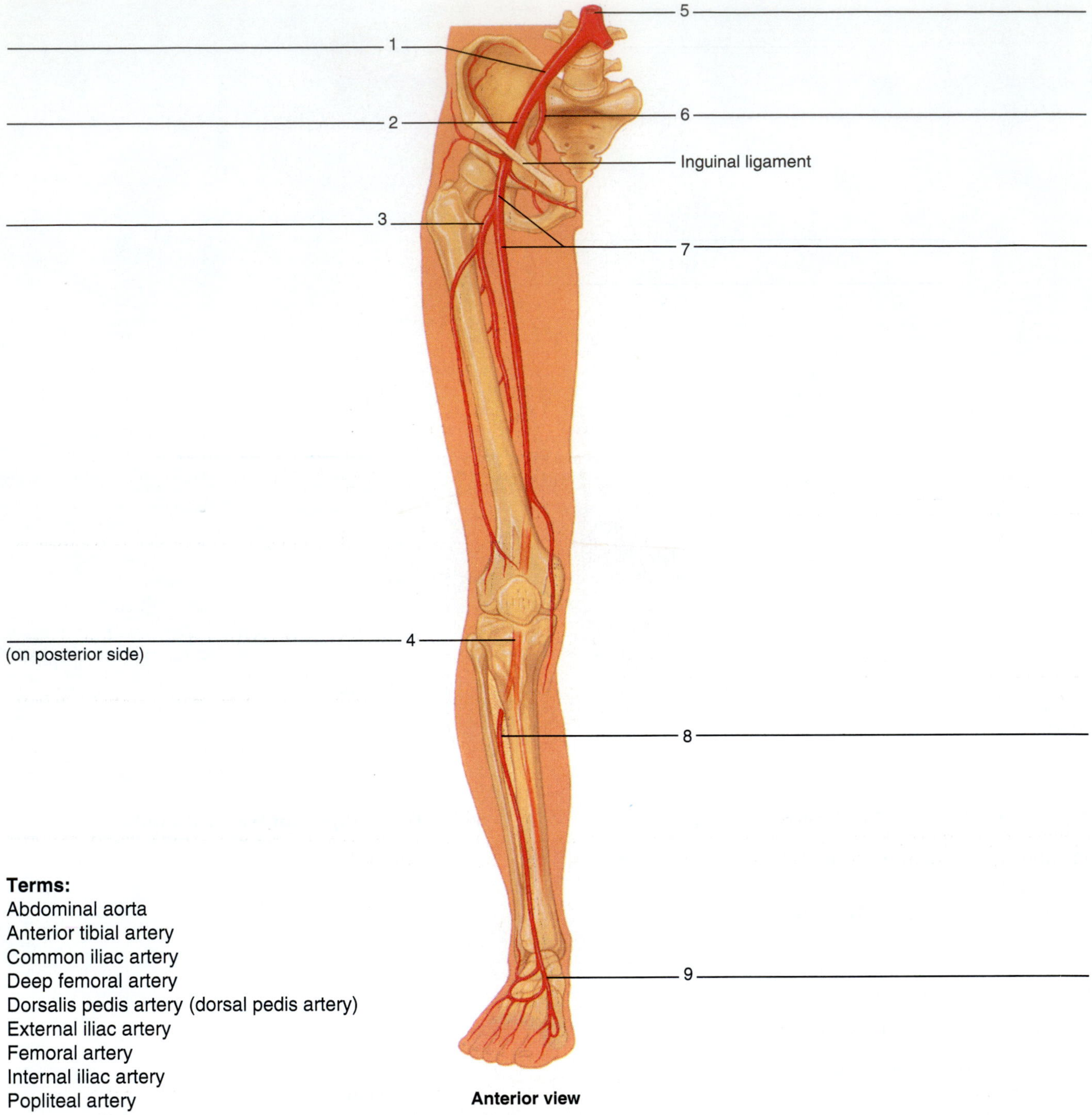

Terms:
Abdominal aorta
Anterior tibial artery
Common iliac artery
Deep femoral artery
Dorsalis pedis artery (dorsal pedis artery)
External iliac artery
Femoral artery
Internal iliac artery
Popliteal artery

Figure 39.6 Label the major veins associated with the head and neck. (Note that the clavicle has been removed.)

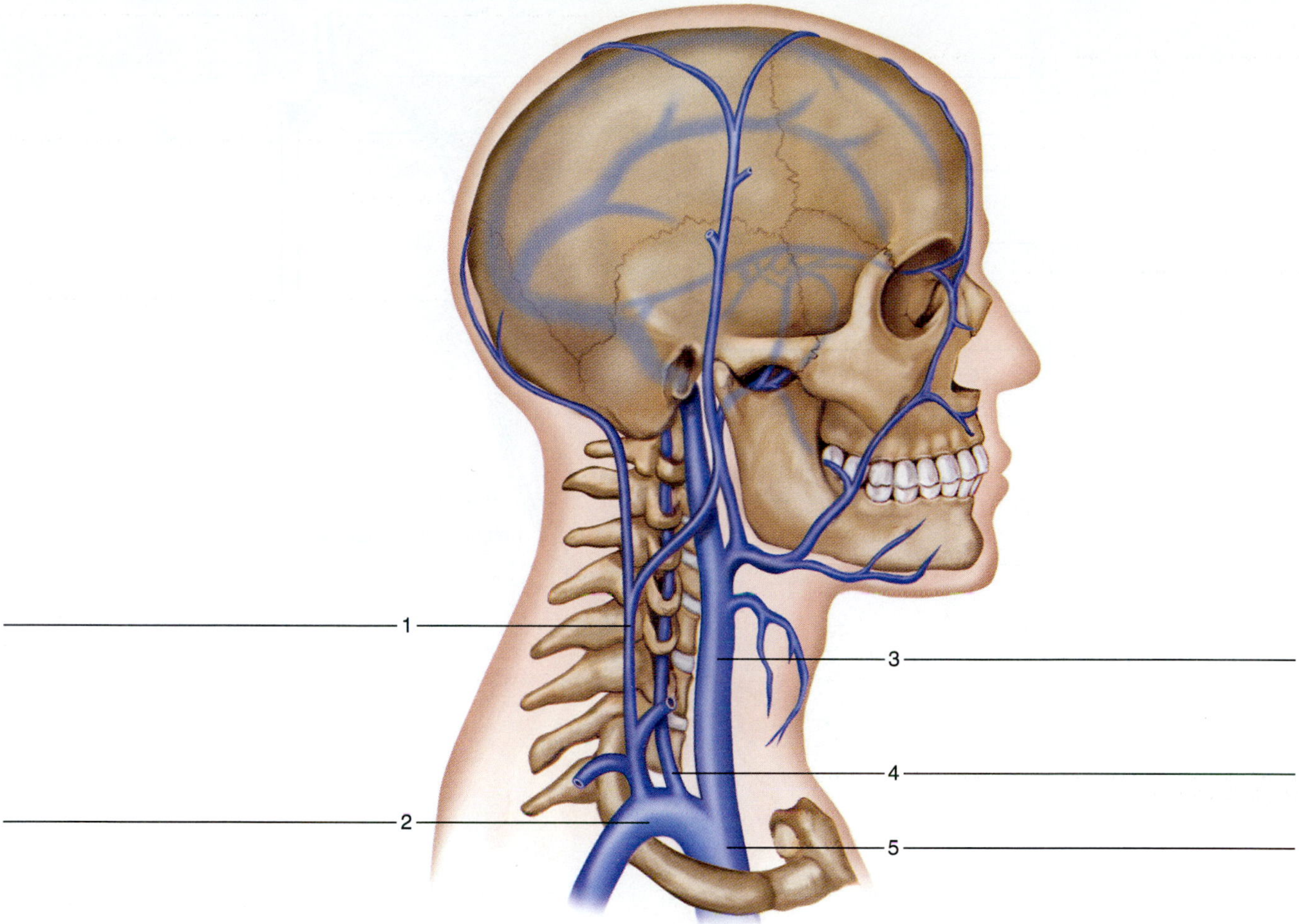

Figure 39.7 Label the veins associated with the thoracic wall, using the terms provided.

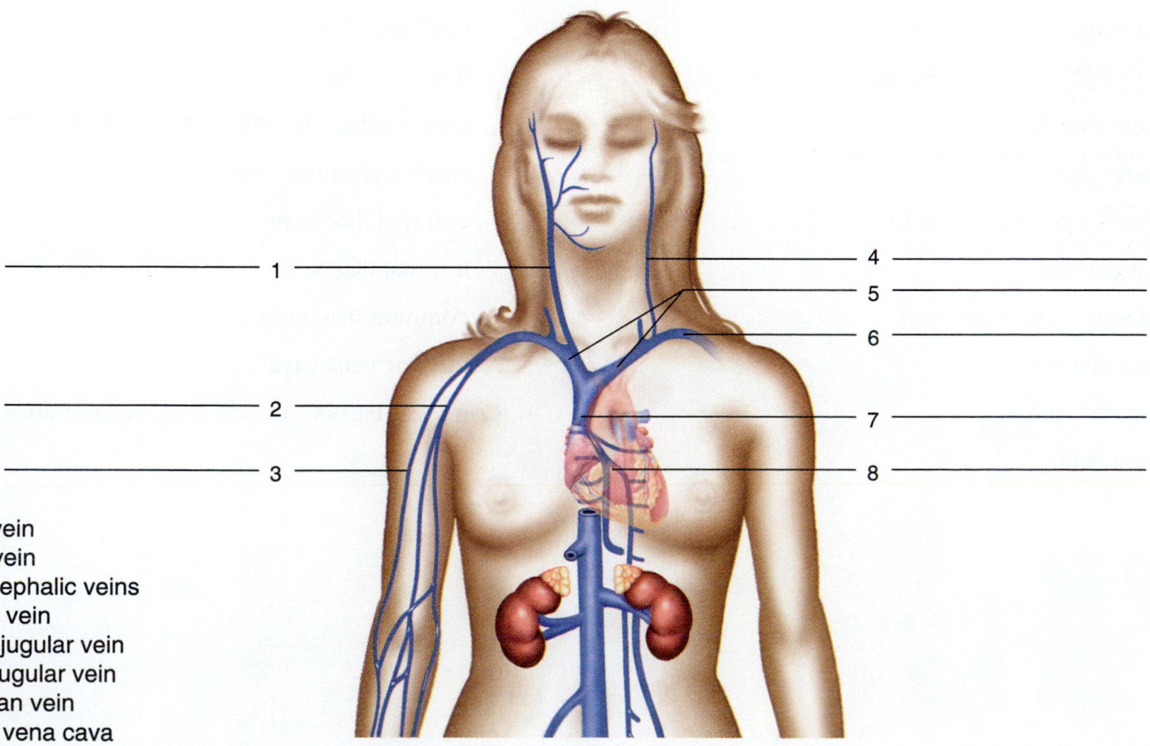

Terms:
Axillary vein
Azygos vein
Brachiocephalic veins
Cephalic vein
External jugular vein
Internal jugular vein
Subclavian vein
Superior vena cava

Figure 39.8 Label the veins of the upper limb and shoulder.

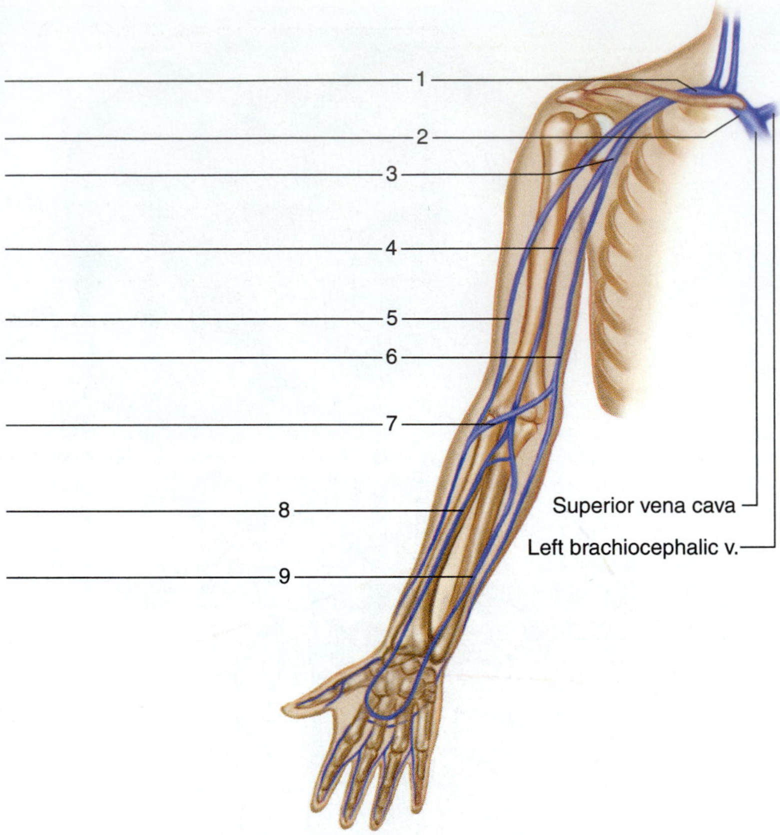

veins of the thoracic wall

- azygos vein
- superior hemiazygos vein
- inferior hemiazygos vein

veins of abdominal viscera

- hepatic portal vein
- gastric vein
- superior mesenteric vein
- splenic vein
- inferior mesenteric vein
- hepatic vein
- gonadal vein
- renal vein

veins from lower limb and pelvis

- anterior tibial vein
- posterior tibial vein
- popliteal vein
- femoral vein
- great saphenous vein
- small saphenous vein
- external iliac vein
- internal iliac vein
- common iliac vein
- inferior vena cava

4. Complete Parts C, D, and E of the laboratory report.

Figure 39.9 Label the veins that drain the abdominal viscera.

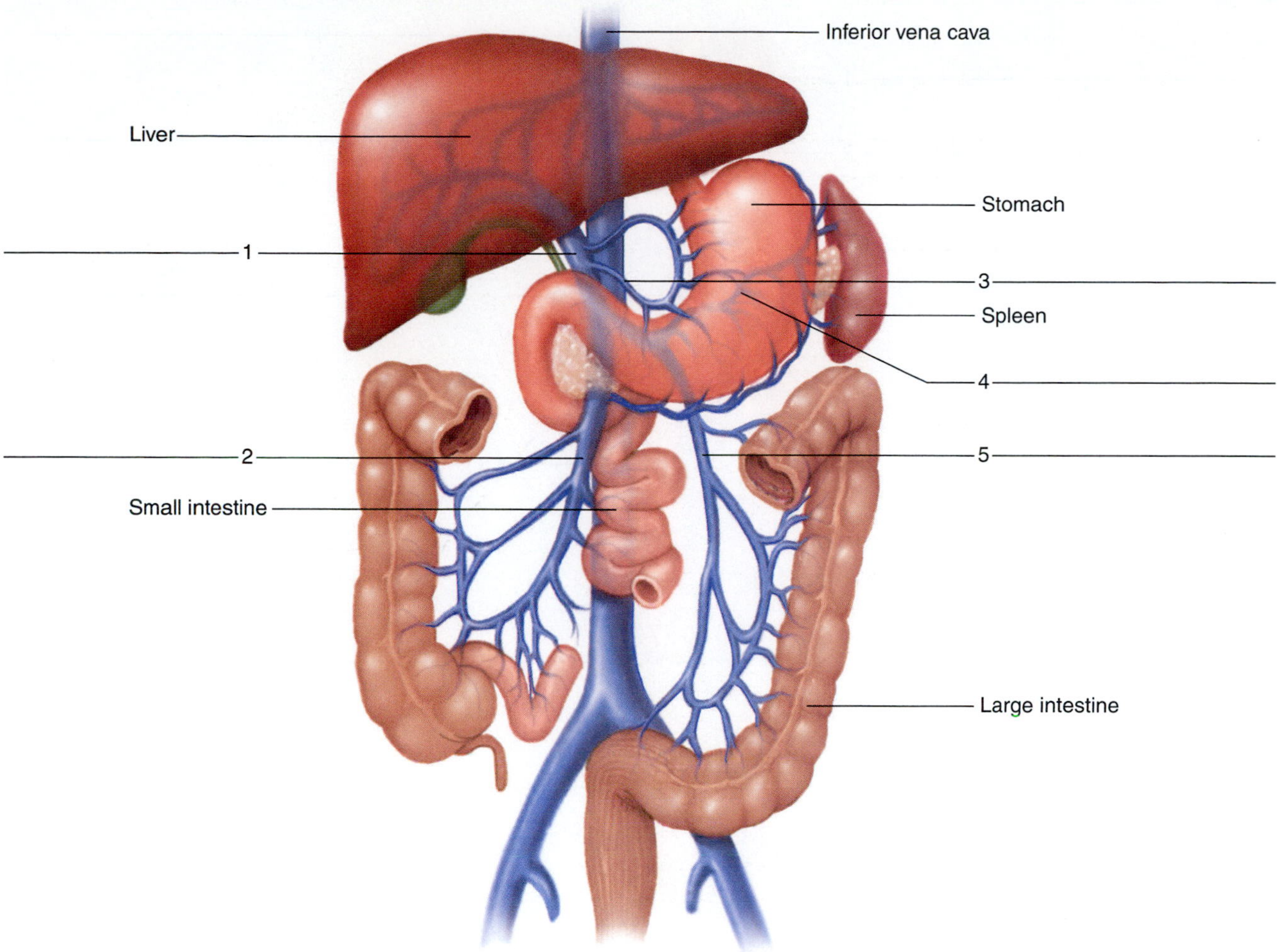

Web Quest

Identify the major arteries and veins, and describe their functions. Observe an animation of the exchanges in a capillary. Search these at www.mhhe.com/shieress8

Figure 39.10 Label the veins of the pelvis and lower limb, using the terms provided.

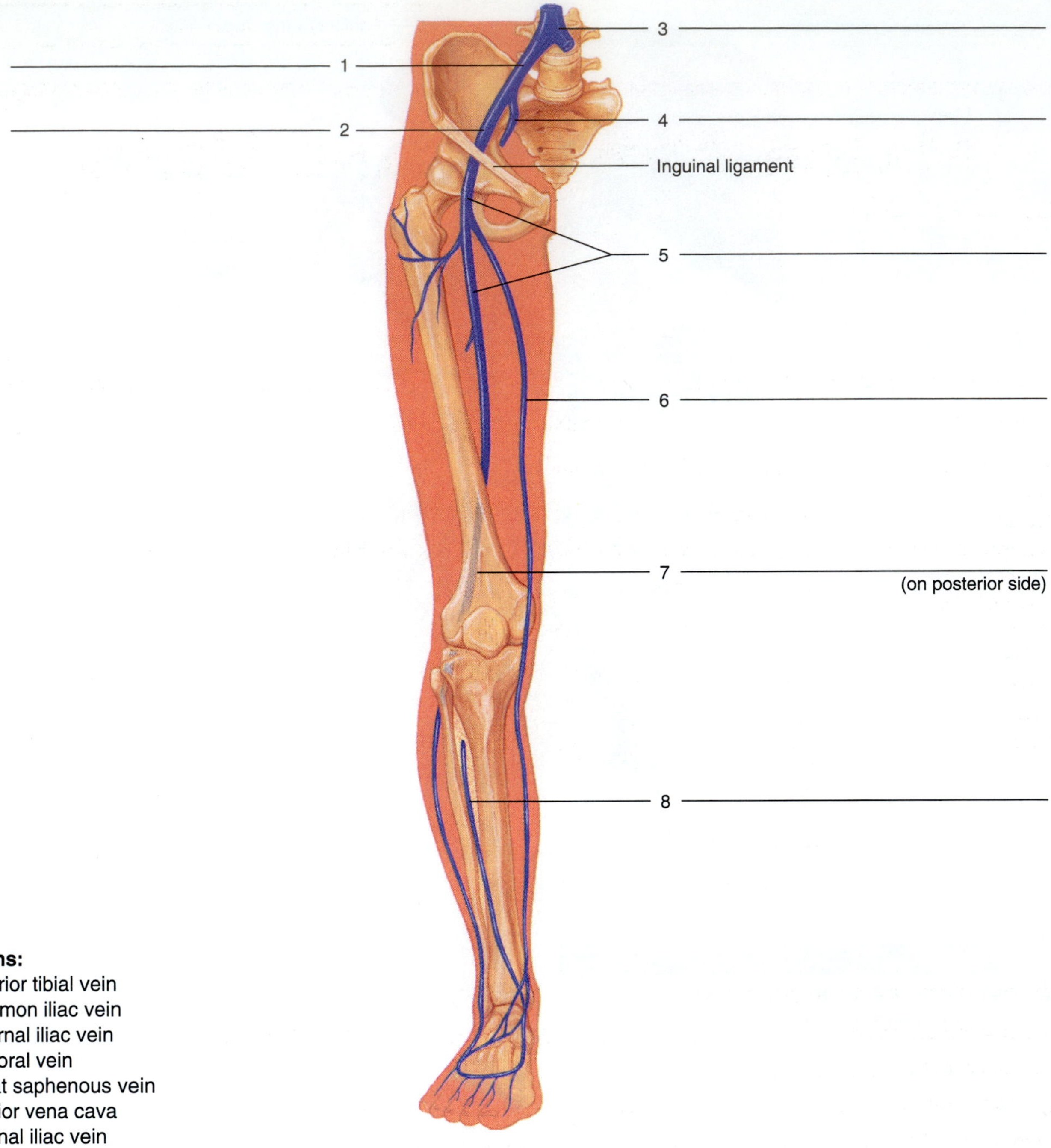

Anterior view

Terms:
Anterior tibial vein
Common iliac vein
External iliac vein
Femoral vein
Great saphenous vein
Inferior vena cava
Internal iliac vein
Popliteal vein

Laboratory Report **39**

Name ____________________
Date ____________________
Section ____________________

MAJOR ARTERIES AND VEINS

PART A

Match the arteries in column A with the descriptions in column B. Place the letter of your choice in the space provided.

Column A	Column B
a. Anterior tibial	_______ 1. Jaw, teeth, and face
b. Brachial	_______ 2. Kidney
c. Celiac	_______ 3. Upper digestive tract, spleen, and liver
d. External carotid	_______ 4. Foot and toes
e. Internal carotid	_______ 5. Gluteal muscles
f. Internal iliac	_______ 6. Biceps muscle
g. Phrenic	_______ 7. Knee joint
h. Popliteal	_______ 8. Adrenal gland
i. Renal	_______ 9. Diaphragm
j. Suprarenal	_______ 10. Brain
k. Ulnar	_______ 11. Forearm muscles

PART B

Provide the name of the missing artery in each of the following sequences:

1. brachiocephalic artery, ______________________________, right axillary artery
2. ascending aorta, ______________________________, thoracic aorta
3. abdominal aorta, ______________________________, diaphragm
4. abdominal aorta, ______________________________, ovary
5. aortic arch, ______________________________, left external carotid artery
6. axillary artery, ______________________________, radial artery
7. common iliac artery, ______________________________, femoral artery

PART C

Match each vein in column A with the vein it drains into from column B. Place the letter of your choice in the space provided.

Column A		Column B
a. Anterior tibial	_______	1. Popliteal
b. Basilic	_______	2. Axillary
c. Brachiocephalic	_______	3. Inferior vena cava
d. Common iliac	_______	4. Subclavian
e. External jugular	_______	5. Brachial
f. Femoral	_______	6. Superior vena cava
g. Popliteal	_______	7. Femoral
h. Radial	_______	8. External iliac

PART D

Provide the name of the missing vein or veins in each of the following sequences:

1. subclavian vein, ____, superior vena cava
2. posterior tibial vein, ____, femoral vein
3. internal iliac vein, ____, inferior vena cava
4. medial cubital vein, ____, axillary vein
5. kidney, ____, inferior vena cava
6. femoral vein, ____, common iliac vein
7. liver, ____, inferior vena cava

PART E

Label the major arteries and veins indicated in figure 39.11.

Figure 39.11 Label the major arteries and veins of the systemic and pulmonary circuits, using the terms provided.

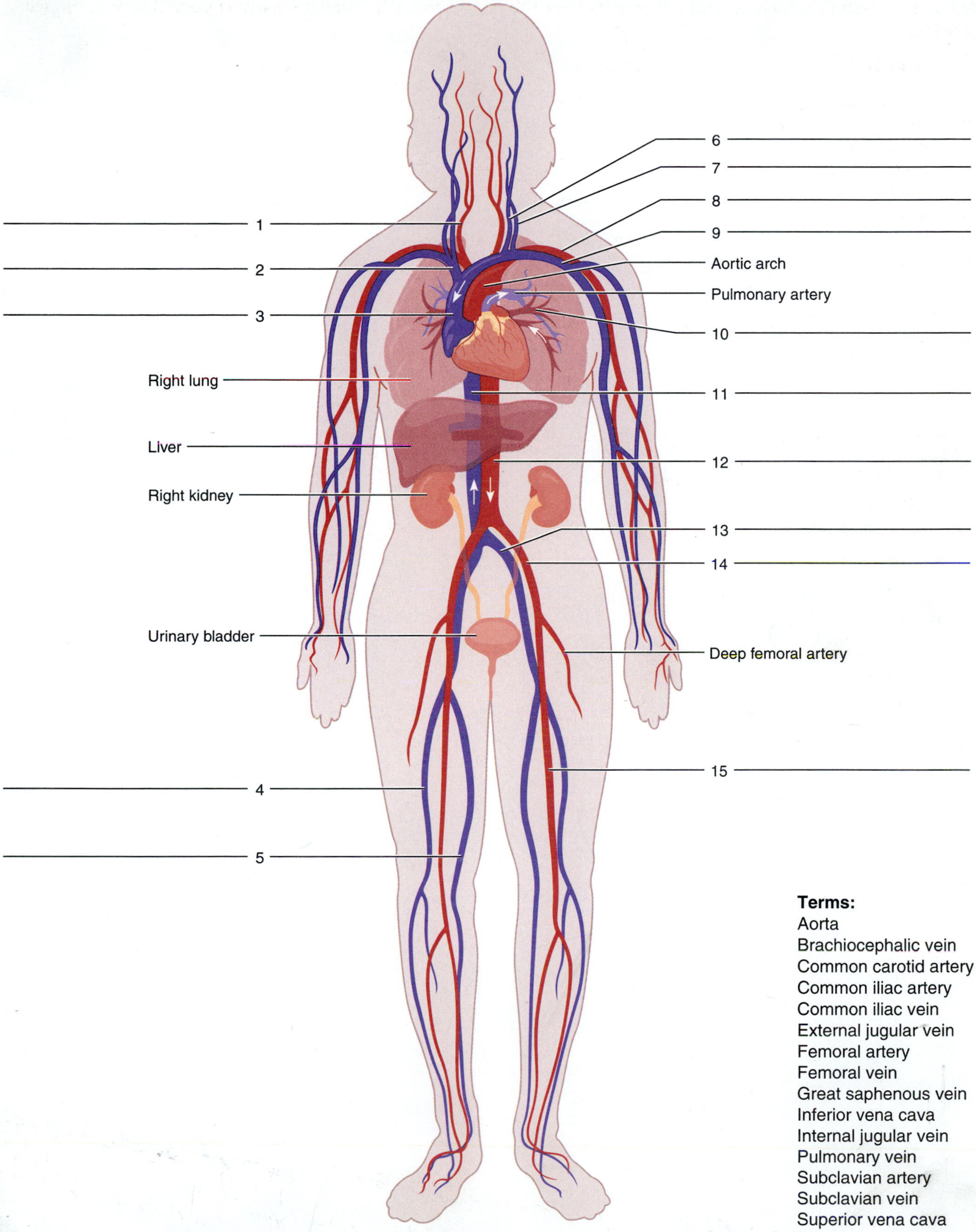

Terms:

Aorta
Brachiocephalic vein
Common carotid artery
Common iliac artery
Common iliac vein
External jugular vein
Femoral artery
Femoral vein
Great saphenous vein
Inferior vena cava
Internal jugular vein
Pulmonary vein
Subclavian artery
Subclavian vein
Superior vena cava

Laboratory Exercise 40

Lymphatic System

MATERIALS NEEDED

Textbook
Human torso
Anatomical chart of the lymphatic system
Compound microscope
Prepared microscope slides:
  Lymph node section
  Human thymus section
  Human spleen section

The lymphatic system is closely associated with the cardiovascular system and includes a network of capillaries and vessels that assist in the circulation of body fluids. These lymphatic capillaries and vessels provide pathways through which excess fluid can be transported away from intercellular spaces within tissues and returned to the bloodstream.

The organs of the lymphatic system also help to protect the tissues against infections by filtering particles from lymph and by supporting the activities of lymphocytes that furnish immunity against specific disease-causing agents.

PURPOSE OF THE EXERCISE

To review the structure of the lymphatic system and to observe the microscopic structure of a lymph node, thymus, and spleen.

LEARNING OBJECTIVES

After completing this exercise, you should be able to

1. locate and identify the major lymphatic pathways in an anatomical chart or model;
2. locate and identify the major chains of lymph nodes in an anatomical chart or model;
3. describe the structure of a lymph node;
4. identify the major microscopic structures of a lymph node, thymus, and spleen.

PROCEDURE A—LYMPHATIC PATHWAYS

1. Review the section entitled "Lymphatic Pathways" in chapter 14 of the textbook.
2. As a review activity, label figure 40.1.
3. Complete Part A of Laboratory Report 40.
4. Observe the human torso and the anatomical chart of the lymphatic system, and locate the following features:

   lymphatic vessels

   thoracic duct

   right lymphatic duct

   lymph nodes

   lymphatic trunks

   subclavian veins

   jugular veins

PROCEDURE B—LYMPH NODES

1. Review the section entitled "Lymph Nodes" in chapter 14 of the textbook.
2. As a review activity, label figure 40.2.
3. Complete Part B of the laboratory report.
4. Observe the anatomical chart of the lymphatic system and the human torso and locate the clusters of lymph nodes in the following regions:

   cervical region

   axillary region

   inguinal region

   pelvic cavity

   abdominal cavity

   thoracic cavity

5. Palpate the lymph nodes in your cervical region. They are located along the lower border of the mandible and between the mandible and the sternocleidomastoid muscle. They feel like small, firm lumps.

Figure 40.1 Label the diagram by placing the correct numbers in the spaces provided.

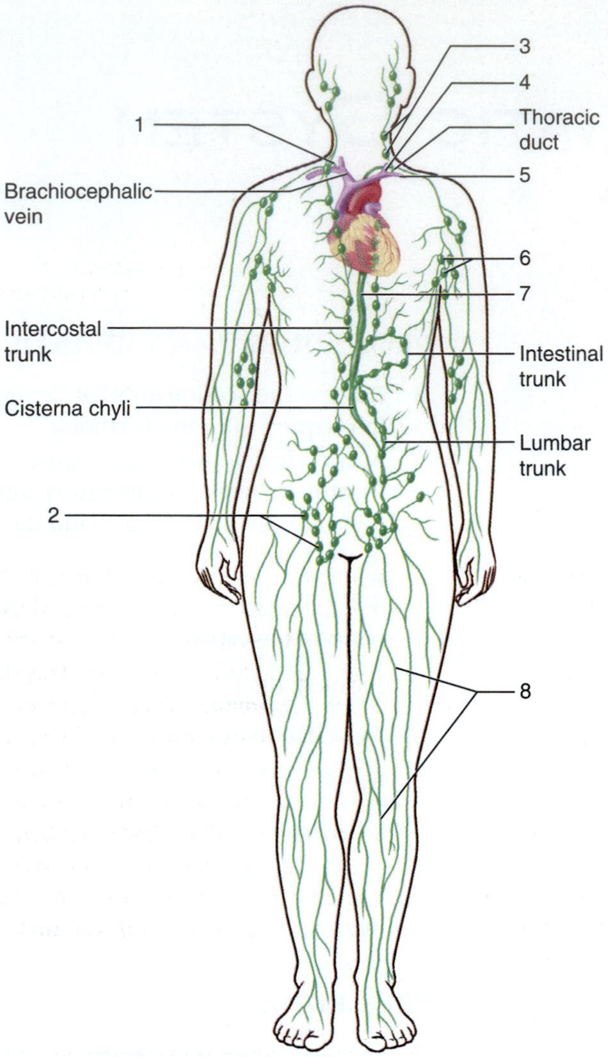

_____ Axillary lymph nodes
_____ Cervical lymph nodes
_____ Inguinal lymph nodes
_____ Jugular vein
_____ Lymphatic vessels
_____ Right lymphatic duct (collecting duct)
_____ Subclavian vein
_____ Thoracic duct (collecting duct)

Figure 40.2 Label this diagram of a lymph node by placing the correct numbers in the spaces provided.

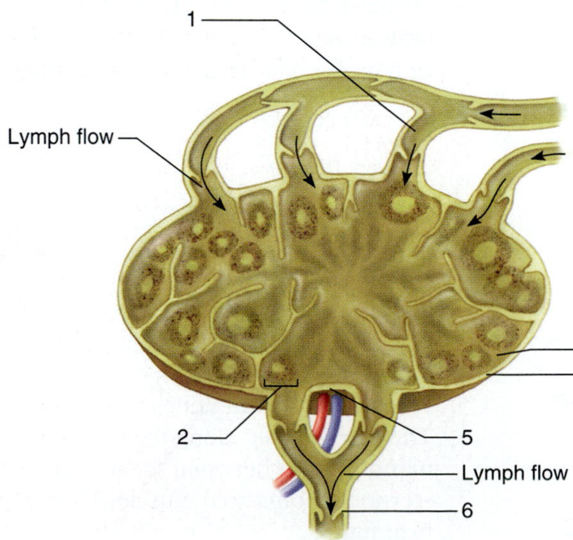

_____ Afferent lymphatic vessel
_____ Capsule
_____ Efferent lymphatic vessel
_____ Hilum
_____ Nodule
_____ Sinus

Figure 40.3 A section of the thymus gland (10× micrograph enlarged to 20×).

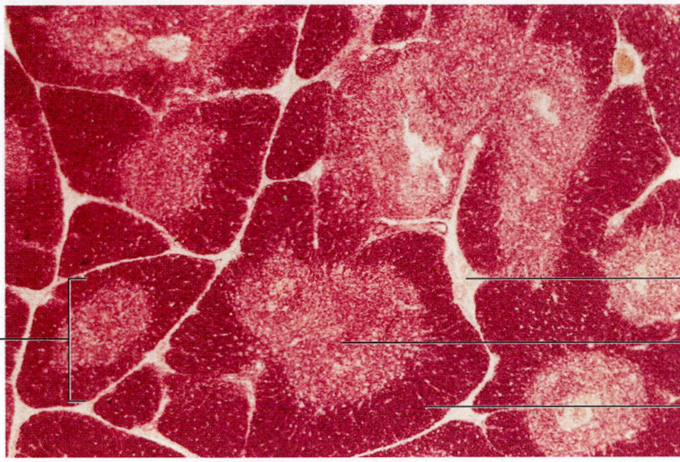

6. Obtain a prepared microscope slide of a lymph node and observe it, using low-power magnification. Identify the *capsule* that surrounds the node and is mainly composed of collagenous fibers, the *lymph nodules* that appear as dense masses near the surface of the node, and the *lymph sinus* that appears as a narrow space between the nodules and the capsule.
7. Using high-power magnification, examine a nodule within the lymph node. Note that the nodule contains densely packed *lymphocytes.*
8. Prepare a labeled sketch of a representative section of a lymph node in Part D of the laboratory report.

PROCEDURE C—THYMUS AND SPLEEN

1. Review the section entitled "Thymus and Spleen" in chapter 14 of the textbook.
2. Locate the thymus and spleen in the anatomical chart of the lymphatic system and on the human torso.
3. Complete Part C of the laboratory report.
4. Obtain a prepared microscope slide of human thymus and observe it, using low-power magnification (fig. 40.3). Note how the thymus is subdivided into *lobules* by *septa* of connective tissue that contain blood vessels. Identify the *capsule* of loose connective tissue that surrounds the thymus; the outer *cortex* of a lobule that is composed of densely packed cells and is deeply stained; and the inner *medulla* of a lobule that is composed of more loosely packed lymphocytes and epithelial cells and is lightly stained.
5. Examine the cortex tissue of a lobule using high-power magnification. The cells of the cortex are composed of densely packed *lymphocytes* among some epithelial cells and macrophages. Some of these cortical cells may be undergoing mitosis, so their chromosomes may be visible.
6. Prepare a labeled sketch of a representative section of the thymus in Part D of the laboratory report.
7. Obtain a prepared slide of the human spleen and observe it, using low-power magnification (fig. 40.4). Identify the *capsule* of dense connective tissue that surrounds the spleen. Note that the tissues of the spleen include circular *nodules* of *white* (in unstained tissue) *pulp* that are enclosed in a matrix of *red pulp.*

Figure 40.4 Micrograph of a section of the spleen (15×).

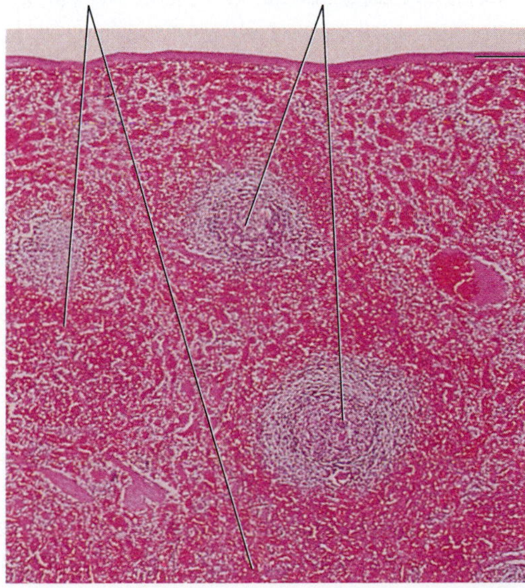

8. Using high-power magnification, examine a nodule of white pulp. The cells of this area are mainly *lymphocytes.* Also, there may be an arteriole centrally located in the nodule.
9. Prepare a labeled sketch of a representative section of the spleen in Part D of the laboratory report.

Web Quest

Locate the six major areas of lymph nodes, and identify the components of the lymphatic system. Search these at www.mhhe.com/shieress8

Laboratory Report **40**

Name ______________________________

Date ______________________________

Section ______________________________

Lymphatic System

PART A

Complete the following statements:

1. Lymphatic pathways begin as ______________________ that merge to form lymphatic vessels.
2. The wall of a lymphatic capillary consists of a single layer of ______________________ cells.
3. Once tissue (interstitial) fluid is inside a lymph capillary, the fluid is called ______________ .
4. Lymphatic vessels have walls that are similar to those of ______________ .
5. Lymphatic vessels contain ______________ that help prevent the backflow of lymph.
6. Lymphatic vessels usually lead to ______________ that filter the fluid being transported.
7. The ______________ is the larger and longer of the two lymphatic collecting ducts.

PART B

Complete the following statements:

1. Lymph nodes contain large numbers of white blood cells called ______________ and macrophages that fight invading microorganisms.
2. The indented region of a bean-shaped lymph node is called the ______________ .
3. ______________ that contain germinal centers are the structural units of a lymph node.
4. The spaces within a lymph node are called ______________________ through which lymph circulates.
5. Lymph enters a node through a(n) ______________ lymphatic vessel.
6. The partially encapsulated lymph nodes in the pharynx are called ______________ .
7. The aggregations of lymph nodules found within the mucosal lining of the small intestine are called ______________________ .

PART C

Complete the following statements:

1. The ____________________ gland is located in the mediastinum, anterior to the aortic arch.
2. The thymus is composed of lymphatic tissue, which is subdivided into ____________________ .
3. The lymphocytes that leave the thymus function in providing ____________________ .
4. The hormones secreted by the thymus are called ____________________ .
5. The ________________ is the largest lymphatic organ.
6. The sinuses within the spleen contain ____________________ .
7. The tiny islands (nodules) of tissue within the spleen that contain many lymphocytes comprise the ________________________ .
8. The ________________________ of the spleen contain large numbers of red blood cells, lymphocytes, and macrophages.
9. ____________________ within the spleen function to engulf and destroy foreign particles and cellular debris.

PART D

Lymph node sketch (_____×)	Thymus sketch (_____×)
Spleen sketch (_____×)	

Laboratory Exercise 41

Organs of the Digestive System

MATERIALS NEEDED

Textbook
Human torso
Skull with teeth
Teeth, sectioned
Tooth model, sectioned
Paper cup
Compound microscope
Prepared microscope slides of the following:
- Parotid gland (salivary gland)
- Esophagus
- Stomach (fundus)
- Small intestine (jejunum)
- Large intestine

Virtual Physiology Lab: 8, Digestion of Fat

The digestive system includes the organs associated with the alimentary canal and several accessory structures. The alimentary canal, which is a muscular tube, passes through the body from the opening of the mouth to the anus. It includes the mouth, pharynx, esophagus, stomach, small intestine, and large intestine. The canal is adapted to move substances throughout its length. It is specialized in various regions to store, digest, and absorb food materials and to eliminate the residues. The accessory organs, which include the salivary glands, liver, gallbladder, and pancreas, secrete products into the alimentary canal that aid digestive functions.

PURPOSE OF THE EXERCISE

To review the structure and function of the digestive organs and to examine the tissues of these organs microscopically.

LEARNING OBJECTIVES

After completing this exercise, you should be able to

1. locate the major digestive organs;
2. describe the functions of these organs;
3. recognize tissue sections of these organs;
4. identify the major features of each tissue section.

PROCEDURE A—MOUTH AND SALIVARY GLANDS

1. Review the sections entitled "Mouth" and "Salivary Glands" in chapter 15 of the textbook.
2. As a review activity, label figures 41.1, 41.2, and 41.3.
3. Examine the mouth of the torso and a skull. Locate the following structures:

oral cavity

vestibule

tongue

- frenulum
- papillae

lingual tonsils

palate

- hard palate
- soft palate
- uvula

palatine tonsils

pharyngeal tonsils (adenoids)

gums (gingivae)

Figure 41.1 Label the major features of the mouth.

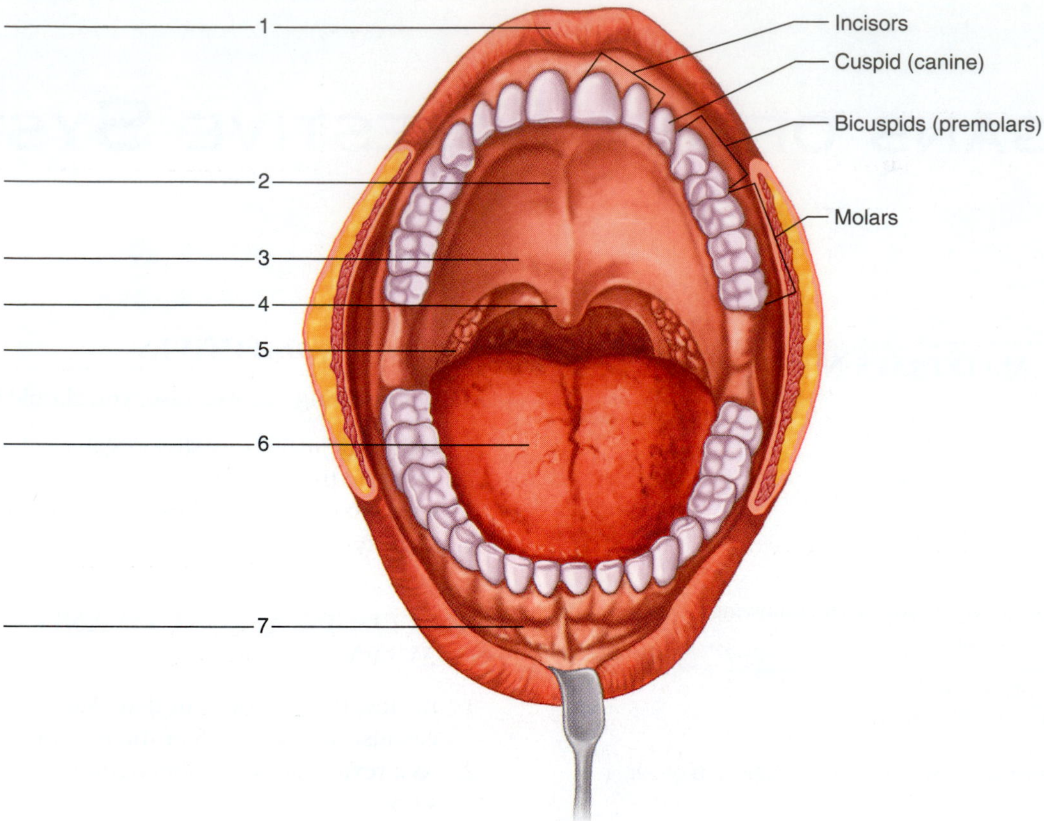

Figure 41.2 Label the features associated with the major salivary glands.

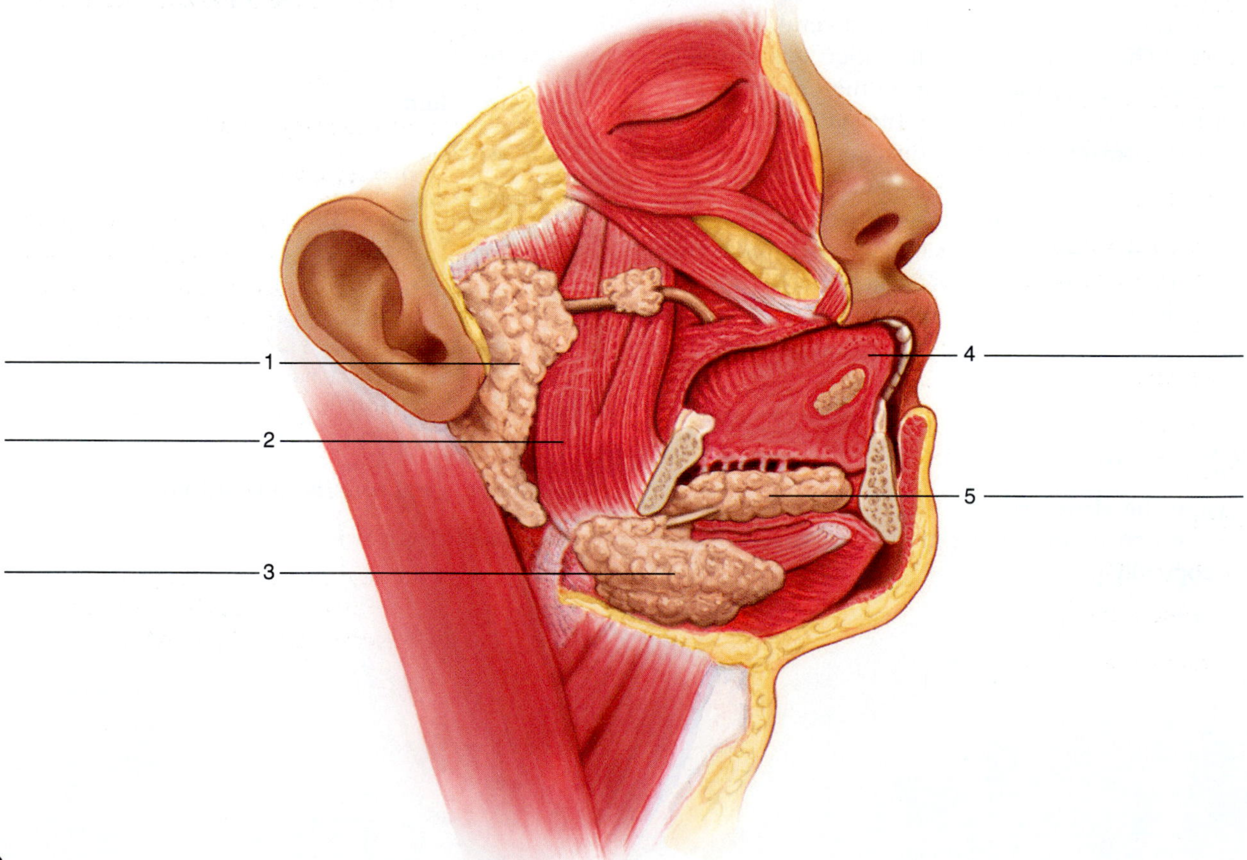

Figure 41.3 Label the features of this cuspid tooth.

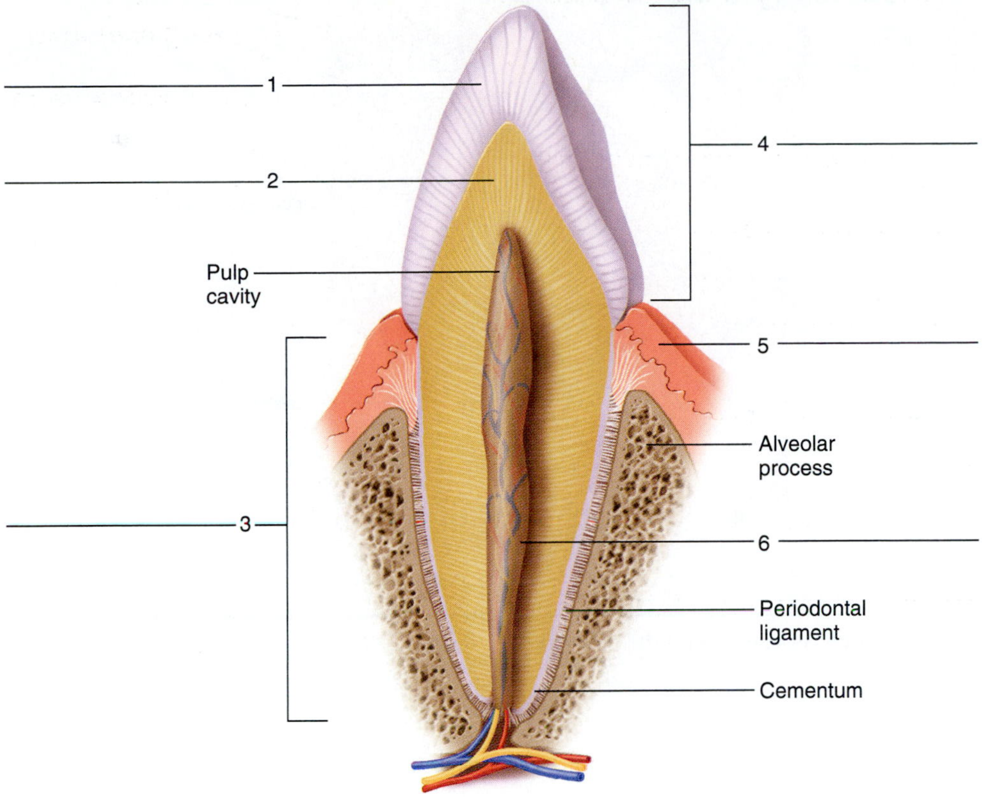

teeth

incisors

cuspids (canines)

bicuspids (premolars)

molars

4. Examine a sectioned tooth and a tooth model. Locate the following features:

crown

enamel

dentin

neck

root

pulp cavity

cementum

root canal

5. Observe the head of the torso and locate the following:

parotid salivary gland

submandibular salivary gland

sublingual salivary gland

6. Examine a microscopic section of a parotid gland, using low- and high-power magnification. Note the numerous glandular cells arranged in clusters around small ducts. Also note a larger secretory duct surrounded by lightly stained cuboidal epithelial cells (fig. 41.4).
7. Complete Part A of Laboratory Report 41.

PROCEDURE B—PHARYNX AND ESOPHAGUS

1. Review the section entitled "Pharynx and Esophagus" in chapter 15 of the textbook.
2. As a review activity, label figure 41.5.

Figure 41.4 Micrograph of the parotid salivary gland (300×).

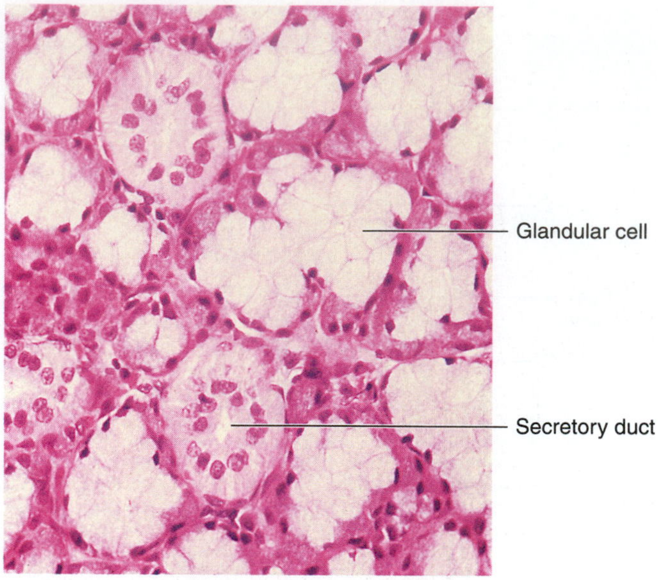

Figure 41.5 Label the features associated with the pharynx.

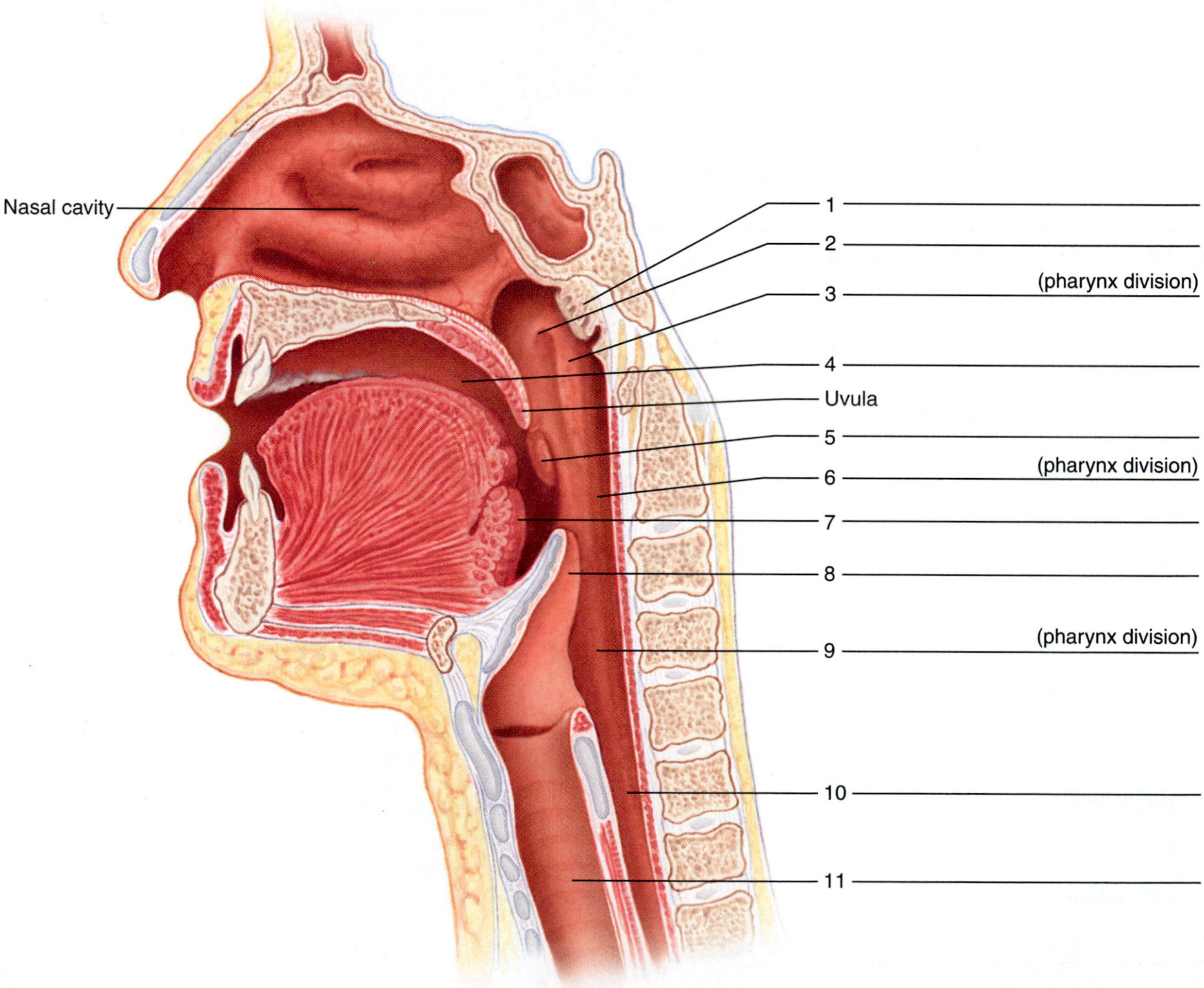

3. Observe the torso and locate the following features:

   pharynx

   nasopharynx

   oropharynx

   laryngopharynx

   esophagus

   lower esophageal sphincter (cardiac sphincter)

4. Have your partner take a drink from a cup of water. Carefully watch the movements in the anterior region of the neck. What steps in the swallowing process did you observe?

   ____

   ____

   ____

   ____

   ____

5. Examine a microscopic section of the esophagus wall, using low-power magnification. Note that the inner lining is composed of stratified squamous epithelium and that there are layers of muscle tissue in the wall. Locate some mucous glands in the submucosa. They appear as clusters of lightly stained cells.
6. Complete Part B of the laboratory report.

PROCEDURE C—THE STOMACH

1. Review the section entitled "Stomach" in chapter 15 of the textbook.
2. As a review activity, label figures 41.6 and 41.7.
3. Observe the torso and locate the following features of the stomach:

   cardiac region

   fundic region

   body region

   pyloric region

   pyloric canal

   rugae

   pyloric sphincter (valve)

Figure 41.6 Label the major regions of the stomach and associated structures.

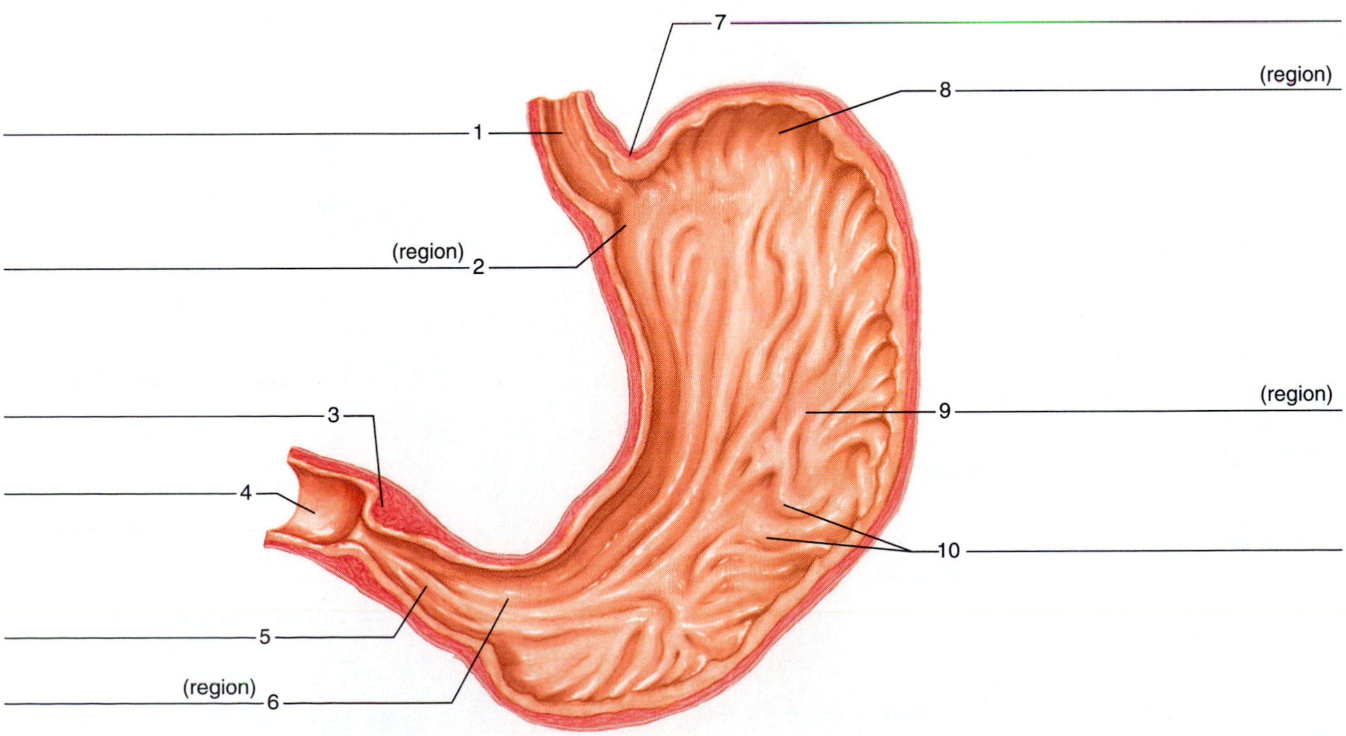

Figure 41.7 Label the lining of the stomach by placing the correct numbers in the spaces provided.

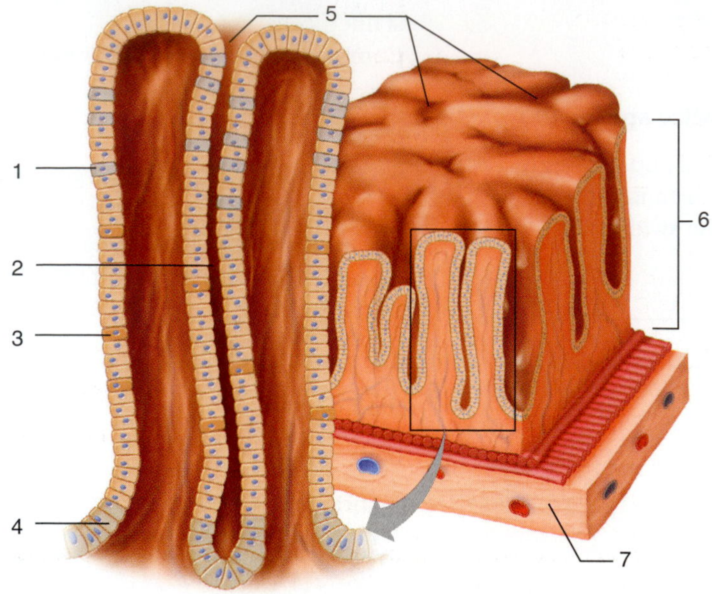

_____ Chief cell
_____ Gastric gland
_____ Gastric pits
_____ Mucosa
_____ Mucous cell
_____ Parietal cell
_____ Submucosa

Figure 41.8 Micrograph of the mucosa of the stomach wall (50× micrograph enlarged to 100×).

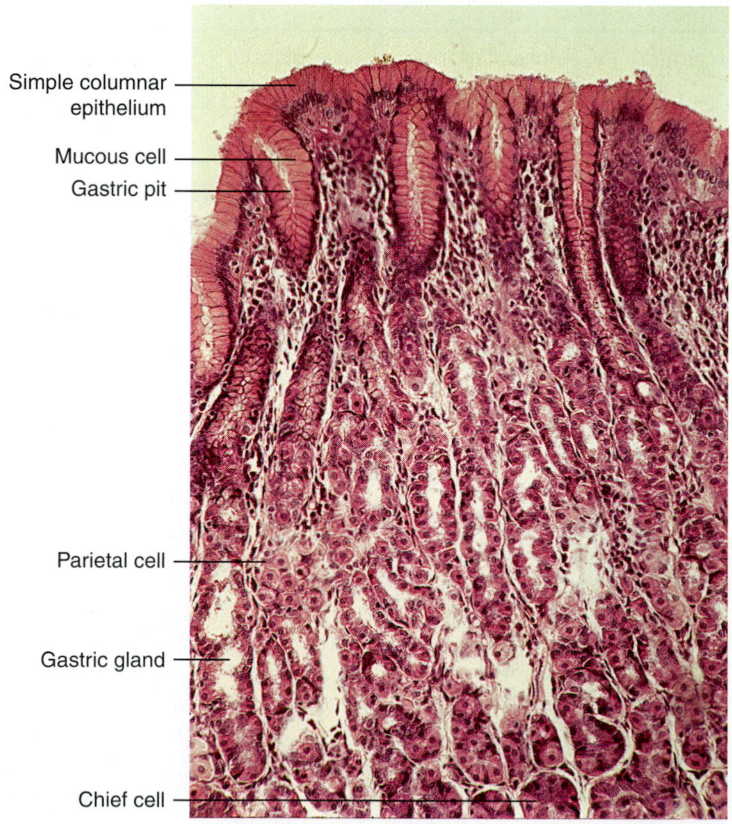

4. Examine a microscopic section of the stomach wall, using low-power magnification (fig. 41.8). Note how the inner lining of simple columnar epithelium dips inward to form gastric pits. The gastric glands are tubular structures that open into the gastric pits. Near the deep ends of these glands, you should be able to locate some intensely stained (bluish) chief cells and some lightly stained (pinkish) parietal cells. What are the functions of these cells? ______________________________
____
____
____
____

5. Complete Part C of the laboratory report.

PROCEDURE D—PANCREAS AND LIVER

1. Review the sections entitled "Pancreas" and "Liver" in chapter 15 of the textbook.
2. As a review activity, label figure 41.9.
3. Observe the torso and locate the following structures:

   pancreas

   pancreatic duct

   liver

   gallbladder

   hepatic ducts

   cystic duct

   common bile duct

   hepatopancreatic sphincter (sphincter of Oddi)

4. Complete Part D of the laboratory report.

Figure 41.9 Label the features associated with the liver and pancreas.

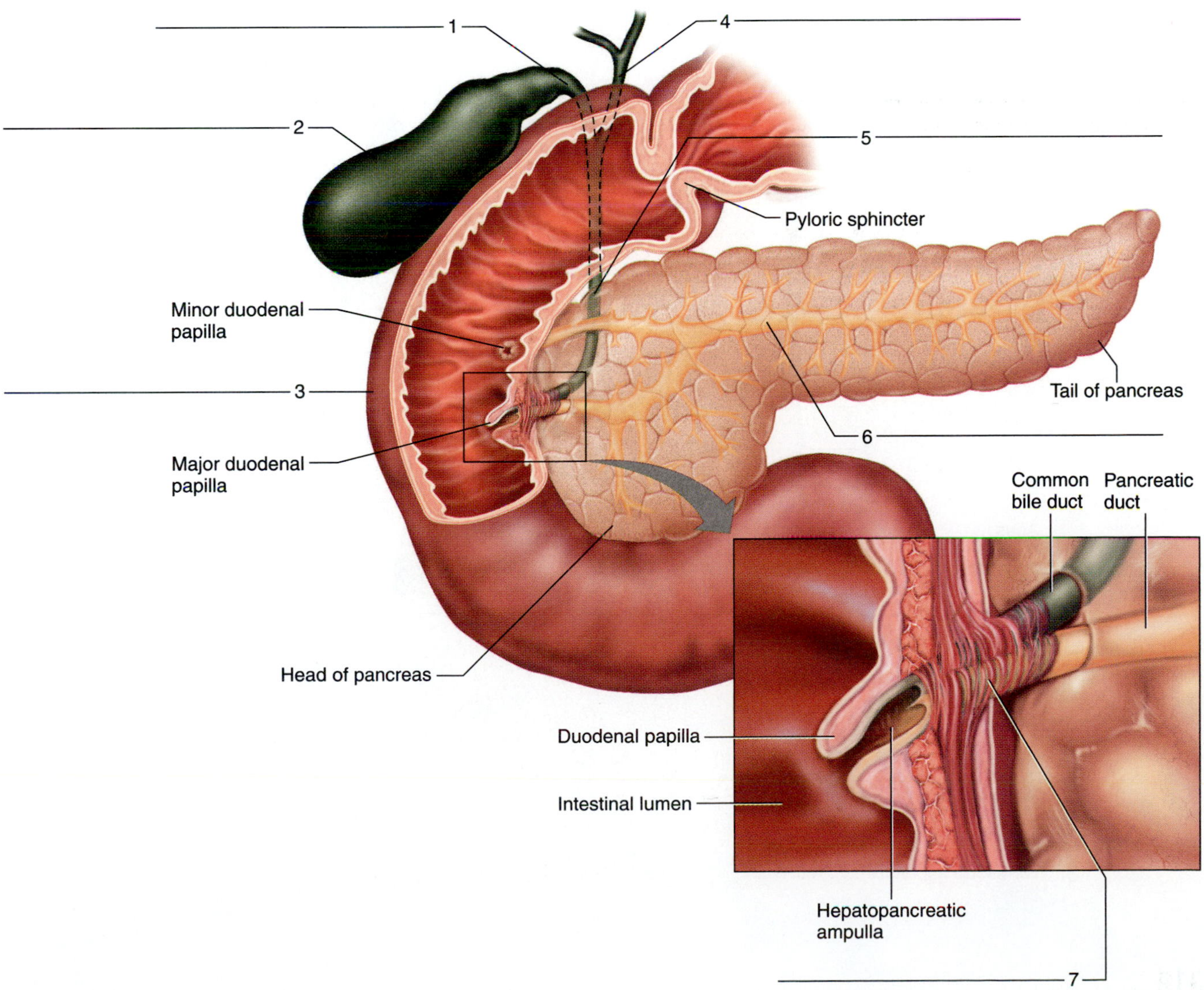

PROCEDURE E—SMALL AND LARGE INTESTINES

1. Review the sections entitled "Small Intestine" and "Large Intestine" in chapter 15 of the textbook.
2. As a review activity, label figure 41.10.
3. Observe the torso and locate each of the following features:

small intestine

duodenum

jejunum

ileum

mesentery

ileocecal sphincter (valve)

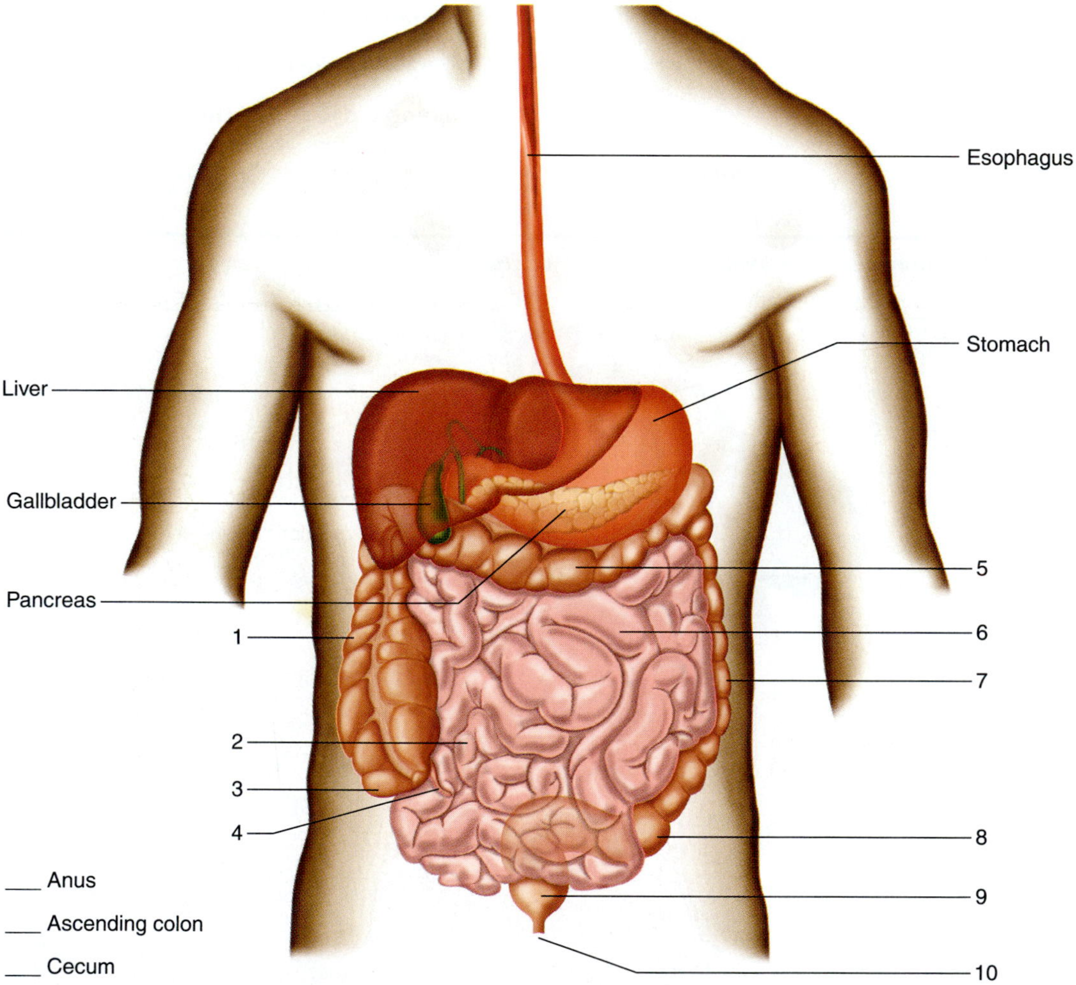

Figure 41.10 Label this diagram of the small and large intestines by placing the correct numbers in the spaces provided.

___ Anus
___ Ascending colon
___ Cecum
___ Descending colon
___ Ileum
___ Jejunum
___ Rectum
___ Sigmoid colon
___ Transverse colon
___ Vermiform appendix

large intestine

cecum

vermiform appendix

ascending colon

transverse colon

descending colon

sigmoid colon

rectum

anal canal

anal columns

anal sphincter muscles

internal anal sphincter

external anal sphincter

anus

4. Using low-power magnification, examine a microscopic section of the small intestine wall. Identify the mucosa, submucosa, muscular layer, and serosa. Note the villi that extend into the lumen of the tube. Study a single villus, using high-power magnification. Note the core of connective tissue and the covering of simple columnar epithelium that contains some lightly stained goblet cells (fig. 41.11). What is the function of these villi?

____

____

____

____

Figure 41.11 Micrograph of the small intestine wall (40×).

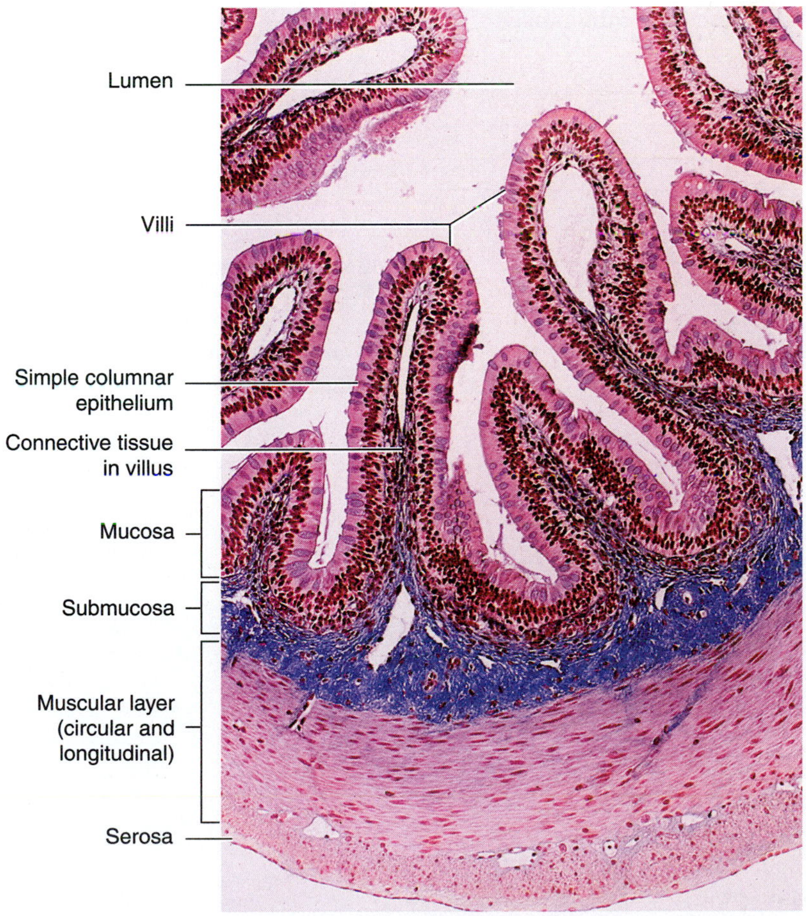

Figure 41.12 Micrograph of the large intestine wall (64×).

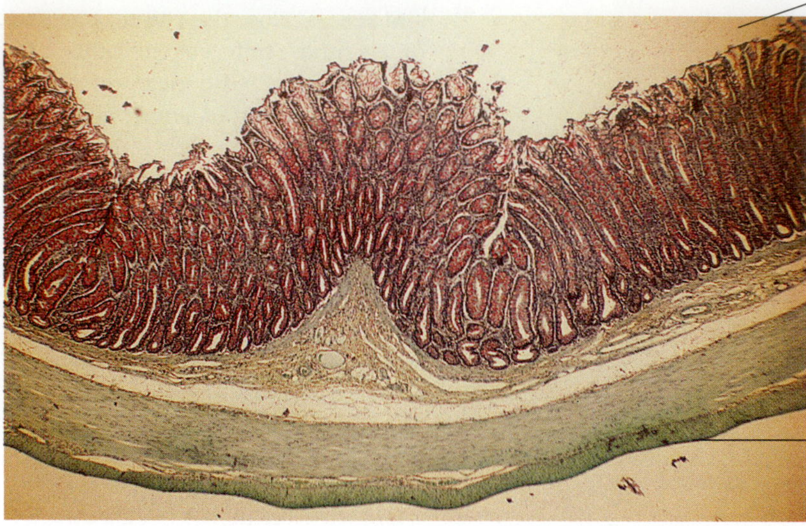

5. Examine a microscopic section of the large intestine wall. Note the lack of villi. Locate the four layers of the wall. Also note the tubular mucous glands that open on the surface of the inner lining and the numerous lightly stained goblet cells (fig. 41.12). What is the function of the mucus secreted by these glands?

____

____

____

____

6. Complete Part E of the laboratory report.

Critical Thinking Application

How is the structure of the small intestine better adapted for absorption than the large intestine?

____

____

____

____

Web Quest

Summarize the functions of the organs of the digestive system. Search these at www.mhhe.com/shieress8

Laboratory Report **41**

Name ______________________

Date ______________________

Section ______________________

Organs of the Digestive System

PART A

Match the terms in column A with the descriptions in column B. Place the letter of your choice in the space provided.

Column A

a. Adenoids (pharyngeal tonsils)
b. Amylase
c. Crown
d. Dentin
e. Frenulum
f. Incisor
g. Molar
h. Oral cavity
i. Palate
j. Papillae
k. Periodontal ligament
l. Serous cell
m. Sublingual gland
n. Uvula
o. Vestibule

Column B

_______ 1. Bonelike substance beneath tooth enamel
_______ 2. Smallest of major salivary glands
_______ 3. Tooth specialized for grinding
_______ 4. Chamber between tongue and palate
_______ 5. Projections on tongue surface
_______ 6. Cone-shaped projection of soft palate
_______ 7. Secretes the digestive enzymes in saliva
_______ 8. Attaches tooth to jaw
_______ 9. Chisel-shaped tooth
_______ 10. Roof of oral cavity
_______ 11. Space between the teeth, cheeks, and lips
_______ 12. Anchors tongue to floor of mouth
_______ 13. Lymphatic tissue in posterior wall of pharynx near auditory tubes
_______ 14. Portion of tooth projecting beyond gum
_______ 15. Splits starch into disaccharides

PART B

Complete the following:

1. The part of the pharynx superior to the soft palate is called the ________________ .
2. The middle part of the pharynx is called the ________________ .
3. The inferior portion of the pharynx is called the ________________ .
4. List the six major actions in the swallowing reflex. ________________
5. ________________ is the main secretion of the esophagus.
6. The esophagus is about ________________ cm long.
7. Summarize the functions of the esophagus. ________________

PART C

Complete the following:

1. Name the four regions of the stomach. ________________
2. Name the valve that prevents the regurgitation of food from the small intestine back into the stomach. ________________
3. Name three types of secretory cells found in gastric glands. ________________
4. Name the gastric cells that secrete digestive enzymes. ________________
5. Name the gastric cells that secrete hydrochloric acid. ________________
6. Name the most important digestive enzyme secreted in the stomach. ________________
7. Name the substance needed for the efficient absorption of vitamin B_{12}. ________________
8. Name the hormone that regulates gastric secretions. ________________
9. Summarize the functions of the stomach. ________________

PART D

Match the terms in column A with the descriptions in column B. Place the letter of your choice in the space provided.

Column A		Column B
a. Amylase	_______	1. Activates protein-digesting enzyme
b. Bile salts	_______	2. Causes emulsification of fats
c. Cholecystokinin	_______	3. Carries on phagocytosis in liver
d. Enterokinase	_______	4. Nucleic acid-digesting enzyme
e. Kupffer cell	_______	5. Carbohydrate-digesting enzyme
f. Lipase	_______	6. Fat-digesting enzyme
g. Nuclease	_______	7. Protein-digesting enzyme
h. Secretin	_______	8. Stimulates the gallbladder to release bile
i. Trypsin	_______	9. Stimulates pancreas to secrete fluids high in bicarbonate ions

PART E

Complete the following:

1. Name the three portions of the small intestine. ______________________________

   ____
2. Describe the function of the mesentery. ______________________________

   ____

   ____
3. Name the lymphatic capillary found in an intestinal villus. ______________________________
4. Name the tubular glands found between the bases of the intestinal villi. ______________________________
5. Name five digestive enzymes secreted by the intestinal glands. ______________________________

   ____
6. Name the valve that controls the movement of material between the small and large intestines. ______________

   ____
7. Name the small projection that contains lymphatic tissue attached to the cecum. ______________________________
8. Summarize the functions of the small intestine. ______________________________

   ____

   ____

   ____
9. Summarize the functions of the large intestine. ______________________________

   ____

   ____

   ____

Action of a Digestive Enzyme

MATERIALS NEEDED

0.5% amylase solution*
Beakers (50 and 500 mL)
Distilled water
Funnel
Pipets (1 and 10 mL)
Pipet rubber bulbs
0.5% starch solution
Graduated cylinder (10 mL)
Test tubes
Test-tube clamps
Wax marker
Iodine-potassium-iodide solution
Medicine dropper
Ice
Water bath, 37°C (98.6°F)
Porcelain test plate
Benedict's solution
Hot plates
Test-tube rack
Thermometer

*The amylase must be free of sugar. See Appendix 1.

Virtual Physiology Lab: 10, Enzyme Characteristics

SAFETY

- Wear safety glasses when working with acids and when heating test tubes.
- Use test-tube clamps when handling hot test tubes.
- If an open flame is used for heating the test solutions, keep clothes and hair away from the flame.
- Use only a mechanical pipetting device (never your mouth). Use pipets with rubber bulbs or dropping pipets.

The digestive enzyme in salivary secretions is called *amylase.* This enzyme splits starch molecules into sugar (disaccharide) molecules, which is the first step in the digestion of complex carbohydrates.

As in the case of other enzymes, amylase is a protein whose activity is affected by exposure to certain environmental factors, including excessive heat, radiation, electricity, and certain chemicals.

PURPOSE OF THE EXERCISE

To investigate the action of amylase and the effect of heat on its enzymatic activity.

LEARNING OBJECTIVES

After completing this exercise, you should be able to

1. describe the action of amylase;
2. test a solution for the presence of starch or the presence of sugar;
3. test the effects of varying temperatures on the activity of amylase.

PROCEDURE A—AMYLASE ACTIVITY

1. Mark three clean test tubes as *tubes 1, 2, 3,* and prepare the tubes as follows:

   Tube 1: Add 6 mL of amylase solution.

   Tube 2: Add 6 mL of starch solution.

   Tube 3: Add 5 mL of starch solution and 1 mL of amylase solution.

2. Shake the tubes well to mix the contents, and place them in a warm water bath, 37°C (98.6°F), for 10 minutes.
3. At the end of the 10 minutes, test the contents of each tube for the presence of starch. To do this, follow these steps:
   a. Place 1 mL of the solution to be tested in a depression of a porcelain test plate.
   b. Next add two drops of iodine-potassium-iodide solution, and note the color of the mixture. If the solution becomes blue-black, starch is present.
   c. Record the results in Part A of Laboratory Report 42.

4. Test the contents of each tube for the presence of sugar (disaccharides in this instance). To do this, follow these steps:
   a. Place 1 mL of the solution to be tested in a clean test tube.
   b. Add 1 mL of Benedict's solution.
   c. Place the test tube with a test-tube clamp in a beaker of boiling water for 2 minutes.
   d. Note the color of the liquid. If the solution becomes green, yellow, orange, or red, sugar is present. Blue indicates a negative test, whereas green indicates a positive test with the least amount of sugar, and red indicates the greatest amount of sugar present.
   e. Record the results in Part A of the laboratory report.
5. Complete Part A of the laboratory report.

PROCEDURE B—EFFECT OF HEAT

1. Mark three clean test tubes as *tubes 4, 5,* and *6.*
2. Add 1 mL of amylase solution to each of the tubes, and expose each solution to a different test temperature for 3 minutes as follows:

   Tube 4: Place in beaker of ice water (about 0°C [32°F]).

   Tube 5: Place in warm water bath (about 37°C [98.6°F]).

   Tube 6: Place in beaker of boiling water (about 100°C [212°F]). Use a test-tube clamp.
3. Add 5 mL of starch solution to each tube, shake to mix the contents, and return the tubes to their respective test temperatures for 10 minutes. It is important that the 5 mL of starch solution added to tube 4 be at ice-water temperature before it is added to the 1 mL of amylase solution.
4. At the end of the 10 minutes, test the contents of each tube for the presence of starch and the presence of sugar by following the directions in Procedure A.
5. Complete Part B of the laboratory report.

OPTIONAL ACTIVITY

Devise an experiment to test the effect of some other environmental factor on amylase activity. For example, you might test the effect of a strong acid by adding a few drops of concentrated hydrochloric acid to a mixture of starch and amylase solutions. Be sure to include a control in your experimental plan. That is, include a tube containing everything except the factor you are testing. Then you will have something with which to compare your results. *Carry out your experiment only if it has been approved by the laboratory instructor.*

Laboratory Report **42**

Name ______________________

Date ______________________

Section ______________________

ACTION OF A DIGESTIVE ENZYME

PART A—AMYLASE ACTIVITY

1. Test results:

Tube		Starch	Sugar
1	Amylase solution		
2	Starch solution		
3	Starch-amylase solution		

2. Complete the following:
   a. Explain the reason for including tube 1 in this experiment.

   ____

   ____

   ____

   b. What is the importance of tube 2?

   ____

   ____

   ____

   c. What do you conclude from the results of this experiment?

   ____

   ____

   ____

PART B—EFFECT OF HEAT

1. Test results:

Tube		Starch	Sugar
4	0°C (32°F)		
5	37°C (98.6°F)		
6	100°C (212°F)		

2. Complete the following:
   a. What do you conclude from the results of this experiment?

   b. If digestion failed to occur in one of the tubes in this experiment, how can you tell if the amylase was destroyed by the factor being tested or if the amylase activity was simply inhibited by the test treatment?

Critical Thinking Application

What test result would occur if the amylase used contained sugar? ______________________ Will your results be valid? Explain your answer.

LABORATORY EXERCISE 43

ORGANS OF THE RESPIRATORY SYSTEM

MATERIALS NEEDED

Textbook
Human skull (sagittal section)
Human torso
Larynx model
Thoracic organs model
Compound microscope
Prepared microscope slides of the following:
  Trachea (cross section)
  Lung, human (normal)

For Demonstrations:

Animal lung with trachea (fresh or preserved)
Prepared microscope slides of the following:
  Lung tissue (smoker)
  Lung tissue (emphysema)

SAFETY

- Wear disposable gloves when working on the fresh or preserved animal lung demonstration.
- Wash your hands before leaving the laboratory.

The organs of the respiratory system include the nose, nasal cavity, sinuses, pharynx, larynx, trachea, bronchial tree, and lungs. They mainly function to process incoming air and to transport it to and from the atmosphere outside the body and the air sacs of the lungs.

In the air sacs, gas exchanges take place between the air and the blood of nearby capillaries. The blood, in turn, transports gases to and from the air sacs and the body cells. This entire process of transporting and exchanging gases between the atmosphere and the body cells is called *respiration.*

PURPOSE OF THE EXERCISE

To review the structure and function of the respiratory organs and to examine the tissues of some of these organs microscopically.

LEARNING OBJECTIVES

After completing this exercise, you should be able to

1. locate the major organs of the respiratory system;
2. describe the functions of these organs;
3. recognize tissue sections of the trachea and lung;
4. identify the major features of these tissue sections.

PROCEDURE A—RESPIRATORY ORGANS

1. Review the section entitled "Organs of the Respiratory System" in chapter 16 of the textbook.
2. As a review activity, label figures 43.1 and 43.2.
3. Examine the sagittal section of the human skull and locate the following features:

nose

nostrils

nasal cavity

nasal septum

nasal conchae

sinuses

maxillary sinus

frontal sinus

ethmoidal sinus

sphenoidal sinus

Figure 43.1 Label the major features of the respiratory system.

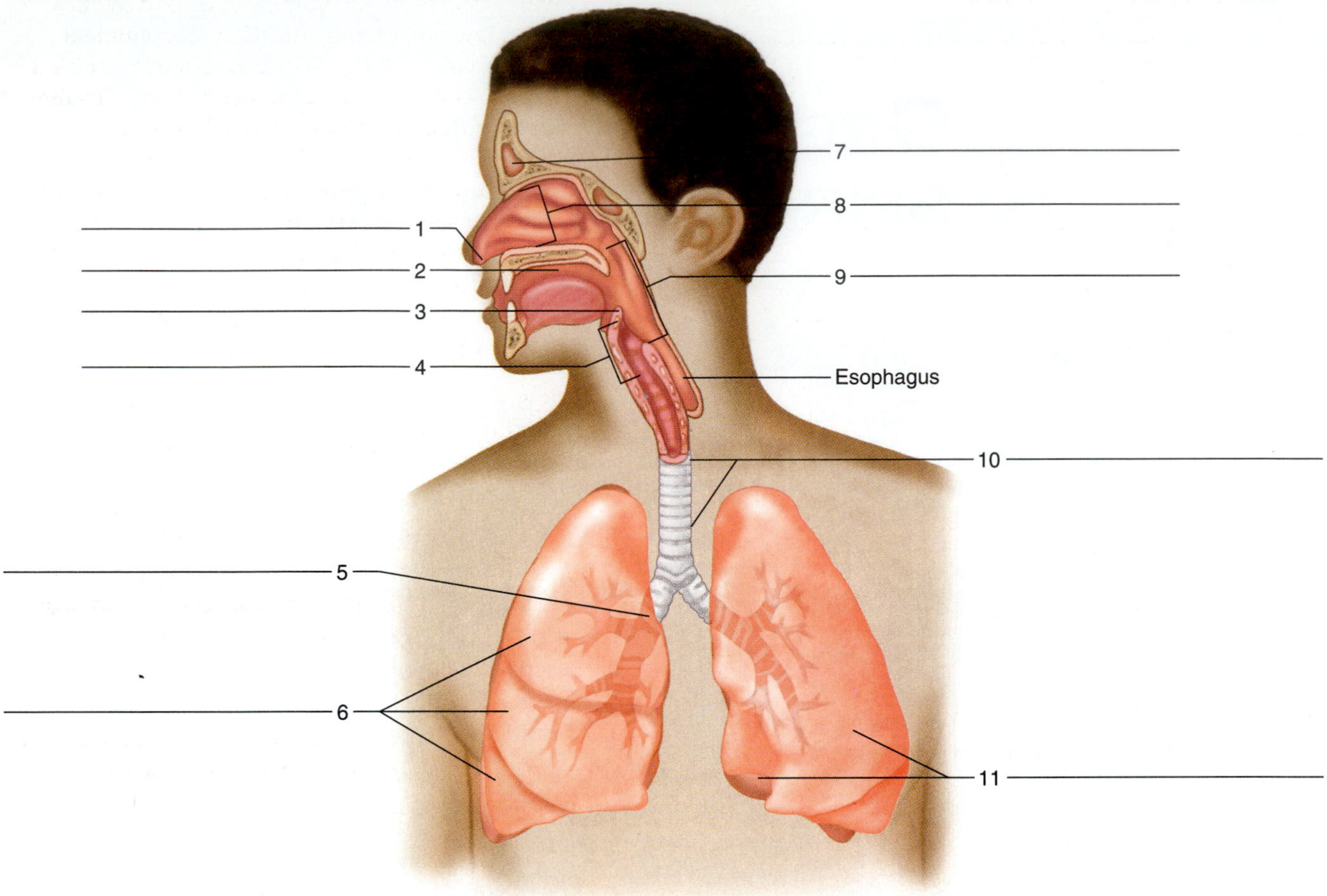

4. Observe the larynx model, the thoracic organs model, and the human torso. Locate the features listed in step 3. Also locate the following:

   pharynx

   nasopharynx

   oropharynx

   laryngopharynx

   larynx (palpate your own larynx)

   vocal cords

   false vocal cords

   true vocal cords

   thyroid cartilage ("Adam's apple")

   cricoid cartilage

   epiglottic cartilage

   trachea (palpate your own trachea)

   bronchi

   primary bronchi

   secondary bronchi

   lung

   lobe

   visceral pleura

   parietal pleura

   pleural cavity

5. Complete Part A of Laboratory Report 43.

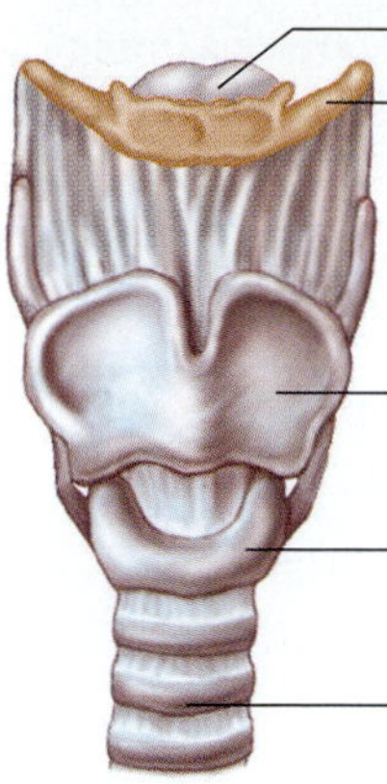

(a)

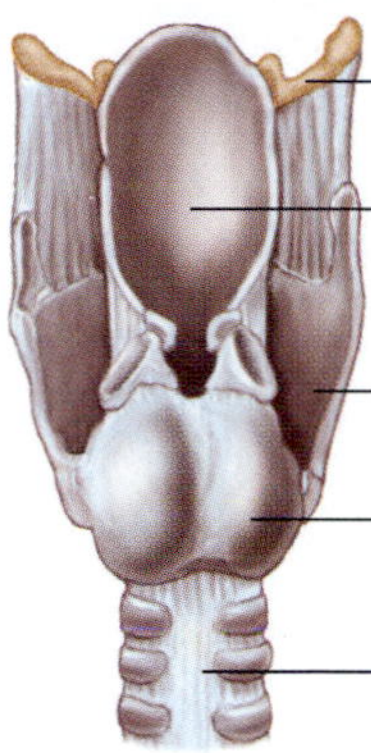

(b)

Figure 43.2 Label the major features of the larynx: (*a*) anterior view; (*b*) posterior view.

DEMONSTRATION

Observe the animal lung and the attached trachea. Identify the larynx, major laryngeal cartilages, trachea, and the incomplete cartilaginous rings of the trachea. Open the larynx and locate the vocal folds. Examine the visceral pleura on the surface of a lung, and squeeze a portion of a lung between your fingers. How do you describe the texture of the lung?

____

____

____

PROCEDURE B—RESPIRATORY TISSUES

1. Obtain a prepared microscope slide of a trachea, and use low-power magnification to examine it. Notice the inner lining of ciliated pseudostratified columnar epithelium and the deep layer of hyaline cartilage, which represents a portion of an incomplete (C-shaped) tracheal ring (fig. 43.3).
2. Use high-power magnification to observe the cilia on the free surface of the epithelial lining. Locate the wineglass-shaped goblet cells, which secrete a protective mucus, in the epithelium.
3. Prepare a labeled sketch of a representative portion of the tracheal wall in Part B of the laboratory report.
4. Obtain a prepared microscope slide of a human lung. Examine it, using low-power magnification, and note the numerous open spaces of the air sacs (alveoli). Look for a bronchiole—a tube with a relatively thick wall and a wavy inner lining. Locate the smooth muscle tissue in the wall of this tube (fig. 43.4). You also may see a section of cartilage as part of the bronchiole wall.

Figure 43.3 Micrograph of a section of the tracheal wall (63×).

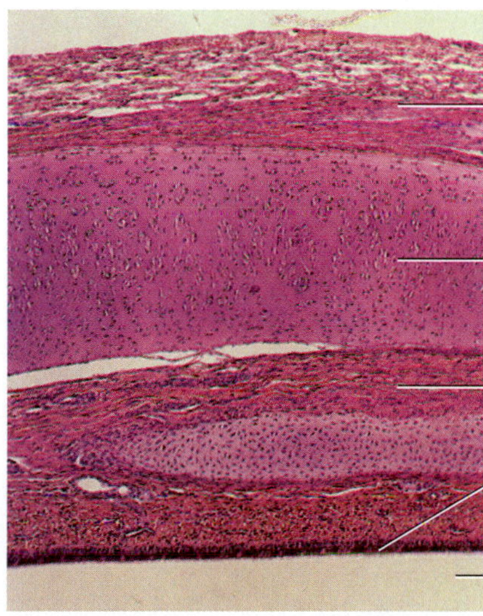

Figure 43.4 Micrograph of human lung tissue (35×).

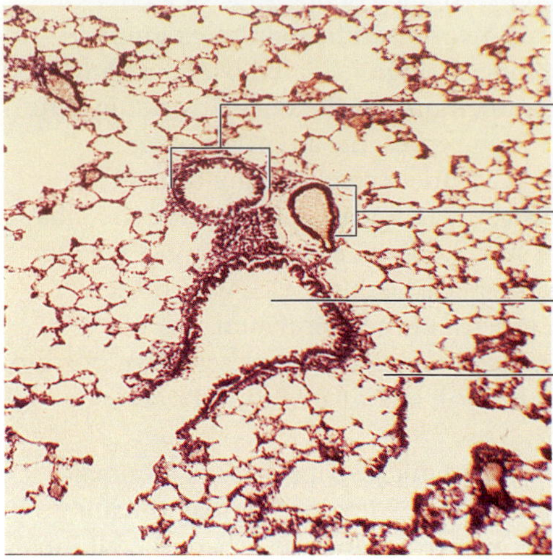

5. Use high-power magnification to examine the alveoli. Note that their walls are composed of simple squamous epithelium. You also may see sections of blood vessels filled with blood cells.
6. Prepare a labeled sketch of a representative portion of the lung in Part B of the laboratory report.
7. Complete Part C of the laboratory report.

DEMONSTRATION

Examine the prepared microscope slides of the lung tissue of a smoker and a person with emphysema, using low-power magnification. How does the smoker's lung tissue compare with that of the normal lung tissue that you examined previously?

___

___

___

How does the emphysema patient's lung tissue compare with the normal lung tissue?

___

___

___

Web Quest

Review the structure and function of the respiratory system at www.mhhe.com/shieress8

Laboratory Report **43**

Name ______________________

Date ______________________

Section ______________________

ORGANS OF THE RESPIRATORY SYSTEM

PART A

Match the terms in column A with the descriptions in column B. Place the letter of your choice in the space provided.

Column A

a. Alveolus
b. Cricoid cartilage
c. Epiglottis
d. Glottis
e. Lung
f. Nasal concha
g. Pharynx
h. Pleural cavity
i. Sinus (paranasal sinus)
j. Vocal cord (true)

Column B

_______ 1. Potential space between visceral and parietal pleurae
_______ 2. Most inferior portion of larynx
_______ 3. Serves as resonant chamber and reduces weight of skull
_______ 4. Microscopic air sacs for gas exchange
_______ 5. Consists of large lobes
_______ 6. Opening between vocal cords
_______ 7. Fold of mucous membrane containing elastic fibers responsible for sounds
_______ 8. Increases surface area of nasal mucous membrane
_______ 9. Passageway for air and food
_______ 10. Partially covers opening of larynx during swallowing

PART B

1. Prepare a labeled sketch of a portion of the tracheal wall.

2. Prepare a labeled sketch of a portion of lung tissue.

PART C

Complete the following:

1. What is the function of the mucus secreted by the goblet cells? ______________________________
____
____

2. Describe the function of the cilia in the respiratory tubes. ______________________________
____
____

3. How is breathing affected if the smooth muscle of the bronchial tree relaxes? ______________________________
____
____

Critical Thinking Application

Why are the alveolar walls so thin?

____
____
____
____

BREATHING AND RESPIRATORY VOLUMES AND CAPACITIES

MATERIALS NEEDED

Textbook
Spirometer, handheld (dry portable)
70% alcohol
Cotton
Disposable mouthpieces
Meterstick

For Demonstration:

Lung function model
Virtual Physiology Lab: 6, Pulmonary Function
Virtual Physiology Lab: 7, Respiration and Exercise

SAFETY

- Clean the spirometer with cotton moistened with 70% alcohol before each use.
- Place a new disposable mouthpiece on the stem of the spirometer before each use.
- Dispose of the cotton and mouthpieces according to your laboratory instructor.

Breathing involves the movement of air from outside the body through the bronchial tree and into the alveoli and the reversal of this air movement. These movements are caused by changes in the size of the thoracic cavity that result from skeletal muscle contractions and from the elastic recoil of stretched tissues.

The volumes of air that move in and out of the lungs during various phases of breathing are called *respiratory air volumes* and *capacities.* These volumes can be measured by using an instrument called a *spirometer.* However, the values obtained vary with a person's age, sex, height, and weight. Various respiratory capacities can be calculated by combining two or more of the respiratory volumes.

PURPOSE OF THE EXERCISE

To review the mechanisms of breathing and to measure or calculate certain respiratory air volumes and respiratory capacities.

LEARNING OBJECTIVES

After completing this exercise, you should be able to

1. describe the mechanisms responsible for inspiration and expiration;
2. define the respiratory air volumes and respiratory capacities;
3. measure or calculate the respiratory air volumes and capacities.

PROCEDURE A—BREATHING MECHANISMS

1. Review the sections entitled "Inspiration" and "Expiration" in chapter 16 of the textbook.
2. Complete Part A of Laboratory Report 44.

DEMONSTRATION

Observe the mechanical lung function model. Note that it consists of a heavy glass bell jar with a rubber sheeting tied over its wide open end (fig. 44.1). Its narrow upper opening is plugged with a rubber stopper through which a glass Y tube is passed. Small rubber balloons are fastened to the arms of the Y.

What happens to the balloons when the rubber sheeting is pulled downward? ______________________

____

____

What happens when the sheeting is pushed upward?

____

____

How do you explain these changes? ______________

____

____

____

What part of the respiratory system is represented by the rubber sheeting? ______________________

by the bell jar? ______________________

by the Y tube? ______________________

by the balloons? ______________________

Figure 44.1 A lung function model.

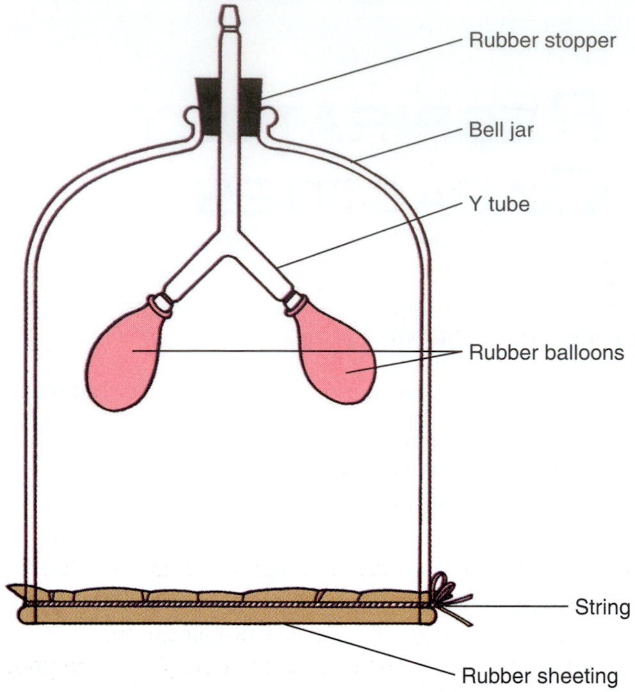

Figure 44.2 A handheld spirometer can be used to measure respiratory air volumes.

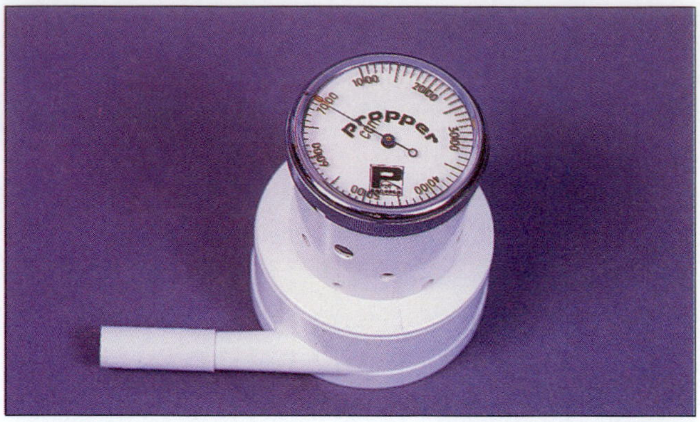

WARNING

If the subject begins to feel dizzy or light-headed while performing Procedure B, stop the exercise and breathe normally.

PROCEDURE B—RESPIRATORY AIR VOLUMES AND CAPACITIES

1. Review the section entitled "Respiratory Air Volumes and Capacities" in chapter 16 of the textbook.
2. Complete Part B of the laboratory report.
3. Obtain a handheld spirometer. Note that the needle can be set at zero by rotating the adjustable dial. Before using the instrument, clean it with cotton moistened with 70% alcohol and place a new disposable mouthpiece over its stem. The instrument should be held with the dial upward, and air should be blown into the disposable mouthpiece. Movement of the needle indicates the air volume that leaves the lungs (fig. 44.2).
4. *Tidal volume* is the volume of air that enters (or leaves) the lungs during a *respiratory cycle* (one inspiration plus the following expiration). *Resting tidal volume* is the volume of air that enters (or leaves) the lungs during normal, quiet breathing (fig. 44.3). To measure this volume, follow these steps:
   a. Sit quietly for a few moments.
   b. Position the spirometer dial so that the needle points to zero.
   c. Place the mouthpiece between your lips and exhale three ordinary expirations into it after inhaling through the nose each time. *Do not force air out of your lungs; exhale normally.*
   d. Divide the total value indicated by the needle by 3, and record this amount as your resting tidal volume on the table in Part C of the laboratory report.
5. *Expiratory reserve volume* is the volume of air in addition to the tidal volume that leaves the lungs during forced expiration. To measure this volume, follow these steps:
   a. Breathe normally for a few moments. Set the needle to zero.
   b. At the end of an ordinary expiration, place the mouthpiece between your lips and exhale all of the air you can force from your lungs through the spirometer.
   c. Record the results as your expiratory reserve volume in Part C.
6. *Vital capacity* is the maximum volume of air that can be exhaled after taking the deepest breath possible. To measure this volume, follow these steps:
   a. Breathe normally for a few moments. Set the needle at zero.
   b. Breathe in and out deeply a couple of times, then take the deepest breath possible.
   c. Place the mouthpiece between your lips and exhale all the air out of your lungs, slowly and forcefully.

Figure 44.3 Graphic representation of respiratory volumes and capacities.

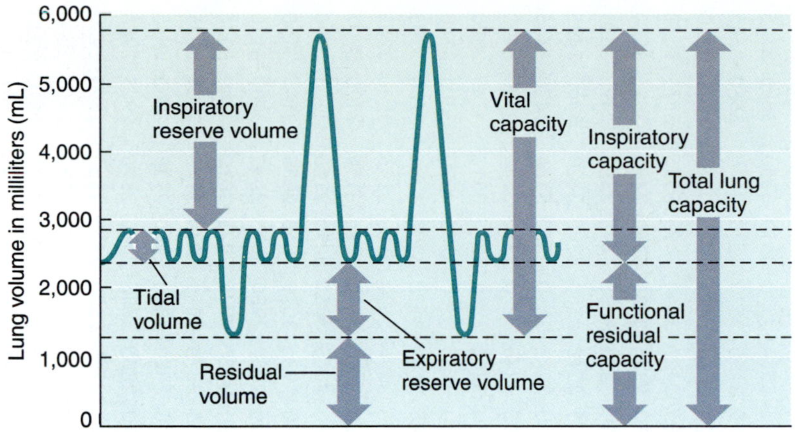

d. Record the value as your vital capacity in Part C. Compare your result with that expected for a person of your sex, age, and height as listed in tables 44.1 and 44.2. Use the meterstick to determine your height in centimeters if necessary or multiply your height in inches times 2.54 to calculate your height in centimeters. Considerable individual variations from the expected will be noted due to parameters other than sex, age, and height, which could include physical shape, health, medications, and others.

Critical Thinking Application

It can be noted from the data in tables 44.1 and 44.2 that vital capacities gradually decrease with age. Propose an explanation for this normal correlation.

____

____

____

____

7. *Inspiratory reserve volume* (IRV) is the volume of air in addition to the tidal volume that enters the lungs during forced inspiration. Calculate your inspiratory reserve volume by subtracting your tidal volume (TV) and your expiratory reserve volume (ERV) from your vital capacity (VC):

$$IRV = VC - (TV + ERV)$$

8. *Inspiratory capacity* (IC) is the maximum volume of air a person can inhale following exhalation of the tidal volume. Calculate your inspiratory capacity by adding your tidal volume (TV) and your inspiratory reserve volume (IRV):

$$IC = TV + IRV$$

9. *Functional residual capacity* (FRC) is the volume of air that remains in the lungs following exhalation of the tidal volume. Calculate your functional residual capacity (FRC) by adding your expiratory reserve volume (ERV) and your residual volume (RV), which you can assume is 1,200 mL:

$$FRC = ERV + 1{,}200$$

10. Complete Part C of the laboratory report.

OPTIONAL ACTIVITY

Determine your *minute respiratory volume.* To do this, follow these steps:

1. Sit quietly for a while, and then to establish your breathing rate, count the number of times you breathe in 1 minute. This might be inaccurate because conscious awareness of breathing rate can alter the results. You might ask a laboratory partner to record your breathing rate at some time when you are not expecting it to be recorded.
2. Calculate your minute respiratory volume by multiplying your breathing rate by your tidal volume:

__________ × __________ = ________________

(breathing rate) (tidal volume) (minute respiratory volume)

3. This value indicates the total volume of air that moves into your respiratory passages during each minute of ordinary breathing.

Table 44.1 Predicted Vital Capacities (in Milliliters) for Females

	Height in Centimeters																								
Age	**146**	**148**	**150**	**152**	**154**	**156**	**158**	**160**	**162**	**164**	**166**	**168**	**170**	**172**	**174**	**176**	**178**	**180**	**182**	**184**	**186**	**188**	**190**	**192**	**194**
16	2950	2990	3030	3070	3110	3150	3190	3230	3270	3310	3350	3390	3430	3470	3510	3550	3590	3630	3670	3715	3755	3800	3840	3880	3920
17	2935	2975	3015	3055	3095	3135	3175	3215	3255	3295	3335	3375	3415	3455	3495	3535	3575	3615	3655	3695	3740	3780	3820	3860	3900
18	2920	2960	3000	3040	3080	3120	3160	3200	3240	3280	3320	3360	3400	3440	3480	3520	3560	3600	3640	3680	3720	3760	3800	3840	3880
20	2890	2930	2970	3010	3050	3090	3130	3170	3210	3250	3290	3330	3370	3410	3450	3490	3525	356+5	3605	3645	3695	3720	3760	3800	3840
22	2860	2900	2940	2980	3020	3060	3095	3135	3175	3215	3255	3290	3330	3370	3410	3450	3490	3530	3570	3610	3650	3685	3725	3765	3800
24	2830	2870	2910	2950	2985	3025	3065	3100	3140	3180	3220	3260	3300	3335	3375	3415	3455	3490	3530	3570	3610	3650	3685	3725	3765
26	2800	2840	2880	2920	2960	3000	3035	3070	3110	3150	3190	3230	3265	3300	3340	3380	3420	3455	3495	3530	3570	3610	3650	3685	3725
28	2775	2810	2850	2890	2930	2965	3000	3040	3070	3115	3155	3190	3230	3270	3305	3345	3380	3420	3460	3495	3535	3570	3610	3650	3685
30	2745	2780	2820	2860	2895	2935	2970	3010	3045	3085	3120	3160	3195	3235	3270	3310	3345	3385	3420	3460	3495	3535	3570	3610	3645
32	2715	2750	2790	2825	2865	2900	2940	2975	3015	3050	3090	3125	3160	3200	3235	3275	3310	3350	3385	3425	3460	3495	3535	3570	3610
34	2685	2725	2760	2795	2835	2870	2910	2945	2980	3020	3055	3090	3130	3165	3200	3240	3275	3310	3350	3385	3425	3460	3495	3535	3570
36	2655	2695	2730	2765	2805	2840	2875	2910	2950	2985	3020	3060	3095	3130	3165	3205	3240	3275	3310	3350	3385	3420	3460	3495	3530
38	2630	2665	2700	2735	2770	2810	2845	2880	2915	2950	2990	3025	3060	3095	3130	3170	3205	3240	3275	3310	3350	3385	3420	3455	3490
40	2600	2635	2670	2705	2740	2775	2810	2850	2885	2920	2955	2990	3025	3060	3095	3135	3170	3205	3240	3275	3310	3345	3380	3420	3455
42	2570	2605	2640	2675	2710	2745	2780	2815	2850	2885	2920	2955	2990	3025	3060	3100	3135	3170	3205	3240	3275	3310	3345	3380	3415
44	2540	2575	2610	2645	2680	2715	2750	2785	2820	2855	2890	2925	2960	2995	3030	3060	3095	3130	3165	3200	3235	3270	3305	3340	3375
46	2510	2545	2580	2615	2650	2685	2715	2750	2785	2820	2855	2890	2925	2960	2995	3030	3060	3095	3130	3165	3200	3235	3270	3305	3340
48	2480	2515	2550	2585	2620	2650	2685	2715	2750	2785	2820	2855	2890	2925	2960	2995	3030	3060	3095	3130	3160	3195	3230	3265	3300
50	2455	2485	2520	2555	2590	2625	2655	2690	2720	2755	2785	2820	2855	2890	2925	2955	2990	3025	3060	3090	3125	3155	3190	3225	3260
52	2425	2455	2490	2525	2555	2590	2625	2655	2690	2720	2755	2790	2820	2855	2890	2925	2955	2990	3020	3055	3090	3125	3155	3190	3220
54	2395	2425	2460	2495	2530	2560	2590	2625	2655	2690	2720	2755	2790	2820	2855	2885	2920	2950	2985	3020	3050	3085	3115	3150	3180
56	2365	2400	2430	2460	2495	2525	2560	2590	2625	2655	2690	2720	2755	2790	2820	2855	2885	2920	2950	2980	3015	3045	3080	3110	3145
58	2335	2370	2400	2430	2460	2495	2525	2560	2590	2625	2655	2690	2720	2750	2785	2815	2850	2880	2920	2945	2975	3010	3040	3075	3105
60	2305	2340	2370	2400	2430	2460	2495	2525	2560	2590	2625	2655	2685	2720	2750	2780	2810	2845	2875	2915	2940	2970	3000	3035	3065
62	2280	2310	2340	2370	2405	2435	2465	2495	2525	2560	2590	2620	2655	2685	2715	2745	2775	2810	2840	2870	2900	2935	2965	2995	3025
64	2250	2280	2310	2340	2370	2400	2430	2465	2495	2525	2555	2585	2620	2650	2680	2710	2740	2770	2805	2835	2865	2895	2925	2955	2990
66	2220	2250	2280	2310	2340	2370	2400	2430	2460	2495	2525	2555	2585	2615	2645	2675	2705	2735	2765	2800	2825	2860	2890	2920	2950
68	2190	2220	2250	2280	2310	2340	2370	2400	2430	2460	2490	2520	2550	2580	2610	2640	2670	2700	2730	2760	2795	2820	2850	2880	2910
70	2160	2190	2220	2250	2280	2310	2340	2370	2400	2425	2455	2485	2515	2545	2575	2605	2635	2665	2695	2725	2755	2780	2810	2840	2870
72	2130	2160	2190	2220	2250	2280	2310	2335	2365	2395	2425	2455	2480	2510	2540	2570	2600	2630	2660	2685	2715	2745	2775	2805	2830
74	2100	2130	2160	2190	2220	2245	2275	2305	2335	2360	2390	2420	2450	2475	2505	2535	2565	2590	2620	2650	2680	2710	2740	2765	2795

From E. DeF. Baldwin and E. W. Richards, Jr., "Pulmonary Insufficiency 1, Physiologic Classification, Clinical Methods of Analysis, Standard Values in Normal Subjects" in *Medicine* 27:243, © by William & Wilkins. Used by permission.

Table 44.2 Predicted Vital Capacities (in Milliliters) for Males

Age	Height in Centimeters																								
	146	148	150	152	154	156	158	160	162	164	166	168	170	172	174	176	178	180	182	184	186	188	190	192	194
16	3765	3820	3870	3920	3975	4025	4075	4130	4180	4230	4285	4335	4385	4440	4490	4540	4590	4645	4595	4745	4800	4850	4900	4955	5005
18	3740	3790	3840	3890	3940	3995	4045	4095	4145	4200	4250	4300	4350	4405	4455	4505	4555	4610	4660	4710	4760	4815	4865	4915	4965
20	3710	3760	3810	3860	3910	3960	4015	4065	4115	4165	4215	4265	4320	4370	4420	4470	4520	4570	4625	4675	4725	4775	4825	4875	4930
22	3680	3730	3780	3830	3880	3930	3980	4030	4080	4135	4185	4235	4285	4335	4385	4435	4485	4535	4585	4635	4685	4735	4790	4840	4890
24	3635	3685	3735	3785	3835	3885	3935	3985	4035	4085	4135	4185	4235	4285	4330	4380	4430	4480	4530	4580	4630	4680	4730	4780	4830
26	3605	3655	3705	3755	3805	3855	3905	3955	4000	4050	4100	4150	4200	4250	4300	4350	4395	4445	4495	4545	4595	4645	4695	4740	4790
28	3575	3625	3675	3725	3775	3820	3870	3920	3970	4020	4070	4115	4165	4215	4265	4310	4360	4410	4460	4510	4555	4605	4655	4705	4755
30	3550	3595	3645	3695	3740	3790	3840	3890	3935	3985	4035	4080	4130	4180	4230	4275	4325	4375	4425	4470	4520	4570	4615	4665	4715
32	3520	3565	3615	3665	3710	3760	3810	3855	3905	3950	4000	4050	4095	4145	4195	4240	4290	4340	4385	4435	4485	4530	4580	4625	4675
34	3475	3525	3570	3620	3665	3715	3760	3810	3855	3905	3950	4000	4045	4095	4140	4190	4225	4285	4330	4380	4425	4475	4520	4570	4615
36	3445	3495	3540	3585	3635	3680	3730	3775	3825	3870	3920	3965	4010	4060	4105	4155	4200	4250	4295	4340	4390	4435	4485	4530	4580
38	3415	3465	3510	3555	3605	3650	3695	3745	3790	3840	3885	3930	3980	4025	4070	4120	4165	4210	4260	4305	4350	4400	4445	4495	4540
40	3385	3435	3480	3525	3575	3620	3665	3710	3760	3805	3850	3900	3945	3990	4035	4085	4130	4175	4220	4270	4315	4360	4410	4455	4500
42	3360	3405	3450	3495	3540	3590	3635	3680	3725	3770	3820	3865	3910	3955	4000	4050	4095	4140	4185	4230	4280	4325	4370	4415	4460
44	3315	3360	3405	3450	3495	3540	3585	3630	3675	3725	3770	3815	3860	3905	3950	3995	4040	4085	4130	4175	4220	4270	4315	4360	4405
46	3285	3330	3375	3420	3465	3510	3555	3600	3645	3690	3735	3780	3825	3870	3915	3960	4005	4050	4095	4140	4185	4230	4275	4320	4365
48	3255	3300	3345	3390	3435	3480	3525	3570	3615	3655	3700	3745	3790	3835	3880	3925	3970	4015	4060	4105	4150	4190	4235	4280	4325
50	3210	3255	3300	3345	3390	3430	3475	3520	3565	3610	3650	3695	3740	3785	3830	3870	3915	3960	4005	4050	4090	4135	4180	4225	4270
52	3185	3225	3270	3315	3355	3400	3445	3490	3530	3575	3620	3660	3705	3750	3795	3835	3880	3925	3970	4010	4055	4100	4140	4185	4230
54	3155	3195	3240	3285	3325	3370	3415	3455	3500	3540	3585	3630	3670	3715	3760	3800	3845	3890	3930	3975	4020	4060	4105	4145	4190
56	3125	3165	3210	3255	3295	3340	3380	3425	3465	3510	3550	3595	3640	3680	3725	3765	3810	3850	3895	3940	3980	4025	4065	4110	4150
58	3080	3125	3165	3210	3250	3290	3335	3375	3420	3460	3500	3545	3585	3630	3670	3715	3755	3800	3840	3880	3925	3965	4010	4050	4095
60	3050	3095	3135	3175	3220	3260	3300	3345	3385	3430	3470	3500	3555	3595	3635	3680	3720	3760	3805	3845	3885	3930	3970	4015	4055
62	3020	3060	3110	3150	3190	3230	3270	3310	3350	3390	3440	3480	3520	3560	3600	3640	3680	3730	3770	3810	3850	3890	3930	3970	4020
64	2990	3030	3080	3120	3160	3200	3240	3280	3320	3360	3400	3440	3490	3530	3570	3610	3650	3690	3730	3770	3810	3850	3900	3940	3980
66	2950	2990	3030	3070	3110	3150	3190	3230	3270	3310	3350	3390	3430	3470	3510	3550	3600	3640	3680	3720	3760	3800	3840	3880	3920
68	2920	2960	3000	3040	3080	3120	3160	3200	3240	3280	3320	3360	3400	3440	3480	3520	3560	3600	3640	3680	3720	3760	3800	3840	3880
70	2890	2930	2970	3010	3050	3090	3130	3170	3210	3250	3290	3330	3370	3410	3450	3480	3520	3560	3600	3640	3680	3720	3760	3800	3840
72	2860	2900	2940	2980	3020	3060	3100	3140	3180	3210	3250	3290	3330	3370	3410	3450	3490	3530	3570	3610	3650	3680	3720	3760	3800
74	2820	2860	2900	2930	2970	3010	3050	3090	3130	3170	3200	3240	3280	3320	3360	3400	3440	3470	3510	3550	3590	3630	3670	3710	3740

From E. DeF. Baldwin and E. W. Richards, Jr., "Pulmonary Insufficiency 1, Physiologic Classification, Clinical Methods of Analysis, Standard Values in Normal Subjects" in *Medicine* 27:243, © by William & Wilkins. Used by permission.

Laboratory Report **44**

Name ______________________________

Date ______________________________

Section ______________________________

Breathing and Respiratory Volumes and Capacities

PART A

Complete the following statements:

1. Breathing can also be called ______________________________ .
2. The weight of air causes a force called ______________________ pressure.
3. The weight of air at sea level is sufficient to support a column of mercury within a tube ______________________ mm high.
4. If the pressure inside the lungs decreases, outside air is pushed into the airways by ______________________ .
5. Nerve impulses are carried to the diaphragm by the ______________________________ nerves.
6. When the diaphragm contracts, the size of the thoracic cavity ______________________________ .
7. The ribs are raised by the contraction of the ______________________________ muscles, which increases the size of the thoracic cavity.
8. Only a thin film of lubricating serous fluid separates the parietal pleura from the ______________________ of a lung.
9. A mixture of lipoproteins, called ______________________________ , acts to reduce the tendency of alveoli to collapse.
10. The force responsible for normal expiration comes from ______________________________ of tissues and from surface tension.
11. Muscles that help to force out more than the normal volume of air by pulling the ribs downward and inward include the ______________________________ .
12. The diaphragm can be forced to move higher than normal by the contraction of the ______________________ muscles.

PART B

Match the air volumes in column A with the definitions in column B. Place the letter of your choice in the space provided.

Column A

a. Expiratory reserve volume
b. Functional residual capacity
c. Inspiratory capacity
d. Inspiratory reserve volume
e. Residual volume
f. Tidal volume
g. Total lung capacity
h. Vital capacity

Column B

_______ 1. Volume in addition to tidal volume that leaves the lungs during forced expiration

_______ 2. Vital capacity plus residual volume

_______ 3. Volume that remains in lungs after the most forceful expiration

_______ 4. Volume that enters or leaves lungs during a respiratory cycle

_______ 5. Volume in addition to tidal volume that enters lungs during forced inspiration

_______ 6. Maximum volume a person can exhale after taking the deepest possible breath

_______ 7. Maximum volume a person can inhale following exhalation of the tidal volume

_______ 8. Volume of air remaining in the lungs following exhalation of the tidal volume

PART C

1. Test results for respiratory air volumes and capacities:

Respiratory Volume or Capacity	Expected Value* (approximate)	Test Result	Percent of Expected Value (test result/expected value ×100)
Tidal volume (resting) (TV)	500 mL		
Expiratory reserve volume (ERV)	1,100 mL		
Vital capacity (VC)	4,600 mL (or enter yours from the table)		
Inspiratory reserve volume (IRV)	3,000 mL		
Inspiratory capacity (IC)	3,500 mL		
Functional residual capacity (FRC)	2,300 mL		

**The values listed are most characteristic for a tall, young adult.*

2. Complete the following:
   a. How do your test results compare with the expected values?

   ____

   ____

   ____

   ____

   b. How does your vital capacity compare with the average value for a person of your sex, age, and height?

   ____

   ____

   c. What measurement in addition to vital capacity is needed before you can calculate your total lung capacity?

   ____

   ____

3. If your experimental results are considerably different than the predicted vital capacities, propose reasons for the differences. As you write this paragraph, consider factors such as smoking, exercise, respiratory disorders, and medications. (Your instructor might have you make some class correlations from class data.)

   ____

   ____

   ____

   ____

   ____

Structure of the Kidney

MATERIALS NEEDED

Textbook
Human torso
Kidney model
Preserved pig (or sheep) kidney
Dissecting tray
Dissecting instruments
Long knife
Compound microscope
Prepared microscope slide of a kidney section

SAFETY

- Wear disposable gloves when working on the kidney dissection.
- Dispose of the kidney and gloves as instructed.
- Wash the dissecting tray and instruments as instructed.
- Wash your laboratory table.
- Wash your hands before leaving the laboratory.

The two kidneys are the primary organs of the urinary system. They are located in the upper quadrants of the abdominal cavity, against the posterior wall and behind the parietal peritoneum. They perform a variety of complex activities that lead to the production of urine.

The other organs of the urinary system include the ureters, which transport urine away from the kidneys; the urinary bladder, which stores urine; and the urethra, which conveys urine to the outside of the body.

PURPOSE OF THE EXERCISE

To review the structure of the kidney, to dissect a kidney, and to observe the major structures of a nephron microscopically.

LEARNING OBJECTIVES

After completing this exercise, you should be able to

1. describe the location of the kidneys;
2. locate and identify the major structures of a kidney;
3. identify the major structures of a nephron;
4. trace the path of filtrate through a renal tubule;
5. trace the path of blood through the renal blood vessels.

PROCEDURE A—KIDNEY STRUCTURE

1. Review the section entitled "Kidney Structure" in chapter 17 of the textbook.
2. As a review activity, label figure 45.1.
3. Complete Part A of Laboratory Report 45.
4. Observe the human torso and the kidney model. Locate the following:

kidneys

ureters

urinary bladder

urethra

renal sinus

renal pelvis

major calyces

minor calyces

renal papillae

renal medulla

renal pyramids

renal cortex

renal columns

nephrons

Figure 45.1 Label the major structures in the longitudinal section of a kidney.

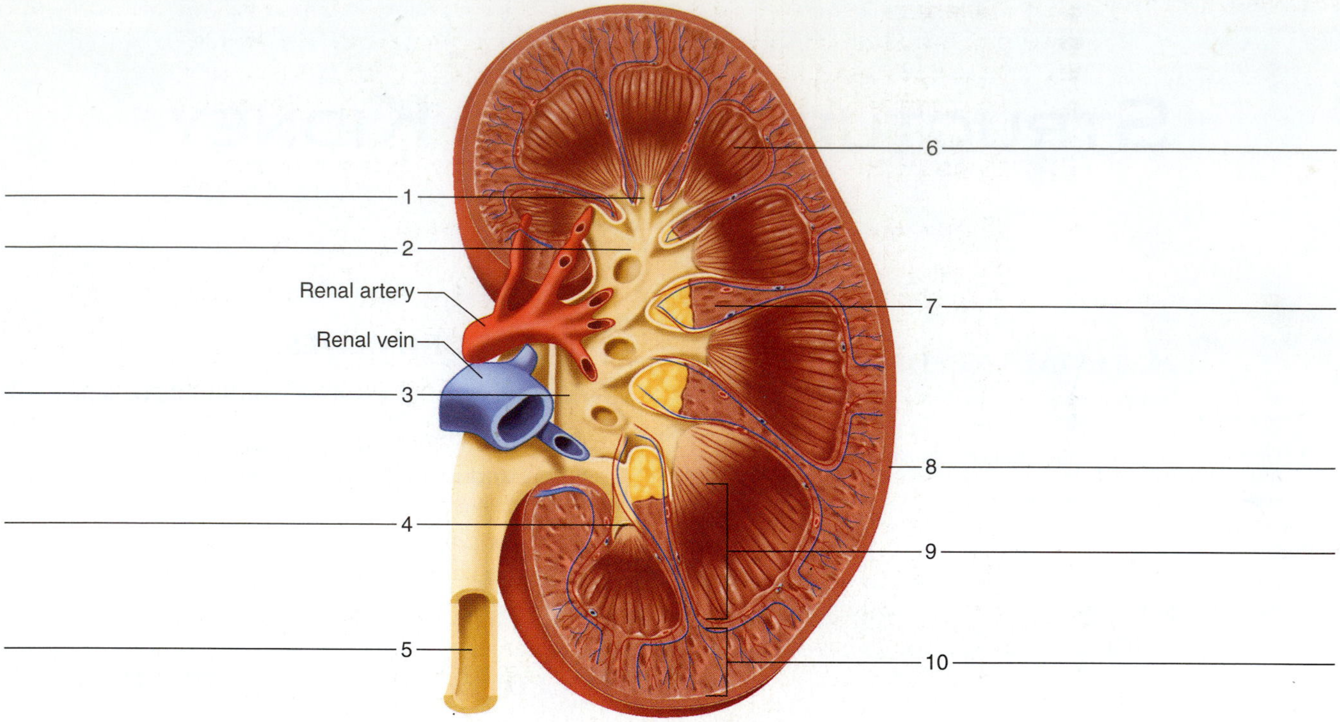

5. To observe the structure of a kidney:
   a. Obtain a preserved pig or sheep kidney and rinse it with water to remove as much of the preserving fluid as possible.
   b. Carefully remove any adipose tissue from the surface of the specimen.
   c. Locate the following features:

      renal capsule

      hilum

      renal artery

      renal vein

      ureter

   d. Use a long knife to cut the kidney in half longitudinally along the frontal plane, beginning on the convex border.
   e. Rinse the interior of the kidney with water, and using figure 45.2 as a reference, locate the following:

      renal pelvis

         major calyces

         minor calyces

      renal cortex

         renal columns (extensions of renal cortex between renal pyramids)

      renal medulla

         renal pyramids

PROCEDURE B—RENAL BLOOD VESSELS AND NEPHRONS

1. Review the sections entitled "Renal Blood Vessels" and "Nephrons" in chapter 17 of the textbook.
2. As a review activity, label figure 45.3.
3. Complete Part B of the laboratory report.
4. Obtain a microscope slide of a kidney section and examine it, using low-power magnification. Locate the *renal capsule, renal cortex* (which appears somewhat granular and may be more darkly stained than the other renal tissues), and *renal medulla* (fig. 45.4).
5. Examine the renal cortex, using high-power magnification. Locate a *renal corpuscle.* These structures appear as isolated circular areas. Identify the *glomerulus,* which is the capillary cluster inside the corpuscle, and the *glomerular capsule,* which

Figure 45.2 Longitudinal section of a pig kidney that has a triple injection of latex (*red* in the renal artery, *blue* in the renal vein, and *yellow* in the ureter and renal pelvis).

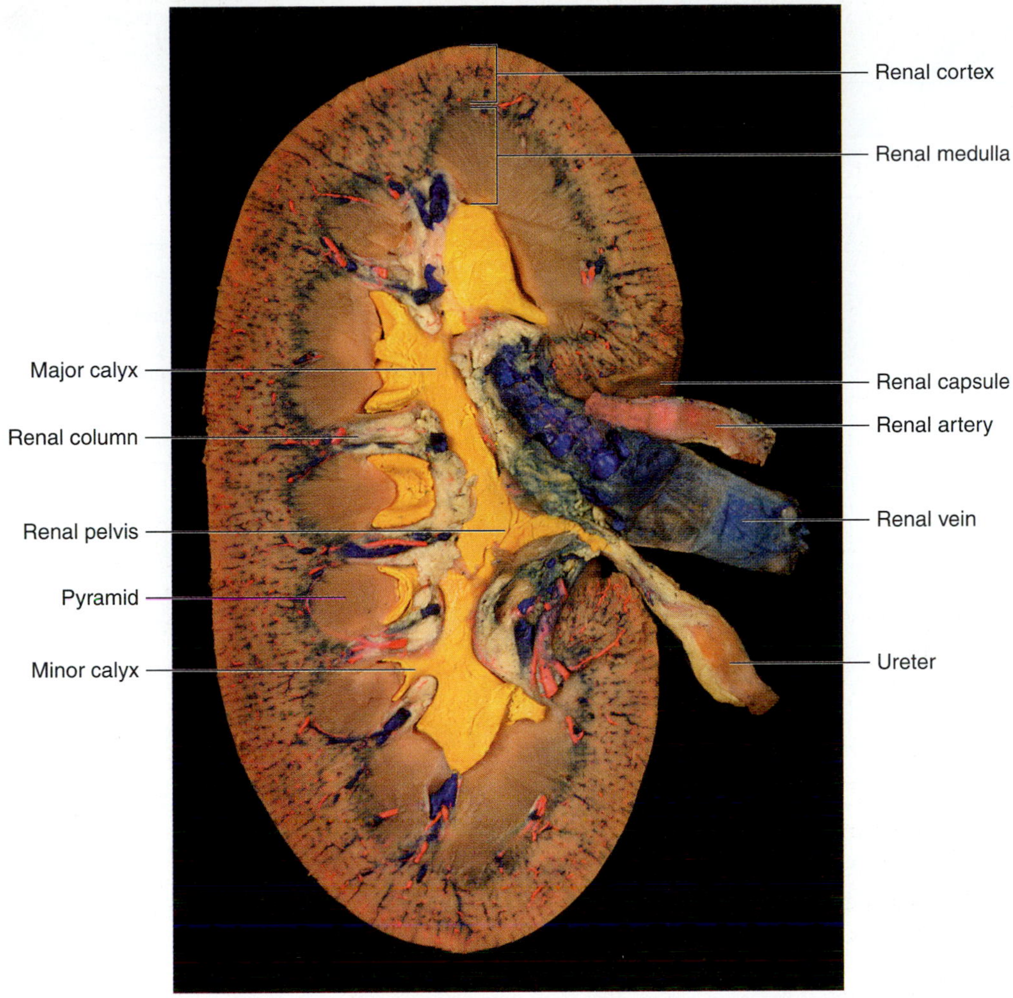

appears as a clear area surrounding the glomerulus. Also note the numerous sections of renal tubules that occupy the spaces between renal corpuscles (fig. 45.4).

6. Prepare a labeled sketch of a representative section of the renal cortex in Part C of the laboratory report.
7. Examine the renal medulla using high-power magnification (fig. 45.4). Identify longitudinal and cross sections of various collecting ducts. Note that these ducts are lined with simple epithelial cells, which vary in shape from squamous to cuboidal.
8. Prepare a labeled sketch of a representative section of renal medulla in Part D of the laboratory report.

Web Quest

Review the structure and function of the urinary system at www.mhhe.com/shieress8

Figure 45.3 Label the major structures of the nephron and the blood vessels associated with it, using the terms provided.

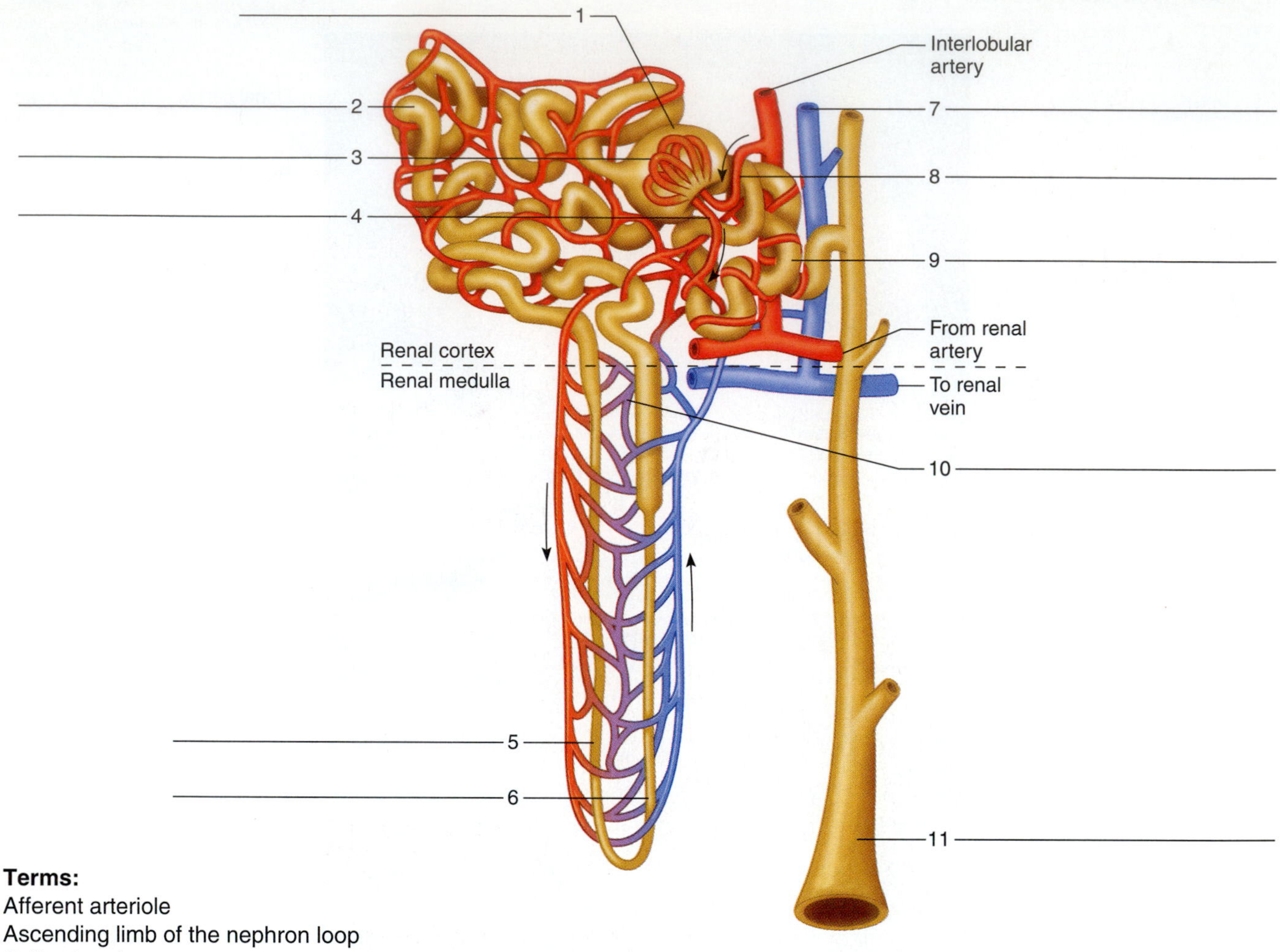

Terms:
Afferent arteriole
Ascending limb of the nephron loop
Collecting duct
Descending limb of the nephron loop
Distal convoluted tubule
Efferent arteriole
Glomerular capsule
Glomerulus
Interlobular vein
Peritubular capillary
Proximal convoluted tubule

Figure 45.4 (*a*) Micrograph of a section of the renal cortex (220×). (*b*) Micrograph of a section of the renal medulla (80× micrograph enlarged to 200×).

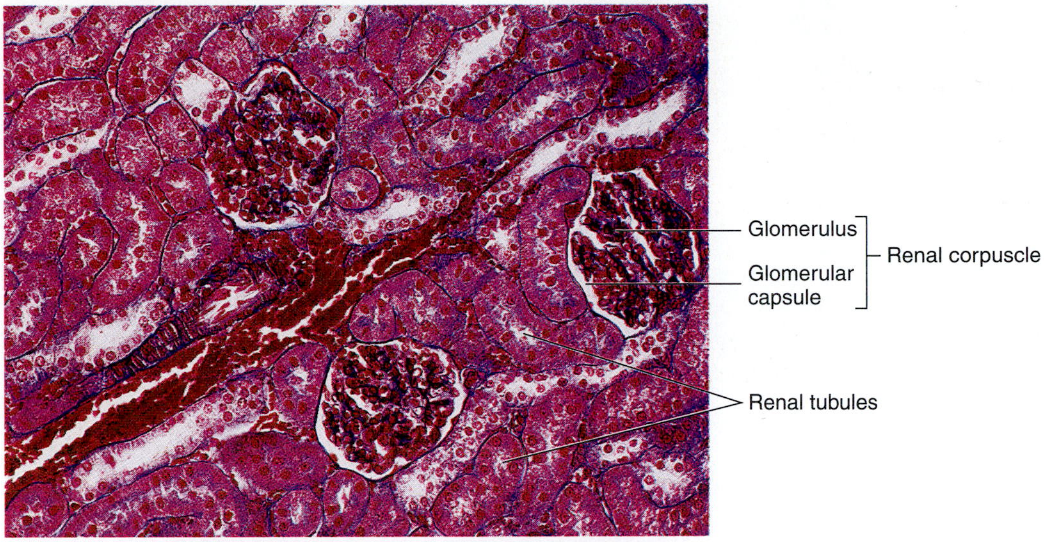

(a)

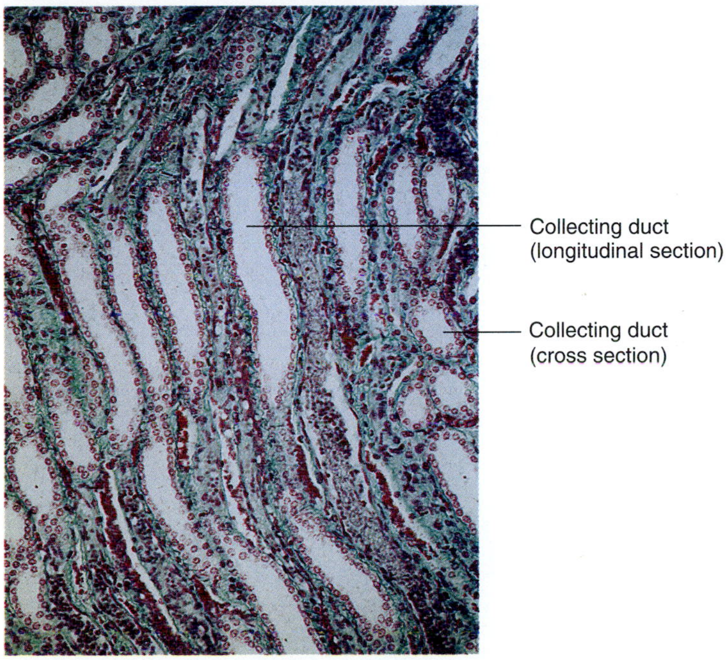

(b)

Laboratory Report **45**

Name ______________________________

Date ______________________________

Section ______________________________

STRUCTURE OF THE KIDNEY

PART A

Match the terms in column A with the descriptions in column B. Place the letter of your choice in the space provided.

Column A		Column B
a. Calyces	_______	1. Shell around renal medulla
b. Hilum	_______	2. Branches of renal pelvis to renal papillae
c. Nephron	_______	3. Conical mass of tissue within renal medulla
d. Renal column	_______	4. Projection with tiny openings into minor calyx
e. Renal cortex	_______	5. Medial depression for blood vessels and ureter to enter kidney chamber
f. Renal papilla	_______	6. Hollow chamber within kidney
g. Renal pelvis	_______	7. Microscopic functional unit of kidney
h. Renal pyramid	_______	8. Cortical tissue between renal pyramids
i. Renal sinus	_______	9. Superior funnel-shaped end of ureter inside renal sinus

PART B

1. Distinguish between a renal corpuscle and a renal tubule. ______________________________

____

____

____

____

2. Number the following structures to indicate their respective positions in relation to the nephron. Assign number 1 to the part attached to the glomerular capsule.

______ ascending limb of the nephron loop

______ collecting duct

______ descending limb of the nephron loop

______ distal convoluted tubule

______ proximal convoluted tubule

______ renal papilla

3. Explain how the blood vessels associated with the renal corpuscle help to maintain a relatively high blood pressure within the glomerulus. ______________________________

4. Define and explain *juxtaglomerular apparatus.* ______________________________

PART C

Prepare a sketch of a representative section of renal cortex. Label the glomerulus, glomerular capsule, and sections of renal tubules.

PART D

Prepare a sketch of a representative section of renal medulla. Label a longitudinal section and a cross section of a collecting duct.

Laboratory Exercise 46

Urinalysis

MATERIALS NEEDED

Disposable urine-collecting container
Paper towel
Urinometer cylinder
Urinometer hydrometer
Laboratory thermometer
pH test paper
Reagent strips (individual or combination strips such as Chemstrip or Multistix) to test for the presence of the following:
- Glucose
- Protein
- Ketones
- Bilirubin
- Hemoglobin/occult blood

Compound microscope
Microscope slide
Coverslip
Centrifuge
Centrifuge tube
Graduated cylinder, 10 mL
Medicine dropper
Sedi-stain
Normal and abnormal simulated urine specimens (optional)

SAFETY

- Wear disposable gloves when working with body fluids.
- Work only with your own urine sample.
- Use an appropriate disinfectant to wash the laboratory table before and after the procedures.
- Place glassware in a disinfectant when finished.
- Dispose of contaminated items as instructed.
- Wash your hands before leaving the laboratory.

Urine is the product of kidney functions, which include the removal of various waste substances from the blood and the maintenance of body fluid and electrolyte balances. Consequently, the composition of urine varies from time to time because of differences in dietary intake and physical activity. Also, the volume of urine produced by the kidneys varies with such factors as fluid intake, environmental temperature, relative humidity, respiratory rate, and body temperature.

An analysis of urine composition and volume often is used to evaluate the functions of the kidneys and other organs. This procedure, called *urinalysis,* involves observing the physical characteristics of a urine sample, testing for the presence of certain organic and inorganic substances, and examining the microscopic solids present in the sample.

PURPOSE OF THE EXERCISE

To perform the observations and tests commonly used to analyze the characteristics and composition of urine.

LEARNING OBJECTIVES

After completing this exercise, you should be able to

1. evaluate the color, transparency, and specific gravity of a urine sample;
2. determine the pH of a urine sample;
3. test a urine sample for the presence of glucose, protein, ketones, bilirubin, and hemoglobin;
4. perform a microscopic study of urine sediment;
5. evaluate the results of these observations and tests.

WARNING

While performing the following tests, you should wear disposable latex gloves so that skin contact with the urine is avoided. Observe all safety procedures listed for this lab. (Normal and abnormal simulated urine specimens could be used instead of real urine for this lab.)

PROCEDURE

1. Proceed to the restroom with a clean disposable container. The first small volume of urine should not be collected because it contains abnormally high levels of microorganisms from the urethra. Collect a midstream sample of about 50 mL of urine. The best collections are the first specimen in the morning or one taken 3 hours after a meal. Refrigerate samples if they are not used immediately.

2. Place a sample of urine in a clean, transparent container. Describe the *color* of the urine. Normal urine varies from light yellow to amber, depending on the presence of urochromes, which are end-product pigments produced during the decomposition of hemoglobin. Dark urine indicates a high concentration of pigments.

   Abnormal urine colors include yellow-brown or green, due to elevated concentrations of bile pigments, and red to dark brown, due to the presence of blood. Certain foods, such as beets or carrots, and various drug substances also may cause color changes in urine, but in such cases the colors have no clinical significance. Enter the results of this and the following tests in Part A of Laboratory Report 46.
3. Evaluate the *transparency* of the urine sample (that is, judge whether the urine is clear, slightly cloudy, or very cloudy). Normal urine is clear enough to see through. You can read newsprint through slightly cloudy urine; you can no longer read newsprint through cloudy urine. Cloudy urine indicates the presence of various substances that may include mucus, bacteria, epithelial cells, fat droplets, or inorganic salts.
4. Determine the *specific gravity* of the urine sample. Specific gravity is the ratio of the weight of something to the weight of an equal volume of pure water. For example, mercury (at 15°C) weighs 13.6 times as much as an equal volume of water; thus, it has a specific gravity of 13.6. Although urine is mostly water, it has substances dissolved in it and is slightly heavier than an equal volume of water. Thus, urine has a specific gravity of more than 1.000. Actually, the specific gravity of normal urine varies from 1.003 to 1.035.

   To determine the specific gravity of a urine sample, follow these steps:

   a. Pour enough urine into a clean urinometer cylinder to fill it about three-fourths full. Any foam that appears should be removed with a paper towel.
   b. Use a laboratory thermometer to measure the temperature of the urine.
   c. Gently place the urinometer hydrometer into the urine, and *make sure that the float is not touching the sides of the cylinder* (fig. 46.1).
   d. Position your eye at the level of the urine surface. Determine which line on the stem of the hydrometer intersects the lowest level of the concave surface (meniscus) of the urine.
   e. Because liquids tend to contract and become denser as they are cooled, or to expand and become less dense as they are heated, it may be necessary to make a temperature correction to obtain an accurate specific gravity measurement. To do this, add 0.001 to the hydrometer reading for each 3 degrees of urine temperature above 25°C or subtract 0.001 for each 3 degrees below 25°C. Enter this calculated value in the table of the laboratory report as the test result.

Figure 46.1 Float the hydrometer in the urine, making sure that it does not touch the sides of the cylinder.

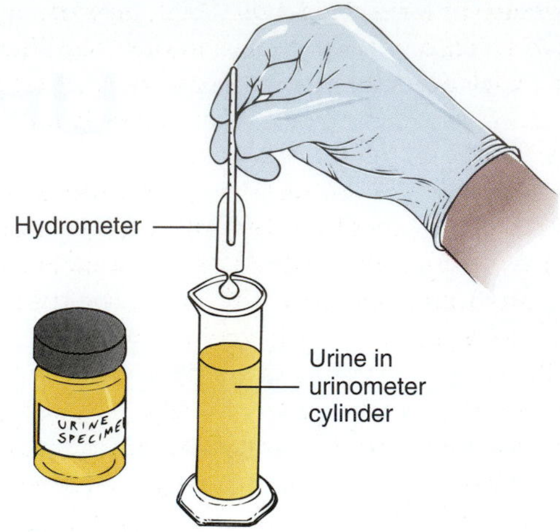

5. Reagent strips can be used to perform a variety of urine tests. In each case, directions for using the strips are found on the strip container. *Be sure to read them.*

   To perform each test, follow these steps:

   a. Obtain a urine sample and the proper reagent strip.
   b. Read the directions on the strip container.
   c. Dip the strip in the urine sample.
   d. Remove the strip at an angle and let it touch the inside rim of the urine container to remove any excess liquid.
   e. Wait for the length of time indicated by the directions on the container before you compare the color of the test strip with the standard color scale on the side of the container. The value or amount represented by the matching color should be used as the test result and recorded in Part A of the laboratory report.

ALTERNATIVE PROCEDURE

If combination reagent test strips (Multistix or Chemstrip) are being utilized, locate the appropriate color chart for each test being evaluated. Wait the designated time for each color reaction before the comparison is made to the standard color scale.

6. Perform the *pH test.* The pH of normal urine varies from 4.6 to 8.0, but most commonly, it is near 6.0 (slightly acidic). The pH of urine may decrease as a result of a diet high in protein, or it may increase with a vegetarian diet. Significant daily variations within the broad normal range are results of concentrations of excesses from variable diets.
7. Perform the *glucose test.* Normally, there is no glucose in urine. However, glucose may appear in the urine temporarily following a meal high in carbohydrates. Glucose also may appear in the urine as a result of uncontrolled diabetes mellitus.
8. Perform the *protein test.* Normally, proteins of large molecular size are not present in urine. However, those of small molecular sizes, such as albumins, may appear in trace amounts, particularly following strenuous exercise. Increased amounts of proteins also may appear as a result of kidney diseases in which the glomeruli are damaged or as a result of high blood pressure.
9. Perform the *ketone test.* Ketones are products of fat metabolism. Usually they are not present in urine. However, they may appear in the urine if the diet fails to provide adequate carbohydrate, as in the case of prolonged fasting or starvation or as a result of insulin deficiency (diabetes mellitus).
10. Perform the *bilirubin test.* Bilirubin, which results from hemoglobin decomposition in the liver, normally is absent in urine. It may appear, however, as a result of liver disorders that cause obstructions of the biliary tract. Urochrome, a normal yellow component of urine, is a result of additional breakdown of bilirubin.
11. Perform the *hemoglobin/occult blood test.* Hemoglobin occurs in the red blood cells, and because such cells normally do not pass into the renal tubules, hemoglobin is not found in normal urine. Its presence in urine usually indicates a disease process, a transfusion reaction, or an injury to the urinary organs.
12. Complete Part A of the laboratory report.
13. A urinalysis usually includes a study of urine sediment—the microscopic solids present in a urine sample. This sediment normally includes mucus, certain crystals, and a variety of cells, such as the epithelial cells that line the urinary tubes and an occasional white blood cell. Other types of solids, such as casts or red blood cells, may indicate a disease or injury if they are present in excess. (Casts are cylindrical masses of cells or other substances that form in the renal tubules and are flushed out by the flow of urine.)

    To observe urine sediment, follow these steps:

    a. Thoroughly stir or shake a urine sample to suspend the sediment, which tends to settle to the bottom of the container.
    b. Pour 10 mL of urine into a clean centrifuge tube and centrifuge it for 5 minutes at a slow speed (1,500 rpm). Be sure to balance the centrifuge with an even number of tubes filled to the same levels.
    c. Carefully decant 9 mL (leave 1 mL) of the liquid from the sediment in the bottom of the centrifuge tube, as directed by your laboratory instructor. Resuspend the 1 mL of sediment.
    d. Use a medicine dropper to remove some of the sediment and place it on a clean microscope slide.
    e. Add a drop of Sedi-stain to the sample, and add a coverslip.
    f. Examine the sediment with low-power (reduce the light when using low power) and high-power magnifications.
    g. Identify the kinds of solids present with the aid of figure 46.2.
    h. In Part B of the laboratory report, make a sketch of each type of sediment that you observed.

14. Complete Part B of the laboratory report.

Figure 46.2 Types of urine sediment. Healthy individuals lack many of these sediments and possess only occasional to trace amounts of others.

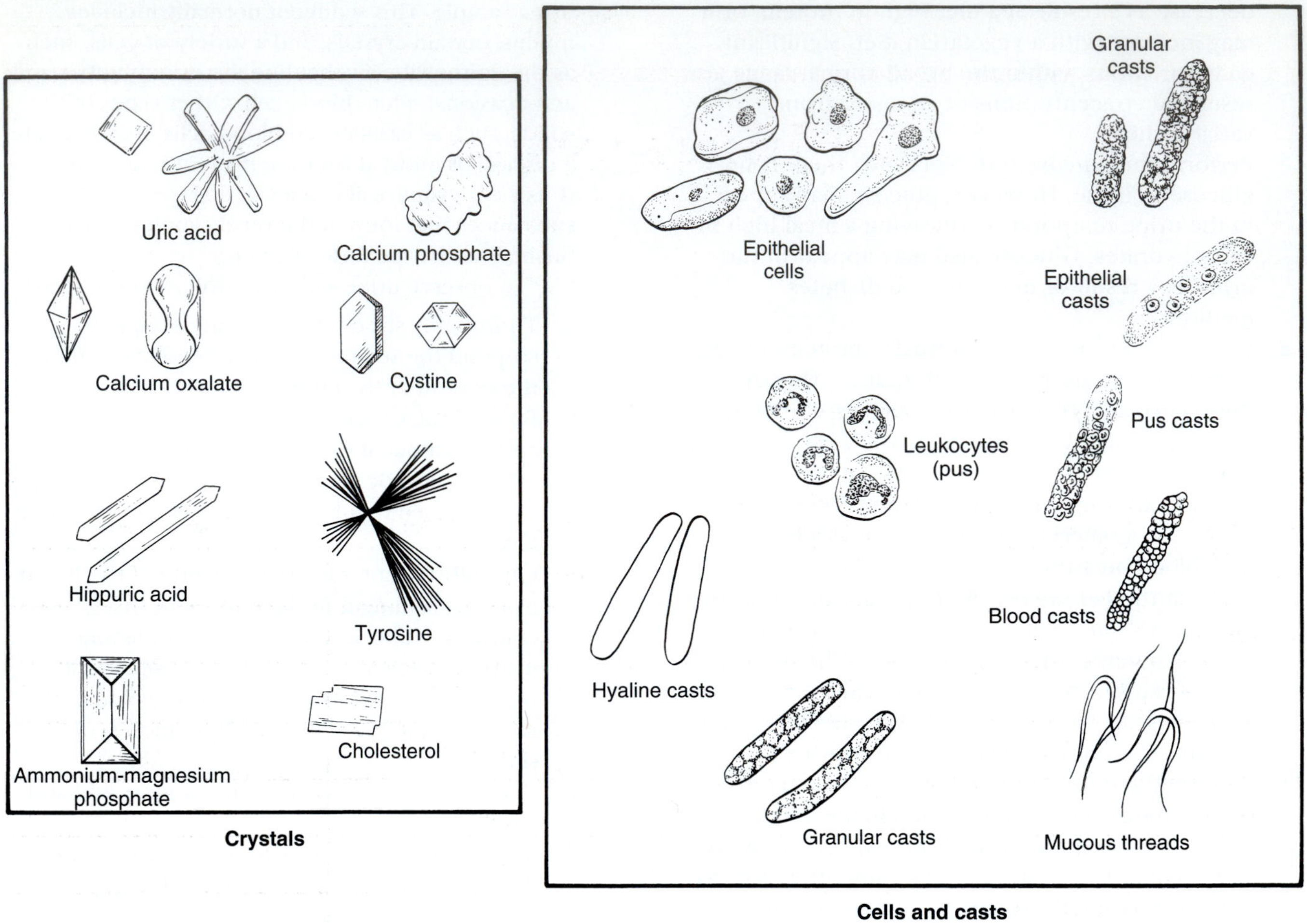

Laboratory Report **46**

Name ______________________

Date ______________________

Section ______________________

URINALYSIS

PART A

1. Enter your observations, test results, and evaluations in the following table.

Urine Characteristics	Observations and Test Results	Normal Values	Evaluations
Color		Light yellow to amber	
Transparency		Clear	
Specific gravity (corrected for temperature)		1.003–1.035	
pH		4.6–8.0	
Glucose		0 (negative)	
Protein		0 to trace	
Ketones		0	
Bilirubin		0	
Hemoglobin/Occult Blood		0	
(Other)			
(Other)			

2. Summarize the results of the urinalysis. ______________________

Critical Thinking Application

Why do you think it is important to refrigerate a urine sample if an analysis cannot be performed immediately after collecting it?

PART B

1. Make a sketch for each type of sediment you observed. Label any from those shown in figure 46.2.

2. Summarize the results of the urine sediment study.

MALE REPRODUCTIVE SYSTEM

MATERIALS NEEDED

Textbook
Human torso
Model of the male reproductive system
Anatomical chart of the male reproductive system
Compound microscope
Prepared microscope slides of the following:
- Testis section
- Epididymis, cross section
- Penis, cross section

The organs of the male reproductive system are specialized to produce and maintain the male sex cells, to transport these cells together with supporting fluids to the female reproductive tract, and to produce and secrete male sex hormones.

These organs include the testes, in which sperm cells and male sex hormones are produced, and sets of internal and external accessory organs. The internal organs include various tubes and glands, whereas the external structures are the scrotum and the penis.

PURPOSE OF THE EXERCISE

To review the structure and functions of the male reproductive organs and to examine some of these organs microscopically.

LEARNING OBJECTIVES

After completing this exercise, you should be able to

1. locate and identify the organs of the male reproductive system;
2. describe the functions of these organs;
3. recognize sections of the testis, epididymis, and penis microscopically;
4. identify the major features of these microscopic sections.

PROCEDURE A—MALE REPRODUCTIVE ORGANS

1. Review the sections entitled "Testes," "Male Internal Accessory Organs," and "Male External Reproductive Organs" in chapter 19 of the textbook.
2. As a review activity, label figures 47.1 and 47.2.
3. Observe the human torso, the model, and anatomic chart of the male reproductive system. Locate the following features:

   testes
   - seminiferous tubules
   - interstitial cells

   epididymis

   vas deferens

   ejaculatory duct

   seminal vesicles

   prostate gland

   bulbourethral glands

   scrotum

   penis
   - corpora cavernosa
   - corpus spongiosum
   - glans penis
   - prepuce
4. Complete Part A of Laboratory Report 47.

Figure 47.1 Label the major structures of the male reproductive system in this sagittal view.

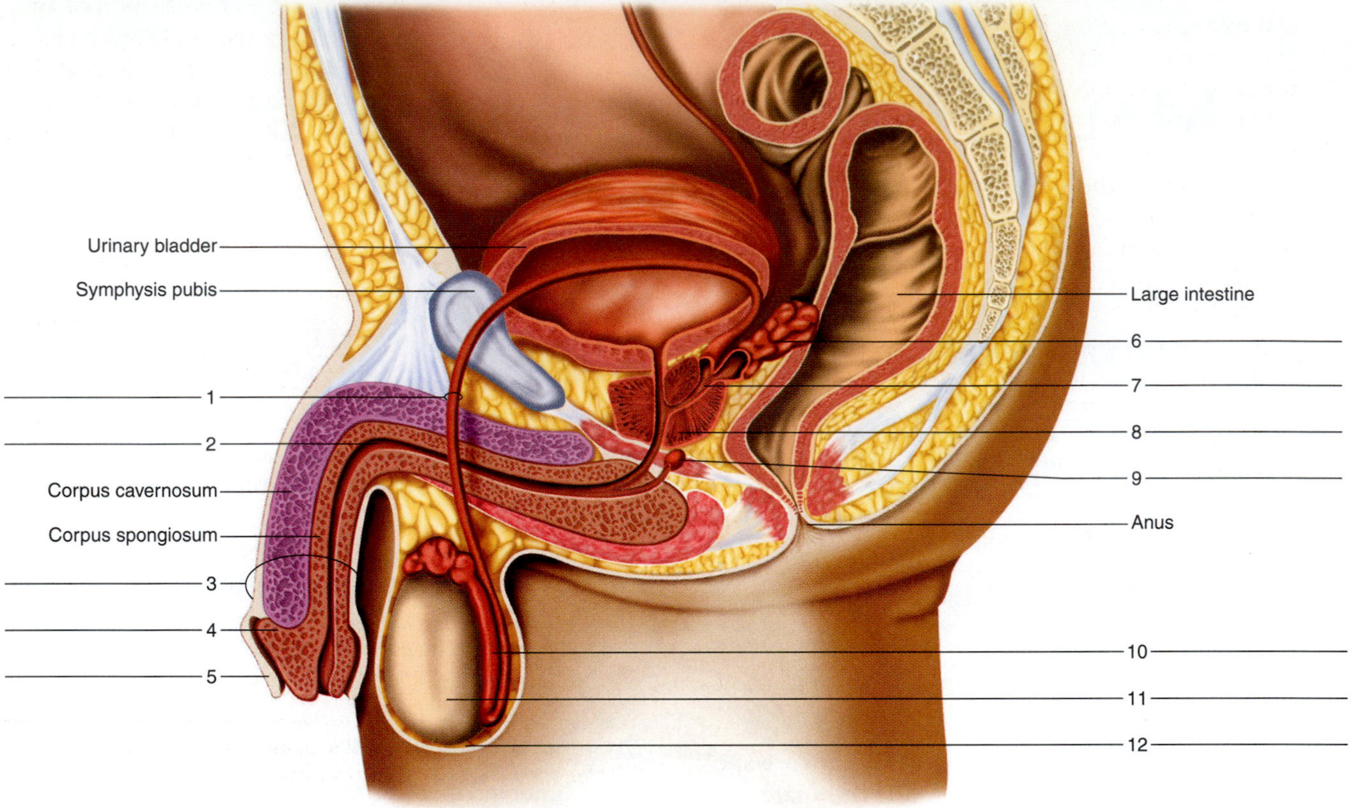

Figure 47.2 Label the diagram of (*a*) the sagittal section of a testis and (*b*) a cross section of a seminiferous tubule by placing the correct numbers in the spaces provided.

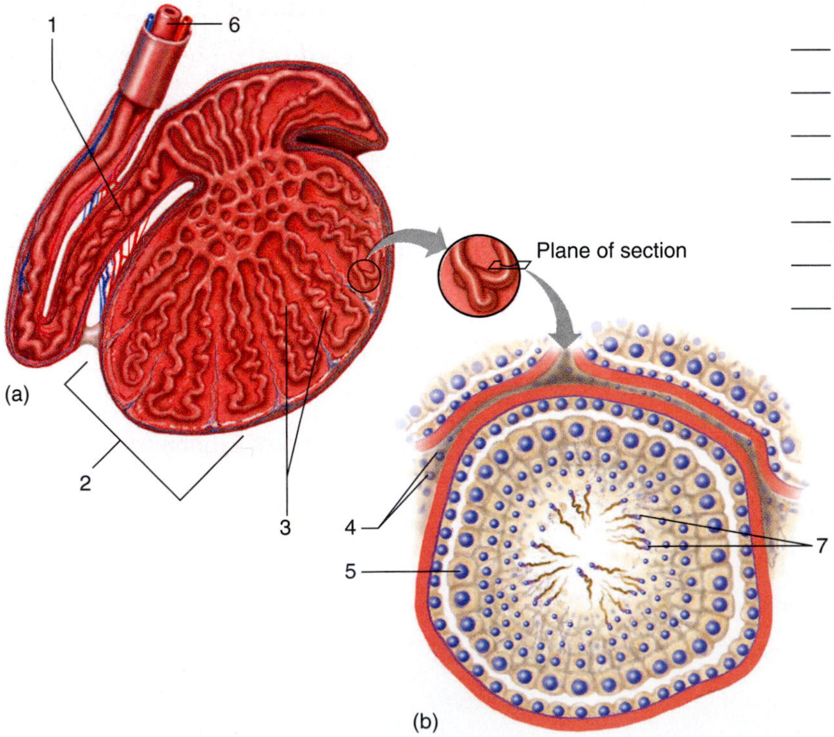

___ Epididymis

___ Epithelial cell (germinal)

___ Interstitial cells

___ Seminiferous tubules

___ Sperm cells (spermatozoa)

___ Testis

___ Vas deferens

PROCEDURE B—MICROSCOPIC ANATOMY

1. Obtain a microscope slide of a human testis section and examine it, using low-power magnification. Locate the thick *fibrous capsule* on the surface and the numerous sections of *seminiferous tubules* inside (fig. 47.3).
2. Focus on some of the seminiferous tubules, using high-power magnification. Locate the *epithelium,* which forms the inner lining of each tube. Within this epithelium, identify some *supporting cells,* which have pale, oval-shaped nuclei, and some *spermatogenic cells,* which have smaller, round nuclei. Near the lumen of the tube, find some darkly stained, elongated heads of developing sperm cells. In the spaces between adjacent seminiferous tubules, locate some isolated *interstitial cells* (fig. 47.4).
3. Prepare a labeled sketch of a representative section of the testis in Part B of the laboratory report.
4. Obtain a microscope slide of a cross section of an *epididymis.* Examine its wall, using high-power magnification. Note the elongated, *pseudostratified columnar epithelial cells,* which comprise most of the inner lining. These cells have nonmotile cilia on their free surfaces (fig. 47.5). Also note the thin layer of smooth muscle and connective tissue surrounding the tube.
5. Prepare a labeled sketch of the epididymis wall in Part B of the laboratory report.
6. Obtain a microscope slide of a *penis* cross section and examine it with low-power magnification (fig. 47.6). Identify the following features:

   corpora cavernosa

   corpus spongiosum

   urethra

   skin

Figure 47.3 Micrograph of a human testis (1.7×).

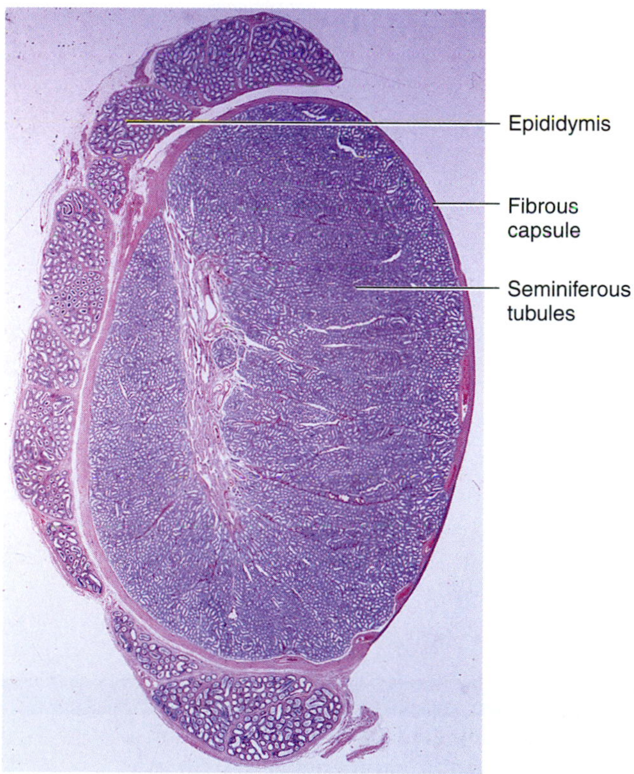

Figure 47.4 Micrograph of seminiferous tubules (50× micrograph enlarged to 135×).

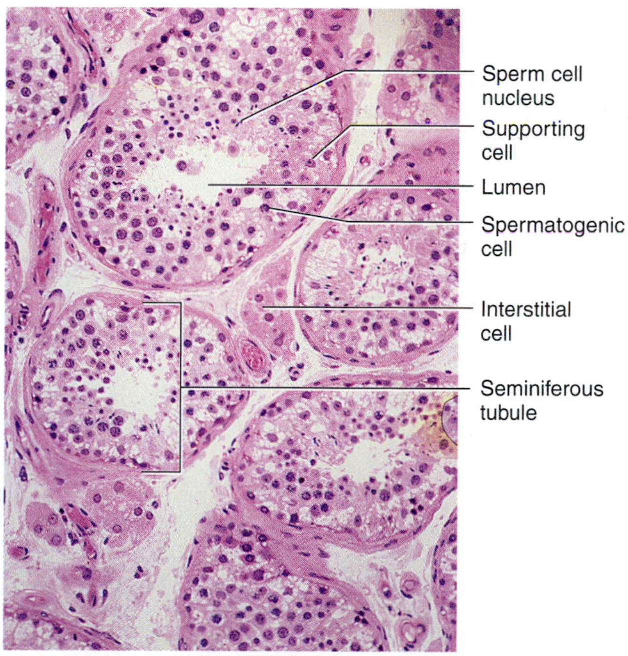

Figure 47.5 Micrograph of a cross section of a human epididymis (50× micrograph enlarged to 145×).

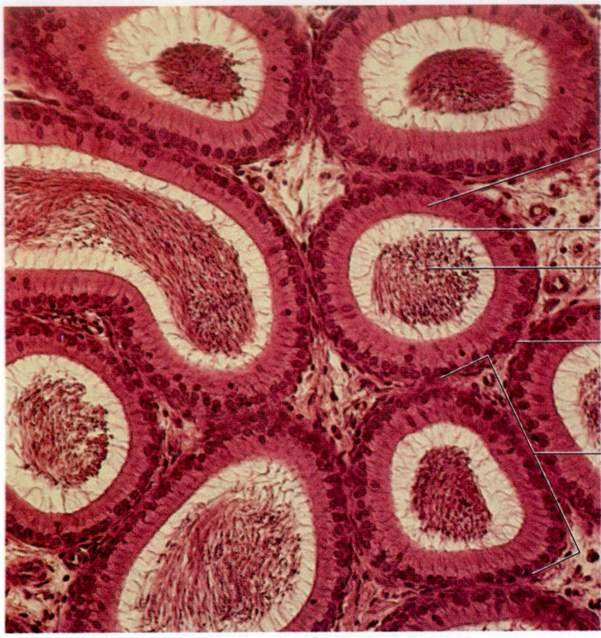

Figure 47.6 Micrograph of a cross section of the body of the penis (5×).

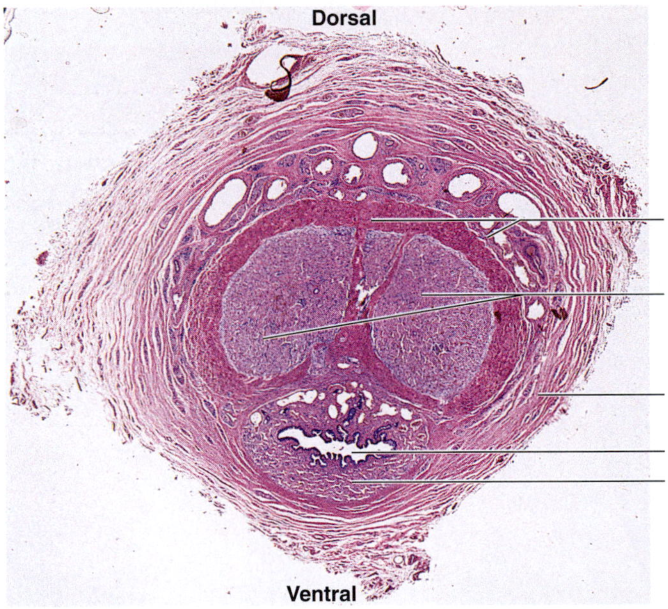

7. Prepare a labeled sketch of a penis cross section in Part B of the laboratory report.
8. Complete Part B of the laboratory report.

Web Quest

Review the structures and functions of the male reproductive system at www.mhhe.com/shieress8

Laboratory Report **47**

Name ______________________

Date ______________________

Section ______________________

MALE REPRODUCTIVE SYSTEM

PART A

Complete the following statements:

1. Connective tissue subdivides a testis into many lobules, which contain ______________________ .
2. The ______________________ is a highly coiled tube on the outer surface of the testis.
3. The ______________________ cells of the epithelium that lines the seminiferous tubules give rise to sperm cells.
4. Undifferentiated sperm cells in the male embryo are called ______________________ .
5. ______________________ is the process by which sperm cells are formed.
6. The number of chromosomes normally present in a human sperm cell is ______________________ .
7. The anterior end of a sperm head, called the ______________________ , contains enzymes that aid in the penetration of an egg cell at the time of fertilization.
8. Sperm cells undergo maturation while they are stored in the ______________________ .
9. The secretion of the seminal vesicles is rich in the monosaccharide called ______________________ .
10. Prostate gland secretion helps neutralize seminal fluid because the prostate secretion is ______________________ .
11. The secretion of the ______________________ glands lubricates the end of the penis in preparation for sexual intercourse.
12. The sensitive, cone-shaped end of the penis is called the ______________________ .

PART B

1. Prepare a labeled sketch of a representative section of the testis.

2. Prepare a labeled sketch of a region of the epididymis wall.

3. Prepare a labeled sketch of a penis cross section.

4. Briefly describe the function of each of the following:

a. supporting cells in seminiferous tubules

____

____

b. spermatogenic cells

____

____

c. interstitial cells

____

____

d. epididymis

____

____

e. corpora cavernosa and corpus spongiosum

____

____

LABORATORY EXERCISE 48

FEMALE REPRODUCTIVE SYSTEM

MATERIALS NEEDED

Textbook
Human torso
Model of the female reproductive system
Anatomical chart of the female reproductive system
Compound microscope
Prepared microscope slides of the following:
- Ovary section with maturing follicles
- Uterine tube, cross section
- Uterine wall section

For Demonstration:

Prepared microscope slides of the following:
- Uterine wall, early proliferative phase
- Uterine wall, secretory phase
- Uterine wall, early menstrual phase

The organs of the female reproductive system are specialized to produce and maintain the female sex cells, to transport these cells to the site of fertilization, to provide a favorable environment for a developing offspring, to move the offspring to the outside, and to produce female sex hormones.

These organs include the ovaries, which produce the egg cells and female sex hormones, and sets of internal and external accessory organs. The internal accessory organs include the uterine tubes, uterus, and vagina. The external organs are the labia majora, labia minora, clitoris, and vestibular glands.

PURPOSE OF THE EXERCISE

To review the structure and functions of the female reproductive organs and to examine some of their features microscopically.

LEARNING OBJECTIVES

After completing this exercise, you should be able to

1. locate and identify the organs of the female reproductive system;
2. describe the functions of these organs;
3. recognize sections of the ovary, uterine tube, and uterine wall microscopically;
4. identify the major features of these microscopic sections.

PROCEDURE A—FEMALE REPRODUCTIVE ORGANS

1. Review the sections entitled "Ovaries," "Female Internal Accessory Organs," "Female External Reproductive Organs," and "Mammary Glands" in chapter 19 of the textbook.
2. As a review activity, label figures 48.1, 48.2, and 48.3.
3. Observe the human torso, the model, and anatomic chart of the female reproductive system. Locate the following features:

ovaries

- medulla
- cortex

uterine tubes (oviducts; fallopian tubes)

- infundibulum
- fimbriae

uterus

- body
- cervix
- endometrium
- myometrium
- perimetrium (serous membrane)

Figure 48.1 Label the structures of the female reproductive system in this sagittal view.

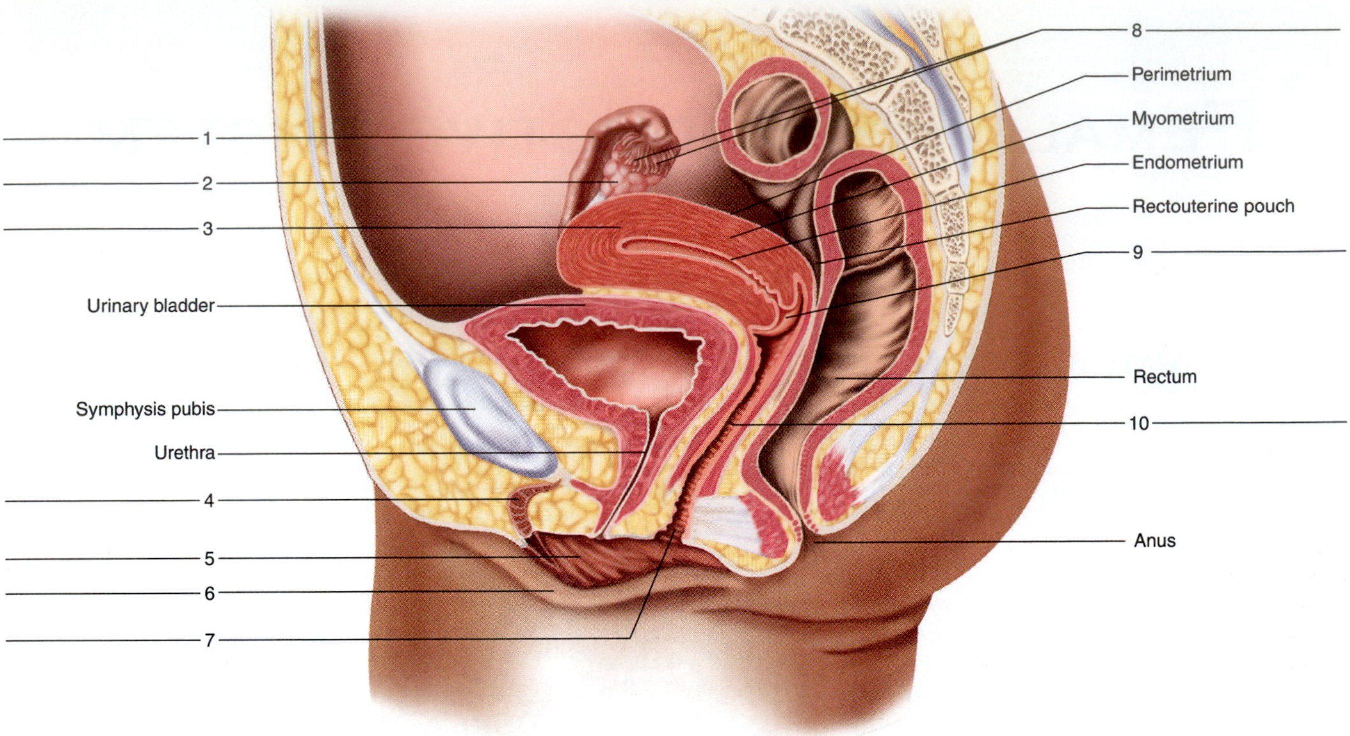

Figure 48.2 Label this ovary by placing the correct numbers in the spaces provided.

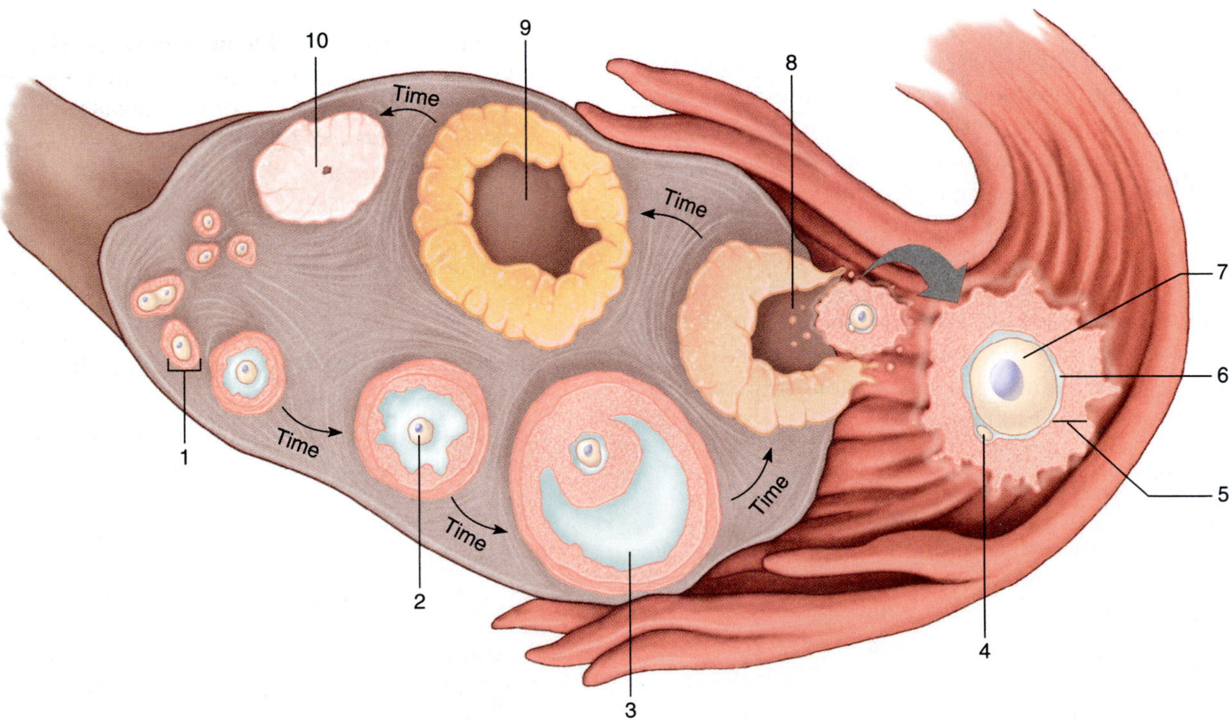

____ Corona radiata
____ Corpus albicans
____ Corpus luteum
____ First polar body
____ Follicular fluid
____ Ovulation site
____ Primary follicle
____ Primary oocyte
____ Secondary oocyte
____ Zona pellucida

Figure 48.3 Label the structures of the breast (anterior view), using the terms provided.

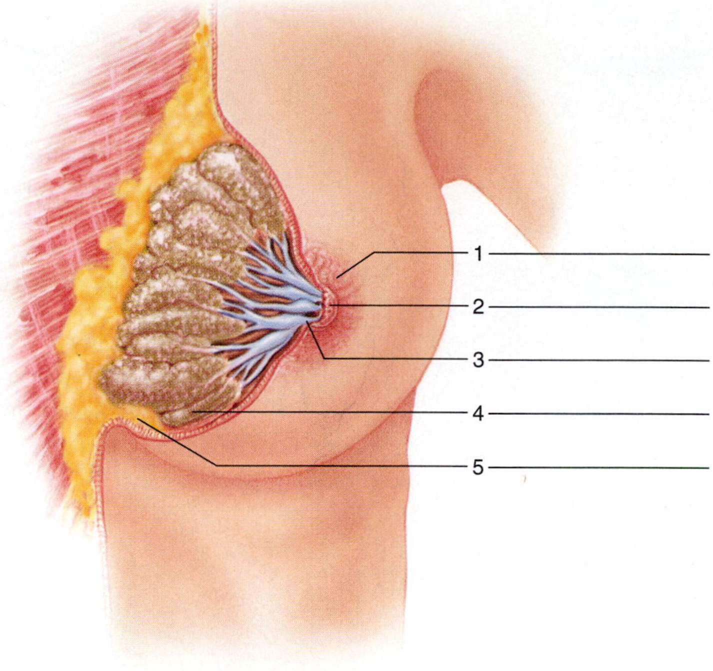

Terms:
Adipose tissue
Alveolar glands
Areola
Lactiferous duct
Nipple

vagina

- vaginal orifice
- hymen
- mucosal layer
- muscular layer
- fibrous layer

vulva (external accessory reproductive organs)

- labia majora
- labia minora
- clitoris
- vestibular glands

breasts

- nipple
- areola
- alveolar glands
- lactiferous ducts
- adipose tissue

4. Complete Part A of Laboratory Report 48.

PROCEDURE B—MICROSCOPIC ANATOMY

1. Obtain a microscope slide of an ovary section with maturing follicles and examine it with low-power magnification (fig. 48.4). Locate the outer layer, or *cortex,* of the ovary, which is composed of densely packed cells, and the inner layer, or *medulla,* which largely consists of loose connective tissue.
2. Focus on the cortex of the ovary, using high-power magnification (fig. 48.5). Note the thin layer of small cuboidal cells on the free surface. These cells comprise the *germinal epithelium.* Also locate some *primordial follicles* just beneath the germinal epithelium. Note that each follicle consists of a single, relatively large *primary oocyte* with a prominent nucleus and a covering of *follicular cells.*
3. Prepare a labeled sketch of the ovarian cortex in Part B of the laboratory report.
4. Use low-power magnification to search the cortex for maturing follicles in various stages of development. Prepare three labeled sketches in Part B of the laboratory report to illustrate the changes that occur in a follicle as it matures.

Figure 48.4 Micrograph of the ovary (30× micrograph enlarged to 80×).

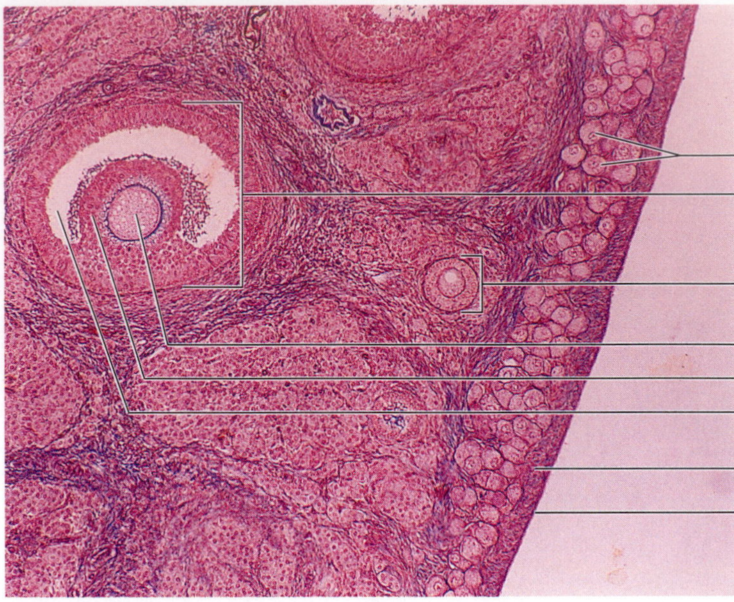

Figure 48.5 Micrograph of the ovarian cortex (100× micrograph enlarged to 200×).

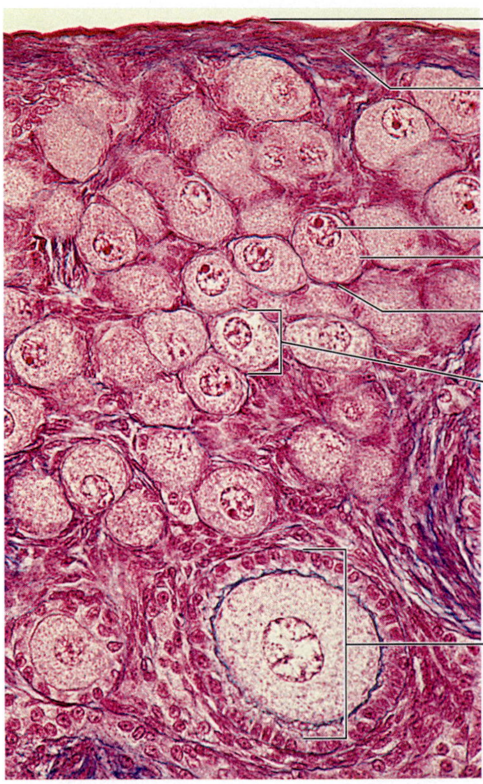

5. Obtain a microscope slide of a cross section of a uterine tube. Examine it, using low-power magnification (fig. 48.6). Note that the shape of the lumen is very irregular.
6. Focus on the inner lining of the uterine tube, using high-power magnification. Note that the lining is composed of *simple columnar epithelium* and that some of the epithelial cells are ciliated on their free surfaces.
7. Prepare a labeled sketch of a representative region of the wall of the uterine tube in Part B of the laboratory report.
8. Obtain a microscope slide of the uterine wall section (fig. 48.7). Examine it, using low-power magnification, and locate the following:

   endometrium

   myometrium

   perimetrium

9. Prepare a labeled sketch of a representative section of the uterine wall in Part B of the laboratory report.
10. Complete Part B of the laboratory report.

Figure 48.6 Micrograph of a cross section of the uterine tube (8×).

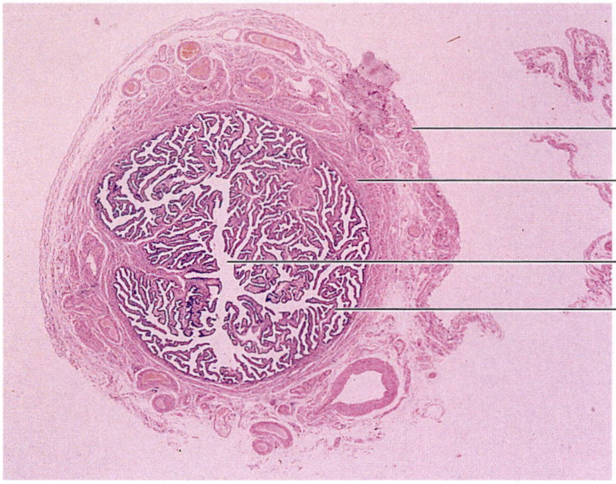

Figure 48.7 Micrograph of a section of the uterine wall (10× micrograph enlarged to 35×).

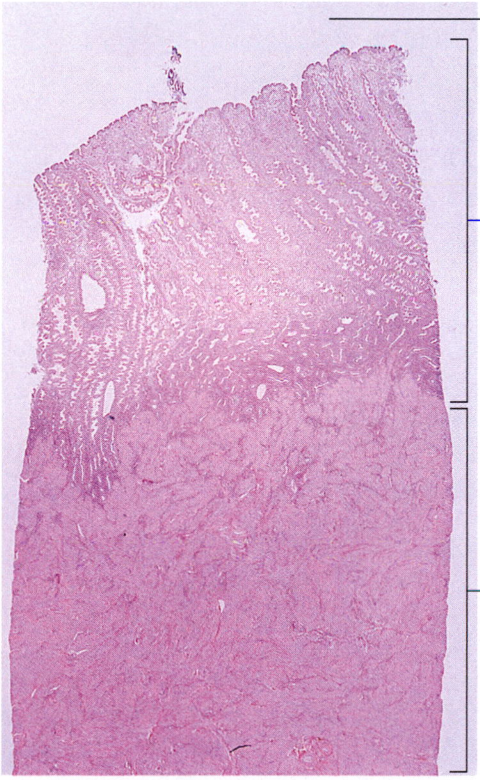

Web Quest

Review the structures and functions of the female reproductive system at www.mhhe.com/shieress8

Web Quest

Examine topics of infertility, fetal development, pregnancy, birth, and many more. Search these at www.mhhe.com/shieress8

DEMONSTRATION

Observe the slides in the demonstration microscopes. Each slide contains a section of uterine mucosa taken during a different phase in the menstrual cycle. In the *early proliferative phase,* note the simple columnar epithelium on the free surface of the mucosa and the numerous sections of tubular uterine glands in the tissues beneath the epithelium. In the *secretory phase,* note that the endometrium is thicker and that the uterine glands appear more extensive and that they are coiled. In the *early menstrual phase,* note that the endometrium is thinner because its surface layer has been lost. Also note that the uterine glands are less apparent and that the spaces between the glands contain many leukocytes. What is the significance of these changes?

____

____

____

____

Laboratory Report **48**

Name ____________________

Date ____________________

Section ____________________

FEMALE REPRODUCTIVE SYSTEM

PART A

Complete the following statements:

1. The ovaries are located in the lateral wall of the ____________________ cavity.
2. The ovarian cortex appears granular because of the presence of ____________________ .
3. A primary oocyte is closely surrounded by flattened epithelial cells called ____________________ cells.
4. When a primary oocyte divides, a secondary oocyte and a(n) ____________________ are produced.
5. Primordial follicles are stimulated to develop into primary follicles by the hormone called ____________________ .
6. ____________________ is the process by which a secondary oocyte is released from the ovary.
7. Uterine tubes are also called ____________________ .
8. The ____________________ is the funnel-shaped expansion at the end of a uterine tube.
9. A portion of the uterus called the ____________________ extends downward into the upper portion of the vagina.
10. The inner mucosal lining of the uterus is called the ____________________ .
11. The myometrium is largely composed of ____________________ tissue.
12. The vaginal orifice is partially closed by a thin membrane called the ____________________ .
13. The group of external accessory organs that surround the openings of the urethra and vagina comprise the ____________________ .
14. The rounded mass of fatty tissue overlying the symphysis pubis of the female is called the ____________________ .
15. The female organ that corresponds to the male penis is the ____________________ .
16. The ____________________ of the female correspond to the bulbourethral glands of the male.

PART B

1. Prepare a labeled sketch of a representative region of the ovarian cortex.

2. Prepare a series of three labeled sketches to illustrate follicular maturation.

3. Prepare a labeled sketch of a representative section of the wall of a uterine tube.

4. Prepare a labeled sketch of a representative section of the uterine wall.

5. Complete each of the following:

   a. Describe the fate of a mature follicle.

   ____

   ____

   ____

   b. Describe the function of the cilia in the lining of the uterine tube.

   ____

   ____

   ____

   c. Briefly describe the changes that occur in the uterine lining during the menstrual cycle. ______________

   ____

   ____

   ____

APPENDIX 1

PREPARATION OF SOLUTIONS

Amylase solution, 0.5%
Place 0.5 g of bacterial amylase in a graduated cylinder or volumetric flask. Add distilled water to the 100-mL level. Stir until dissolved. See the Instructor's Manual for a supplier of amylase that is free of sugar. (Store amylase powder in a freezer until mixing this solution.)

Benedict's solution
Prepared solution is available from various suppliers.

Glucose, 1%
Place 1 g of glucose in a graduated cylinder or volumetric flask. Add distilled water to the 100-mL level. Stir until dissolved.

Iodine-potassium-iodide (IKI solution)
Add 20 g of potassium iodide to 1 L of distilled water, and stir until dissolved. Then add 4.0 g of iodine, and stir again until dissolved. Solution should be stored in a dark stoppered bottle.

Methylene blue
Dissolve 0.3 g of methylene blue powder in 30 mL of 95% ethyl alcohol. In a separate container, dissolve 0.01 g of potassium hydroxide in 100 mL of distilled water. Mix the two solutions. (Prepared solution is available from various suppliers.)

Physiological saline solution
Place 0.9 g of sodium chloride in a graduated cylinder or volumetric flask. Add distilled water to the 100-mL level. Stir until dissolved.

Ringer's solution (frog)
Dissolve the following salts in 1 L of distilled water:

6.50 g sodium chloride
0.20 g sodium bicarbonate
0.14 g potassium chloride
0.12 g calcium chloride

Sodium chloride solutions
1. *0.9% solution.* Place 0.9 g of sodium chloride in a graduated cylinder or volumetric flask. Add distilled water to the 100-mL level. Stir until dissolved.
2. *3.0% solution.* Place 3.0 g of sodium chloride in a graduated cylinder or volumetric flask. Add distilled water to the 100-mL level. Stir until dissolved.

Starch solutions
1. *0.5% solution.* Add 5 g of cornstarch to 1 L of distilled water. Heat until the mixture boils. Cool the liquid, and pour it through a filter. Store the filtrate in a refrigerator.
2. *1.0% solution.* Add 10 g of cornstarch to 1 L of distilled water. Heat until the mixture boils. Cool the liquid, and pour it through a filter. Store the filtrate in a refrigerator.

Wright's stain
Prepared solution is available from various suppliers.

Appendix 2

Assessments of Laboratory Reports

Numerous assessment models can be utilized for laboratory reports. A rubric, which can be used for performance assessments, contains a description of the elements (requirements or criteria) of success to various degrees. The term *rubric* originated from *rubrica terra,* which is Latin for the application of red earth to indicate anything of importance. A rubric used for assessment contains elements for judging student performance, with points awarded for varying degrees of success in meeting the objectives. The content and the quality level necessary to attain certain points are indicated in the rubric. It is effective if the assessment tool is shared with the students before the laboratory exercise is performed.

Following are two sample rubrics that could easily be modified to meet the needs of a specific course. Some of the elements for these sample rubrics may not be necessary for every laboratory exercise. The generalized rubric needs to contain the possible assessment points that correspond to objectives for a specific course. The point value for each element may vary. The specific rubric example contains performance levels for laboratory reports. The elements and the point values could easily be altered to meet the value placed on laboratory reports for a specific course.

ASSESSMENT: GENERALIZED LABORATORY REPORT RUBRIC

Element	Assessment Points Possible	Assessment Points Earned
1. Figures are completely and accurately labeled.	______	______
2. Sketches are accurate, contain proper labels, and are of sufficient detail.	______	______
3. Colored pencils were used extensively to differentiate structures on illustrations.	______	______
4. Matching and fill-in-the-blank answers are completed and accurate.	______	______
5. Short-answer/discussion questions contain complete, thorough, and accurate answers. Some elaboration is evident for some answers.	______	______
6. Data collected are complete, accurately displayed, and contain a valid explanation.	______	______
TOTAL POINTS	______	______

ASSESSMENT: SPECIFIC LABORATORY REPORT RUBRIC

Element	Excellent Performance (4 points)	Proficient Performance (3 points)	Marginal Performance (2 points)	Novice Performance (1 point)	Points Earned
Figure labels	Labels completed with ≥ 90% accuracy.	Labels completed with 80%–89% accuracy.	Labels completed with 70%–79% accuracy.	Labels <70% accurate.	
Sketches	Accurate use of scale, details illustrated, and all structures labeled accurately.	Minor errors in sketches. Missing or inaccurate labels on one or more structures.	Sketch is not realistic. Missing or inaccurate labels on two or more structures.	Several missing or inaccurate labels.	
Matching and fill-in-the-blanks	All completed and accurate.	One to two errors or omissions.	Three to four errors or omissions.	Five or more errors or omissions.	
Short-answer and discussion questions	Answers are complete, valid, and contain some elaboration. No misinterpretations are noted.	Answers are generally complete and valid. Only minor inaccuracies were noted. Minimal elaboration exists.	Marginal answers to the questions and contains inaccurate information.	Many answers are incorrect or fail to address the topic. There may be misinterpretations.	
Data collection and analysis	Data are complete and displayed with a valid interpretation.	Only minor data missing or a slight misinterpretation exists.	Some omissions. Not displayed or interpreted accurately.	Data are incomplete or show serious misinterpretations.	

TOTAL POINTS EARNED ____________

Credits

Photos

Chapter 1
1.1, 1.2: © McGraw-Hill Higher Education, Inc./Jim Womack Photography

Chapter 3
3.1–3.3, 3.6: © McGraw-Hill Higher Education, Inc./Jim Womack Photography

Chapter 4
4.3: Ed Reschke

Chapter 5
5.1: © McGraw-Hill Higher Education, Inc./Jim Womack Photography

Chapter 6
6.1(all), 6.3: Ed Reschke; 6.4a: March of Dimes, Inc.; 6.4b: CNRI/SPL/Photo Researchers, Inc.; 6.5a-d: Ed Reschke

Chapter 7
7.1a,b: Ed Reschke; 7.1c: Manfred Kage/Peter Arnold, Inc.; 7.1d: Ed Reschke; 7.1e: Fred Hossler/Visuals Unlimited; 7.1f: Ed Reschke

Chapter 8
8.1a–e: Ed Reschke; 8.1f: © McGraw-Hill Higher Education, Inc./Carol D. Jacobson, Ph.D. Department of Veterinary Anatomy, Iowa State University; 8.1g: Ed Reschke; 8.1h: © Martin Rotker

Chapter 9
9.1a,b: Ed Reschke; 9.1c: Manfred Kage/Peter Arnold, Inc.; 9.1d: Ed Reschke

Chapter 10
10.3a: Victor B. Eichler; 10.3b: © McGraw-Hill Higher Education, Inc./Carol D. Jacobson, Ph.D. Department of Veterinary Anatomy, Iowa State University; 10.3c,d: Victor B. Eichler

Chapter 11
11.3a,b: © McGraw-Hill Higher Education, Inc./Jim Womack Photography

Chapter 13
13.6–13.10: © McGraw-Hill Higher Education, Inc./Jim Womack Photography

Chapter 14
14.6: © Kent M. Van De Graaff/Utah Valley Hospital, Department of Radiology

Chapter 15
15.6: © Martin Rotker; 15.7: Courtesy Eastman Kodak; 15.8: © Martin Rotker

Chapter 16
16.6–16.8: © Martin Rotker

Chapter 18
18.1: Ed Reschke; 18.4: © H.E. Huxley

Chapter 19
19.3a–c: © McGraw-Hill Higher Education, Inc./Jim Womack Photography

Chapter 20
20.5a-c: © McGraw-Hill Higher Education, Inc./Jim Womack Photography

Chapter 22
22.7a,b: © McGraw-Hill Higher Education, Inc./Jim Womack Photography

Chapter 23
23.3–23.7: Ed Reschke

Chapter 24
24.2a–e: © McGraw-Hill Higher Education, Inc./Jim Womack Photography

Chapter 25
25.3: © Per H. Kjeldsen, University of Michigan/Ann Arbor

Chapter 26
26.5: © Martin Rotker/Photo Researchers

Chapter 27
27.2, 27.6: © McGraw-Hill Higher Education, Inc./Jim Womack Photography

Chapter 28
28.4: BioPhoto Associates/Photo Researchers, Inc.; 28.5, 28.6: © McGraw-Hill Higher Education, Inc./Jim Womack Photography; 28.7: © McGraw-Hill Higher Education, Inc.

Chapter 29
29.4: © McGraw-Hill Higher Education, Inc./Jim Womack Photography; 29.6a: Carroll Weiss/Camera M.D. Studios; 29.8: © McGraw-Hill Higher Education, Inc./Jim Womack Photography

Chapter 30
30.3: © McGraw-Hill Higher Education, Inc./Jim Womack Photography

Chapter 31
31.7: © Biophoto Associates/Photo Researchers; 31.8: Ed Reschke; 31.9: © Biophoto Associates/Photo Researchers; 31.10, 31.11: Ed Reschke

Chapter 33
33.2a–d: © McGraw-Hill Higher Education, Inc./Courtesy of Leica, Inc./James L. Shaffer, photographer; 33.5: Courtesy of Stuart Fox; 33.8: © McGraw-Hill Higher Education, Inc./J. Womack Photography

Chapter 35
35.4, 35.5: © McGraw-Hill Higher Education, Inc./Jim Womack Photography

Chapter 37
37.3: © Biophoto Associates/Photo Researchers

Chapter 40
40.3: John D. Cunningham/Visuals Unlimited; 40.4: Biophoto Associates/Photo Researchers

Chapter 41
41.4: Biophoto Associates/Photo Researchers; 41.8: Ed Reschke; 41.11: Dennis Strete; 41.12: Ed Reschke

Chapter 43
43.3: Ed Reschke; 43.4: Victor B. Eichler

Chapter 44
44.2: © McGraw-Hill Higher Education, Inc./Jim Womack Photography

Chapter 45
45.2: © McGraw-Hill Higher Education, Inc./Jim Womack Photography; 45.4a: Biophoto Associates/Photo Researchers; 45.4b: Manfred Kage/Peter Arnold, Inc.

Chapter 47
47.3: Biophoto Associates/Photo Researchers; 47.4, 47.5: Ed Reschke; 47.6: Michael Peres

Chapter 48
48.4: Ed Reschke/Peter Arnold, Inc; 48.5: Ed Reschke; 48.6, 48.7: Michael Peres

Line Art

Chapter 3
3.4: Courtesy of Swift Instruments, Inc., San Jose, California

Chapter 17
17.2: Original illustration drawn by Ross Martin.

INDEX